T5-AGA-017

Heredity and Infection

Studies in the History of Science, Technology and Medicine
Edited by John Krige, CRHST, Paris, France

Studies in the History of Science, Technology and Medicine aims to stimulate research in the field, concentrating on the twentieth century. It seeks to contribute to our understanding of science, technology and medicine as they are embedded in society, exploring the links between the subjects on the one hand and the cultural, economic, political and institutional contexts of their genesis and development on the other. Within this framework, and while not favouring any particular methodological approach, the series welcomes studies which examine relations between science, technology, medicine and society in new ways e.g. the social construction of technologies, large technical systems.

Other Titles in the Series

Heredity and Infection

The History of Disease Transmission

Edited by

Jean-Paul Gaudillière
CERMES, Paris, France

and

Ilana Löwy

London and New York

First published 2001
by Routledge
11 New Fetter Lane, London EC4P 4EE

Simultaneously published in the USA and Canada
by Routledge
29 West 35th Street, New York, NY 10001

Routledge is an imprint of the Taylor & Francis Group

Typeset by Expo Holdings, Malaysia
Printed and bound in Great Britain by MPG Books Ltd, Bodmin

British Library Cataloguing in Publication Data
A catalogue record for this book is available from the British Library

ISBN: 0–415–27120–7

CONTENTS

LIST OF FIGURES

LIST OF TABLES

LIST OF CONTRIBUTORS

Olga Amsterdamska is a senior lecturer at the Department of Science and Technology Dynamics, University of Amsterdam, The Netherlands

Virginia Berridge is a professor of history at the London School of Hygiene and Tropical Medicine, UK

JoAnne Brown is an historian with the university administration at Johns Hopkins University, Baltimore, USA

Angela N.H. Creager is an assistant professor at the Department of History, Princetown University, USA

Jean-Paul Gaudillière is an historian of biology and medicine at INSERM, Paris, France

Ilana Löwy is an historian of biology and medicine at INSERM, Paris, France

Andrew Mendelsohn is a research scholar at the Max Planek Institute for the History of Science, Berlin, Germany

Paolo Palladino is a lecturer in the history of science and medicine at the University of Lancaster, UK

Patrice Pinell is a senior researcher at INSERM, Paris, France

Jean-Pierre Revillard is a professor of immunopharmacology at INSERM, Paris, France

Jennifer Stanton is a researcher at the Health Policy Unit at the London School of Hygiene and Tropical Medicine, UK

Michael Worboys is a professor of medical history at the Department of History, Sheffield Hallam University, UK

Patrick Zylberman is a researcher at CERMES, Paris, France

INTRODUCTION
HORIZONTAL AND VERTICAL TRANSMISSION OF DISEASES: THE IMPOSSIBLE SEPARATION?

Jean-Paul Gaudillière and Ilana Löwy

TRANSMISSION OF DISEASES AND SCIENTIFIC REVOLUTIONS IN MEDICINE

The concept of 'transmission of human diseases', evokes many things: the transmission of germs, the transmission of blood, the transmission of chemicals, or the transmission of mutated genes. One way to classify patterns and objects of transmission is to distinguish between horizontal transmission, usually mediated by infectious agents, and vertical transmission, commonly understood as the transmission of hereditary traits from parents to offspring.

In the 'dark ages' of medicine, official histories of medicine stress, this distinction between infection and heredity was not understood. Doctors were unable to explain the uneven distribution of human pathologies. A few diseases (such as plague or cholera) appeared suddenly, produced a great number of victims in a short span of time, then disappeared until the next outbreak. Some (such as malaria) were permanently present in specific sites. In some families many individuals suffered from a given pathology due to their weak constitution and bad habits. Other pathologies seemed to be related to specific social conditions or specific occupations. Inheritance, food, living conditions, climate or morality could all hinge on a given pathology.

The mid-nineteenth century debate about the transmission of cholera is a typical example of the controversies triggered by this conjunction of factors. 'Contagionists' believed, roughly, that the disease was transferred through direct contact with a sick person and that the pathology-inducing principles were material entities. They were confronted with 'infectionists' who pointed out that close contact with a sick person often did not induce the disease, that the absence of such contact did not always protect from sickness, and that in numerous cases the appearance of a pathology seemed to depend on specific climatic, geographical and social conditions rather than on a chain of contacts. Heated debates between the 'contagionists' and 'infectionists'—terms which in fact refer to a vast array of non-overlapping points of view—were focused on the question of quarantine: 'infectionists' viewed quarantine as a useless obstacle to the freedom of trade; 'contagionists' believed that quarantine, while imperfect, was the only measure able to stop the progress of an epidemic (Ackernecht, 1967; Delaporte, 1986).

Degeneration was the focus of another nexus of conflicting views about transmission. Debates about the familial distribution of pathological traits emerged in the context of anxieties about the rapid growth of industrial cities with its legacies of poverty, dissolute behavior and criminality. Following the French physician Benedict Morel, degeneration was viewed as a progressive deterioration of minds and bodies, manifested in the accumulation of defects within families whose members, plagued with pathologies such as consumption, nervous breakdown, alcoholism, madness, and mental retardation became less and less suited for life (Pick, 1989). Controversies raged about the causes of degeneration. The process was taken either as the product of social conditions which undermined the regulating power of natural selection or as a manifestation of an internal deterioration of the germinal substance. In spite of contrasting perspectives on the causes of such outcomes, observers shared the feeling that little could be done about familial degeneration since it was difficult to put 'mating in the human race' under medical or social control.

The search for practical ways to stop the spreading of disease did not, however, contradict the belief that human pathologies are induced by multiple causes. Some causes were more frequently associated with a given disease than others, but none was seen as absolutely necessary for the development of a given pathology. And since specialists recognized that it was not possible to reach the putative material causes of disease, and difficult to eliminate poverty, hygiene was perceived as the only efficient place for expert intervention. The improvement of the water supply and of the sewage system, the cleaning of the cities, the isolation of the sick, were viewed as the only possible ways to limit mortality and morbidity (Rosenberg, 1962).

The sad confusion concerning the 'true causes' of transmissible diseases, some historians proposed, disappeared following the 'bacteriological revolution' of the 1880s and 1890s. In a series of elegant studies, Codell Carter traced the origins of the idea that infectious diseases are caused by specific material agents—bacteria—back to the discovery of Koch's bacillus and the changing etiology of tuberculosis. Poverty, poor nutrition, crowded lodgings, fatigue, inherited weakness, all could have contributed to the development of tuberculosis in a given individual, but in the absence of the tuberculosis bacillus even an extremely poor and debilitated individual would not die from tuberculosis. The passage from 'sufficient causes' to 'necessary causes' of disease—that is, from the need to prove that a casual agent can, by itself, produce a pathological condition, to the need to prove that a pathological condition cannot exist without the agent—was a true *gestalt switch* which led to the transformation of bacteriology into a practical discipline (Codell Carter, 1985). The change was not immediate, and historians of medicine have traced the resistance to the new ideas (Maulitz, 1979: Lawrence 1985). Nevertheless, the success of practices originating in the bacteriology laboratory—first antisepsis and asepsis and then anti-diphtheric serum—accelerated the diffusion of the bacteriological gospel among medical practitioners and led to a firm separation between infectious and systemic diseases (Latour, 1987).

A decade later, in the years 1900–1910, studies of heredity took a new turn with the new interest in statistical patterns of vertical transmission, with the redefinition of Mendel's laws of inheritance, and with the elaboration of the concept of mutation. These developments increased the understanding of the role of heredity in pathologies such as hemophilia, Huntington's chorea, etc., a process highlighted by the publication of Garrod's book on the inborn errors of metabolism in 1909 (Garrod, 1909). Mendelian genetics was however not restricted to rare familial diseases. With the notion of predisposition, the new science of inheritance provided promising leads for the origins of the eugenicists favorite targets: tuberculosis, venereal diseases, and mental retardation.

Standard histories of medicine adopt, with slight corrections, the public image of the 'scientific revolution' in medicine which was promoted by the protagonists themselves (Lister, Pasteur, Koch), and which was reinforced by their followers or by the authors of popular works (de Kruif, 1926; Lewis, 1925). According to such histories, the study of infectious diseases was definitively dissociated from the investigation of hereditary ones with the advent of genetics and bacteriology. In other words, the birth of 'scientific medicine' in the late nineteenth century, symbolized by the gradual adoption of the laboratory worker's white coat by the entire medical profession, made possible a clear-cut distinction between hereditary and infectious pathologies (Bynum, 1994, pp. 218–226). The energetic public health campaigns which focused on the eradication of the habit of spitting could be read as a visible sign of this conceptual switch: the new classification of the paradigmatic 'constitutional' disease—tuberculosis—as an infectious one.

This book proposes a different thesis. The case studies gathered in this volume point to the notion that the unproblematic separation between infection and heredity was an exception rather than the rule, and was usually limited to relatively short periods of time or to particular circumstances. While textbooks and official classifications highlighted the difference in nature between hereditary and infectious pathologies, the practices in the laboratory, at the bedside, and in the office of the public health expert, present a very different picture, revealing the persistence of associations between vertical and horizontal transmission of human diseases. In a recent book on the 'Scientific Revolution' of the seventeenth century, Steven Shapin argued that such a revolution did not exist as one single process. What did exist was the uncoordinated introduction of several innovations in scientific practices, often stabilized through their applications outside the scientific sphere (Shapin, 1997). The view of both the 'bacteriological revolution' and the 'genetic revolution' and of their consequences presented in this book recalls the argument advanced by Shapin. It proposes a similar revision of the public image of the 'scientific revolution' in medicine. The development of microbial theories of disease and of new concepts of heredity, we argue, did lead to important changes in the ways scientists, physicians and sanitarians dealt with human disease. But these changes were much less linear, the ruptures more salient, the inter-disciplinary boundaries less well defined, and the differences between national settings more important than the image drawn in *a posteriori* accounts of the scientific turn in the studies of transmissible disease (Geison, 1995; Allen, 1975). Moreover, and this is one of the central topics of this volume, the numerous attempts at the 'purification' of transmission, that is, at the establishment of two mutually exclusive categories of horizontal and vertical transmission, were and are at best a partial successes.

THE INCOMPLETE BACTERIOLOGICAL REVOLUTION

Several chapters in this book examine the 'incomplete scientific revolution' in nineteenth century medicine. They recall that a rigid separation between acquired and inherited traits did not make sense for the nineteenth century physician and show the persistence of constitution and hereditary factors within the bacteriological frame of explanation.

Patrice Pinell's analysis of the neurologists' uses of the term 'degenerative heredity' in France during the second half of the nineteenth century as well as Michael Worboys's investigation of the meaning of the Victorian term 'constitutional proclivities' by tuberculosis (TB) experts lead to the conclusion that these terms did describe an inseparable mixture of inherited and acquired traits: 'constitution' and 'heredity' could change through a lifetime. In the second half of the nineteenth century, Pinell explains, French practitioners employed a general category of 'degeneration' which included pathologies which were linked to poverty, to immorality, to poor habits, and to poor heredity. The innate and acquired elements of these pathologies were interconnected: the induction of inborn defects in children of syphilitic parents was attributed to permanent changes in germinal plasm. For instance, in 1910, a French textbook dedicated to serodiagnosis of syphilis still explained that a syphilitic father can directly transmit the disease to his children without infecting their mother (Gastou et Girault, 1910). Doctors proposed a similar mechanism to explain that alcoholism of the parents may affect their offspring.[1] Poverty, disease and vice, the experts affirmed, deformed the body and the mind and led to 'degenerate' families.

Koch's announcement of the discovery of a microorganism which induced tuberculosis has often been taken as the critical point of the 'bacteriological revolution'. Tuberculosis was the paramount disease of the new etiology, exemplifying Koch's famous postulates: the demonstration that a given pathology is caused by a given microorganism. However, Worboys argues, viewed from Britain, Koch's 1882 announcement was a non-event. In contrast to laboratory work and public health discourses, the practices of the physicians did not change in a significant way following the description of the tuberculosis bacillus. The real change in the way doctors perceived tuberculosis occurred earlier, with the widespread acceptance of the view that tuberculosis is a form of inflammation. The latter approach was based on anatomical and histological observation of tuberculosis lesions, which grounded the conviction that tuberculosis is a specific disease, and to efforts to define the anatomo-pathological characteristics of the disease. The inflammation theory led to the development of new therapeutic approaches based on the use of anti-inflammatory and antiseptic substances to attenuate the morbid symptoms.

The identification of tuberculosis as an inflammation did not undermine its identification as an essentially inherited condition. Doctors were aware of the experimental possibility to inoculate TB, but to explain why people living in the same infectious environment—for instance people living in one single housing unit—may or may not develop the disease, they assumed that in a natural state, tuberculous inflammation can spread only in weakened bodies. Koch's agent was but a minor addition to this theoretical framework. The bacillus was perceived as a factor initiating the inflammatory process. Its role was secondary compared with the fundamental role of 'constitution'. In the 1880s and 1890s, the discovery of the bacillus did not made much impact on clinical understanding, diagnosis, prognosis, and treatment of TB in Britain. Moreover, it had limited effects on preventive practices. Tuberculosis continued to be perceived as a weakly infectious disease which was not dangerous to healthy individuals; as a result, doctors did not insist on the isolation of the sick.

In his analysis of the 'etiologic crisis' of the years 1900–1910, Andrew Mendelsohn insists on the short-lived career of monocausal explanations of infectious diseases. After a brief period of adherence to such a scheme in the late nineteenth century bacteriologists were confronted with the paradoxes of 'natural resistance' and of 'asymptomatic' infection. Repeated failures (and occasional, unpredictable success) to induce cholera in volunteers who drank a concentrated solution of the pathogenic germ led to the conviction that some persons fail to develop infection when in contact with a pathogenic bacterium. Scientists also observed that other individuals, e.g., some of those who were in contact with the typhoid bacillus, became actively infected and contagious without showing any symptoms of the disease. Clearly, contact with even a highly virulent germ could not, by itself, explain the advent of a disease, an observation which might be seen as a confirmation of the statement, attributed to Claude Bernard—'le microbe n'est rien, le terrain est tout'.

The dismissal of simple, monocausal explanations of infectious diseases, Mendelsohn explains, opened the way to the (re)introduction of the notion of variable individuals into pathology. Practitioners and hygienists took into account variability among individuals when treating patients or promoting public health measures. By contrast, nineteenth century treatises on pathology were focused on the variability of diseases, and implicitly assumed that before the outbreak of disease individuals were essentially similar. Bacteriology put an end to this approach, forcing pathologists to recast their theories and to take into account pre-existing differences among individuals. Experts discussing the cases of 'bacteria without disease' agreed rapidly that the 'susceptibility' to a given microorganism has a hereditary and an acquired dimension and that it was practically impossible to separate the two.

JoAnne Brown's chapter presents some unexpected 'side-effects' of the changing etiology of tuberculosis thus illustrating the political dimension of the incomplete but effective bacteriological revolution. In the USA, she explains, blacks were perceived as having a specific, racially-determined susceptibility to tuberculosis and syphilis. Before the advent of bacterial theories, this putative susceptibility was interpreted as a specific constitutional weakness of blacks, which could be attenuated through a general improvement of their living conditions. The description of tuberculosis as an infectious disease led to the perception of blacks as dangerous to the whites, a danger which could best be limited by strict racial segregation. Doctrines of 'social segregationism', developed in the early twentieth century were, Brown explains, strongly inspired by a fear of contamination. One may note that at the same time the development of bacterial theories prompted the British colonial administration in Africa to try to reinforce a rigid segregation between white and black populations. The measure was strongly resisted by white colons who refused to exile their black servants and workers to distant neighborhoods. The description of the role of Koch's bacillus in the induction of tuberculosis, a feature which theoretically could 'democratize' this disease (everybody can be accidentally contaminated by an airborne pathogenic microorganism) led, paradoxically, to the reinforcement of links between TB and race and to the accentuation of the importance of hereditary traits in TB transmission.

THE VITALITY OF HYBRIDS: PERSISTENCE OF DUAL APPROACHES TO TRANSMISSION

Physicians and bacteriologists, Andrew Mendelsohn explains, oscillated in the twentieth century between the perception of constitution as reflecting mainly inherited (germ-plasm) traits (1900–1910), the partly overlapping perception of individual constitution as affected mainly by the environment, (1905–1925) and finally, from the 1920s on, a synthetic view which considered hereditary factors as the base of constitution, but acknowledged the parallel influence of environment.

In France, the blending of heredity, constitution and infection took a peculiar turn due to the emphasis the French physicians placed on the role of adaptation and on the relationship between mothers and infants (Burian *et al.*, 1988; Carol, 1995; Gaudillière, 2000). Rather than building theories of gene transmission, doctors favored theories of family transmission which stressed at the same time the dangers of 'intoxication' of the fetus (thus explaining the pathological descent of parents addicted to alcohol), the dangers of 'infection' of the fetus (thus explaining the transmission of syphilis), and the dangers of 'mutation' in the fetus following the transfer of altered germinal substance. Accordingly doctors interested in hereditary pathologies did not make much use of Mendelian statistics but employed methods, such as the establishment and collection of family trees, which facilitated the incorporation of clinical and environmental elements into the studies of heredity (e.g. poor nutrition of the mother, diseases acquired *in utero* or during the birth, cases of madness in the family, etc.). Strongly committed to natalism, French physicians and medical scientists were not interested in 'genetics' (defined as the study of the laws governing the transmission of inborn traits), but in 'heredity' a much more open-ended concept which included all the possible factors which may influence the quality of offspring. Thus major infectious scourges like syphilis or tuberculosis were kept in line with 'constitutional' diseases.

The views of the French pioneer of studies of 'familial disorders', Apert, discussed by Patrice Pinell, and Patrick Zylberman are in this respect typical. Apert naturalized Mendelian hereditary thinking in France by adapting it to the French tradition of focusing on the 'terrain'. Apert accepted that a pathological condition transmitted in families can occasionally be induced by an 'abrupt mutation', but for him pathologies due to specific changes of the germ-plasm were the exception rather than the rule. The majority of diseases running in families resulted, he proposed, from complex combinations of hereditary and environmental causes. In order to prevent pathological inheritance, a pre-condition for fulfilling the national goal of the production of healthy and robust children, it was important to act on both. Doubting the global efficiency of selection practices, such as contraception or sterilization, Apert thought—like the majority of public health activists of the time—that the solution to the problem of *hérédité morbide* was to promote the well-being of families, and especially the health and the education of mothers. The health politics of René Sand, analyzed in Zylberman's paper, similarly integrated environmental elements into attempts to improve heredity. Faithful to the view, predominant in France in the interwar years, that environmental influences usually outweigh heredity, Sand incorporated elements of 'old hereditarism' in his 'mixed economy' of public health. He saw the promotion of family politics as the most efficient way to improve the quality of future generations, and stressed the central role of the fight against infectious diseases and of the improvement of nutrition.

Mendelsohn's chapter on the aftermath of the bacteriological revolution also provides some insights into the spreading of hybrids between infection and heredity. In the first half of the twentieth century, he explains, bacteriology was firmly connected to medical theories which put the accent on constitution (Lawrence and Weisz, 1998). In France, Bouchard and Tzank focused on the role of the organism in producing the disease (Weisz, 1998). In England, Pearson's studies of tuberculosis were seen as confirming the hereditary character of the disease, while Weissman investigated the possibility to harm germ plasm cells by alcohol. Even Garrod, presented today as the founder of biochemical genetics, published in 1931 a book on *Inborn Factors in Disease*, which attributed an important role to constitutional theories. In the 1910's and 1920's, studies of allergy and delayed hypersensitivity (phenomena in which the reactivity of the body is radically altered by previous contacts with an antigen), and the accumulation of evidence that 'racial' or 'inborn' resistance to infectious diseases was in fact the result of an active immunization in early childhood, led to the widespread notion that the environment shapes 'constitution'. Some experts had chosen, however, to focus on heredity. Thus Ludwik Hirszfeld studied the role of 'constitutional serology' in protection from infection (a theory which was dismissed later) (Keating, 1998). Similarly, Webster (whose work is described in detail in Olga Amsterdamska's paper) investigated the role of inherited factors in resistance to epidemic disease.

The recent debates on AIDS, described in Ilana Löwy's chapter, illustrate the persistence of hybrids and of concerns on the role of constitution in individual susceptibility to infection. In the years 1981–83, AIDS (then GRID-Gay Related Immunodeficiency) was seen as the result of a complicated and multifactorial collapse of the immune system and as an extreme case of increased susceptibility to pathogens. The description of HIV (1984–85) might have transformed AIDS into an 'ordinary' infectious disease. The existence and properties of the virus were not, however, sufficient to explain the differential susceptibility to this disease (that is the existence of individuals who remain HIV-free in spite of the fact that they were exposed to high risk of infection), the differential susceptibility to the effects of HIV infection (the existence of infected individuals who do not develop symptoms of AIDS), and the differences between 'first world' and 'third world' patterns of diffusion of the HIV virus (in the former AIDS was mainly transmitted through male homosexual relations, and contaminated needles employed by intravenous drug users; in the latter AIDS was transmitted mainly through heterosexual relations and through *in utero* transmission).

Individual differences in susceptibility to the virus were usually interpreted in terms of 'inherited resistance' and genetic factors (for instance the lack of receptors for HIV on target cells), while epidemiological differences between industrialized and developing countries were usually attributed to environmental factors (high frequency of other sexually-transmitted infections, poor nutrition and hygiene) without excluding the possibility that hereditary variables contribute to the shaping of individual resistance.

Virginia Berridge and Jennifer Stanton link these debates on the transmission of AIDS to historical issues on 'vertical' transmission of infection by focusing on the role of *in utero* transmission of this disease and of hepatitis B. AIDS and hepatitis B are today defined as infectious diseases. Nevertheless, in the early 1990s, one of the main targets of public health policies was the prevention of *in utero* transmission. Large-scale testing of pregnant women in Western countries aimed at the prevention of the birth of infected babies. Doctors initially advocated selective abortions. Today, they propose drug therapies which reduce the risks of trans-placental passage of the virus. The later method of prevention of vertical transmission of AIDS is, however, feasible only in industrialized countries: its cost prevents its diffusion in developing countries, the major site of vertical HIV transmission. By contrast, the vertical transmission of hepatitis B can be prevented in these countries thanks to the existence of an efficient and cheap anti-hepatitis B vaccine. Hence the efforts to promote mass vaccination campaigns of girls and young women (Muraskin, 1995).

Historically, cancer was another area in which experts evoked the possible role of both horizontal and vertical factors in inducing disease. In this case, the problem was not the role of heredity in the susceptibility to infection, but the search for the causes of a mysterious disease. In the twentieth century this search resulted in cycles of alternating genetic, viral, physiological, and environmental etiologies. Angela Creager and Jean Paul Gaudillière follow the revival of studies on the role of viruses in inducing cancer from the initial conviction that a virus (like the tuberculosis bacillus) is a potential 'irritating agent' at the origin of cancerous transformation of the tissues, to the present day identification of viruses with 'hereditary material'. In the post-WW2 era, with the development of new forms of instrumentation and laboratory work, viruses became macromolecular entities composed of circulating DNA or RNA. As a result the search for 'cancer viruses' was linked to attempts to identify the 'genetic predisposition' to cancer. Ironically, the classical agents of horizontal transmission thus became the basis for vertical transmission. Studies on the putative inheritance of breast cancer in mice led to the description of the 'milk factor'—an agent passed from mothers to suckling newborns—*in fine* redefined as a virus. Studies of such animal viruses were characteristic of cancer research in the United States, as exemplified by the large scale research programs launched by the National Cancer Institute in the 1960s and 1970s (Gaudillière, 1998).

In parallel, studies of family trees in humans were complemented with epidemiological surveys relying on statistical tools of increasing sophistication. This line of investigation, focused on a few rare tumors (such as retinoblastoma), and on two common malignancies, breast cancer and colon cancer, indicated the possibility of inherited susceptibility to cancer. The early recognition that an inherited condition—familiar polyposis—is related to colon cancer did not promote, as Paolo Palladino explains, the experts' interest in genetics. British experts in the treatment of colon followed the hereditary aspect of this pathology (mainly through establishing family trees), but they focused their own work on the modification of the expression of hereditary predisposition by environmental factors, and favored surgical solutions over attempts to introduce eugenics measures.

PURIFICATION AT WORK: MODELIZATION, TECHNIQUES AND FRAMEWORK SHIFTS

Some physicians and many biomedical researchers were bothered by the overlap between horizontal and vertical elements of transmission of diseases, and attempted to unravel the complexities of human pathologies in order to pinpoint a single or 'necessary' cause of a disease whose elimination could lead to a solution of a medical problem. These attempts are often presented in theoretical terms, i.e. applying Koch's postulates when looking for bacteria or applying Mendel's law when looking for a gene. The efforts to separate hereditary and environmental causes of disease were however practical above all. In spite of the intricacies of such processes, they could schematically be gathered under two headings: a) attempts to develop an experimental model which can be controlled more easily than the studied phenomenon, thus reducing the number of variables and making possible the study of their hierarchy; b) attempts at new understanding (re-framing) of a given disease through the application of new methods, techniques and concepts.

A) Modelization

Olga Amsterdamska follows attempts to construct models of epidemics. Epidemics, more than individual susceptibility to infection, are usually perceived as complicated multilayered and multicausal events. Nevertheless, in the 1920s and 1930s, two groups of experts—Topley and Greenwood in London, Webster and his collaborators at the Rockefeller Institute in New-York—attempted to develop an experimental model of an epidemic by testing the spreading of an infection in mice colonies. The approaches elaborated by Webster and Topley and Greenwood were radically different. Webster tried to develop a well-controlled, 'clean' model, with a very limited number of variables. His studies, driven by a logic of laboratory experimentation, led to uncovering what he considered to be the single decisive element in the spread of epidemics—the inherited variability of susceptibility to a given germ. If one assumes that germs have a 'standard virulence', Webster proposed, the dynamics of their propagation in a population is dependent on the number of highly susceptible individuals. Webster's critics contested, however the notion of 'standard virulence' of germs, and saw his well-controlled model as too remote from real-life epidemics. Topley and Greenwood (the latter was a statistician) attempted, by contrast, to study epidemics in 'dirty' conditions in which multiple elements were allowed to vary, and the role of the scientists was to look for quantitative regularities. Their only well-established conclusion was that the dynamics of a given epidemic was influenced mainly by the influx of susceptible individuals to the infected area. Such a finding did not go, however, beyond the reproduction in the laboratory of observations made by epidemiologists in the field.

A similar disjunction between the logic of modelization and the logic of intervention characterized attempts to develop models of genetic disease. Looking at the case of muscular dystrophy, Jean Paul Gaudillière explores the development of a new specialty, medical genetics, in the aftermath of the Second World War. The development of antibiotherapies led to a shift of interest of physicians from infectious to chronic disease, and to increased—and generously funded—efforts to study these diseases in the laboratory. The mass production of mice in the post war era facilitated the development of numerous mutants, which could be used in medicine. The main problem was, however, the adequate linking of data obtained in the laboratory with those provided by the clinicians. The mouse model of muscular dystrophy was successful because it was able to provide such a connection. Clinicians (partly) relied on evidence coming from the research laboratory to interpret their data, and used the 'dystrophic' mice to consolidate the familial distribution and biochemical background of muscular dystrophy. The model had, however, limited impact on therapeutic research. The laboratory model of obesity—the 'obese' mouse—was less successful. This model, first widely employed for laboratory work and educational purposes, lost its pertinence for clinicians in the 1960's when obesity was redefined as a behavioral, social and gender-oriented disease.

B) 'Framework Shifts': Techniques and Concepts

The introduction of new methods can, and occasionally did, shift the balance in favor of vertical or of horizontal transmission. The development, in the 1990s, of genetic tests for susceptibility to familiar forms of colon cancer weakened, Palladino explains, the supremacy of surgeons in the diagnosis and the treatment this disease. The introduction of these tests led to attempts at genetic counseling for families suffering from hereditary polyposis of the colon, and to the perception of this form of cancer as a disease with an important hereditary component. The development of quantitative PCR made possible the quantification of HIV particles in blood and in cells (white blood cells, lymph nodes cells) thus challenging, Löwy proposes, the previous perception of AIDS as a multifactorial disease. The presentation of AIDS as a 'mere infection' was also strengthened by the marketing of a 'winning combination' of drugs which successfully arrested the progress of the disease by blocking viral multiplication. The new, technique-driven, focus on the dynamics of HIV multiplication in the body relegated factors such as the immune status of the infected individual to a secondary status, and reduced the interest in investigating both the environmental and hereditary 'co-factors' of HIV infection.

Creager and Gaudillière focus their study of cancer viruses on the role of experimental arrangements—specific agglomerates of techniques, disciplinary approaches and concepts, which present the studied object to professional and lay gaze. When cancer viruses were mainly studied by pathologists, cancer was seen as a disease of 'asymptomatic carriers'. When cancer researchers turned to the study of macromolecules, the focus of investigations in this area shifted from that of infection coming from the 'outside' to the enemy 'within' (hidden viruses, integrated into chromosomes) and to transfers between generations. Finally, the large-scale use of molecular biology techniques (the visualization of viral nucleic acids and proteins and the study of their interactions within the mammalian cells) built up a genetic image of cancer. Viruses were no more 'pathogens' which induced morbid changes in tissues, or 'mutagens' which induced biochemical modifications in cells, but a cluster of genes, difficult to dissociate from the cell's own genetic material. The development of techniques which allowed a more refined level of analysis thus led to increased blurring of borders between the 'inside' and the 'outside' of organism/cell, between 'causes' and 'effects', and between horizontal and vertical transmission of malignancies.

IMMUNOLOGY AS A PRIVILEGED SITE OF INTERACTIONS BETWEEN VERTICAL AND HORIZONTAL TRANSMISSION OF DISEASES

The close intertwining between vertical and horizontal transmission of diseases, one may argue, is embodied in the history of a single biomedical discipline: immunology. In the early days of bacteriology, diseases were perceived in physiological terms. Natural 'resistance' to diseases was viewed as a problem of nutrition and intoxication. The presence of specific foodstuffs allowed the survival of the microbe in the organism. Alternatively, the permanent presence of toxic chemical substances prevented the multiplication of pathogens. Acquired resistance was somehow more difficult to explain: it was attributed to the 'exhaustion' of a nutrient which was absolutely essential to bacterial survival or to the accumulation of toxic bacterial secretions which ended by killing the bacteria (in analogy with bacterial death in cultures due to overcrowding).

In the late 1880s researchers arrived at the conclusion that the body is not a mere test-tube filled with nutrient broth, and that it plays an active role in fighting infection, either through specialized cells or through the mediation of specific substances in the blood. The latter, named antibodies, focused the attention of immunologists for the next eighty years because they were the active components of the immune sera widely employed in diagnosis and therapy (Lindenman, 1981). From the 1890s on, 'resistance to infection' was perceived as a complicated interaction between 'seed' and 'soil' in which neither the 'seed' nor the 'soil' were passive elements (Silverstein, 1989; Moulin, 1991; Fleck, 1935). Antibodies -and more broadly immune mechanisms- became identified as the main material expression of 'predisposition' to an infectious disease. The status of immune mechanisms in a given moment reflected at the same time the (individualized) hereditary traits, and the (individualized) life history. The accumulation of antibodies and of 'hypersensitivites' was the transcription of the body's previous encounters with pathogens and foreign proteins in the physiological makeup of the organism (Richet, 1911).

The growing separation between 'true hereditary diseases' and other pathological conditions after World War II diminished the role of immunology in studies of the hereditary transmission of pathological conditions. In the 1960s, however, immunology became directly linked with vertical transmission of diseases through studies of a new set of laboratory entities: the histocompatiblity antigens. Studies of histocompatibility (the compatibility of tissues) originated in attempts to transplant tumors from one animal to another. It was found that such grafts (and more generally all tissue grafts) were successful only among genetically identical individuals. In the 1950s new models—'syngeneic' (genetically identical) lines of mice—were crafted for analyzing the genetic background of graft rejection. The result was the birth of a new biomedical domain, immunogenetics.

The language of immunogenetics was a language of macromolecules, genes, and proteins. Histocompatibility genes code for proteins on the cell surface (histocompatibility antigens) that are responsible for graft rejection (Löwy and Gaudillière, 1998). As products of the molecularization of biology, histocompatibility genes remained laboratory curiosities until the early 1960s, when attempts to graft kidneys in humans led to intensified studies of proteins involved in graft rejection. In consequence, histocompatibility antigens were redefined as key elements in self/non-self recognition, and their role in numerous infectious and non-infectious diseases was investigated (Löwy, 1987). Specific histocompatibility antigens were linked with increased likelihood to develop diseases such as diabetes or arthritis (Dausset et Pla, 1985). Transmission of genes which code for histocompatibility antigens was thus simultaneously perceived as an explanation of inborn differences in 'resistance to infectious diseases' and of inborn differences in the probability of developing selected chronic diseases. These histocompatibility antigens and genes paved the way to the wave of recombinant DNA-based studies of pathological genes. On the one hand, they exemplified the birth of molecular genetic predisposition factors. On the other hand, the study of these entities anticipated the sort of correlation analysis and large scale collective surveys which characterize contemporary genomics.

In parallel, the narrow focus on a single group of proteins, the histocompatibility antigens, was enlarged to include studies of numerous other molecules active in immune responses (cellular receptors, cytokines, growth factors) following the development of specific pathologies. These studies led to an increased fluidity of distinctions between different components of immune mechanisms. These developments, summed up in Jean Pierre Revillard's postface, highlight the blurring, at the molecular level, of the limits between the inherited and the acquired elements of immune response. On the molecular or the genomic level, these differences tend to be reduced to differences in the expression and the concentration of active macromolecules.

Here, however, one needs to add a word of caution. The classification of human diseases as either 'hereditary' or 'infectious' frequently generated tensions between different groups of practitioners and/or between different social actors. Historical studies indicate thus that the blurring of the differences between 'heredity' and 'environment' in laboratory investigations did not necessarily affect medical practices nor public health policies. They also point to a symmetrical disjunction, i.e. the blurring of differences between vertical and horizontal transmission of a given disease in public health and medical discourses which contrasts with highly efficient purification work in the laboratory.

Finally, one should add that the knowledge produced by experts may significantly differ from widely circulating images of transmission. The refusal, in the 1980s, to accept HIV-contaminated children in schools, the popularity of books on 'the next plague' and of movies which present heroic scientists fighting a deadly virus, the feeling of guilt of parents of children born with a genetic disease, or the endless speculations about the use of new techniques of human reproduction to produce 'flawless babies', all remind us of the central place of the 'traditional' understanding of vertical and horizontal patterns of transmission in shaping the popular vision of diseases.

The book is divided into four thematic sections: tuberculosis, the 'French exception', experimenting and modeling and clinical practices. This division (certainly not the only one possible) was chosen to favor comparisons but also to highlight the diversity of the solutions chosen by experts who dealt with similar situations.

IN MEMORIAM

This book originated in a conference, organized in May 1996 jointly by INSERM U-158, Paris and the Wellcome Unit for the History of Medicine, University of Cambridge. The conference was supported by INSERM. It was part of a coordinated research project funded by the Franco-British Alliance Program. This volume follows an earlier publication, also the fruit of the INSERM-Wellcome collaboration, *The Molecularization of Biology and Medicine*, (Soraya de Chadeverian and Harmke Kamminga, eds., Harwood, 1998). The collaborative venture was, we believe, highly successful, but unfortunately the patient died. Following a rare surge of cross-channel administrative reshuffling, both the Cambridge Wellcome Unit and INSERM U-158 were closed in 1998.

We would like to thank all the contributors to this volume for their efforts and patience, and to Soraya de Chadeverian and Harmke Kamminga for their advice and help. During the conference Jean Gayon, Anne-Marie Moulin, and Hans-Jörg Rheinberger provided comments which proved critical when revising the papers. John Krige deserves special mention for his judicious remarks, advice and patient translation of the approximate English of some participants into the standard idiom. This enterprise would have got nowhere without the valiant efforts of Marie Collet-Moez and Chantal Celiset from INSERM U 158 who put the manuscript in shape.

NOTE

1. It is interesting to note that although teratogenic effects of various substances absorbed by the mother were well known to embryologists, physicians became alert to the possibility that therapeutics taken by the mother may harm the foetus only following the recent thalidomide affair. Anne Dally, 'Thalidomide: Was the tragedy preventable?' *Lancet*, (1998) 351: 1197–1199.

REFERENCES

E.H. Ackernecht, *Medicine at Paris Hospita* (Baltimore: Johns Hopkins Press, 1967).

G. Allen, *The Life Sciences in the Twentieth Century* (Cambridge: Cambridge University Press, 1975).

R. Burian, J. Gayon, and D. Zallen, 'The Singulare Fate of Genetics in the History of French Biology', *Journal of the History of Biology* (1988) 21: 357–402.

W. Bynum, *Science and Practice of Medicine in the Nineteenth Century* (Cambridge: Cambridge University Press, 1994).

A. Carol, *Histoire de l'eugénisme en France: Les médecins et la procréation* (Paris: Seuil, 1995).

K. Codell Carter, 'Ignatz Semmelweis, Karl Myerhoffer and the rise of germ theory' *Medical History* (1985) 29: 33–53.

K. Codell-Carter, 'Koch's postulates in relation to the work of Jacob Henle and Edwin Klebs', *Medical History* (1985) 29: 353–374.

J. Dausset et M. Pla, *HLA: complexe majeur de l'histocompatibilité chez l'homme* (Paris: Flamarion, 1985).

F. Delaporte, *Disease and Civilisation: The Cholera in France, 1832.* (transl. A. Goldhamer), Cambridge (Mass).: The MIT Press, 1986).

L. Fleck, *Genesis and Development of a Scientific Fact* (Chicago: University of Chicago Press, 1979 (1935)).

A. Garrod, *The Inborn Errors of Metabolism* (London: H. Frowde and Hodder and Stoughton, 1909).

P. Gastou et A. Girault, *Guide pratique dio diagnostique de la syphilis* (Paris: J.B. Baillière et Fils, 1910).

J.P. Gaudillière, 'The Molecularization of Cancer Etiology in the Postwar United States: Instruments, Politics, and Management' in *Molecularizing Biology and Medicine*, S. de Chadarevian and H. Kamminga (eds), (Amsterdam: Harwood Academic Publishers, 1998) pp. 139–170.

J.P. Gaudillière, 'Mendelism und Medicine', *CRAS* (2000) 323: 1117–1126.

G. Geison, *The Private Life of Louis Pasteur* (Princeton: Princeton University Press, 1995).

P. Keating, 'Holistic bacteriology: Ludwik Hirszfeld's doctrine of serogenesis between the two World Wars' in Lawrence and Weisz (des) *Greater than the Parts: Holism in Biomedicine, 1920–1950* (Oxford: Oxford University Press, 1998) pp. 283–302.

P. de Kruif, *Microbe Hunters* (New York: Harcourt & Brace, 1926).

B. Latour, *The Pasteurization of France* (Cambridge (Mass): Harvard University Press, 1987).

C. Lawrence, 'Incommunicable knowledge: Science, technology and the clinical art in Britain, 1850–1914', *Journal of Contemporary History* (1985) 20: 503–520.

C. Lawrence, and G. Weisz (eds.), *Greater than the Parts: Holism in Biomedicine, 1920–1950* (Oxford: Oxford University Press, 1998).

S. Lewis, *Arrowsmith* (New York: The New American Library of Classic Literature, 1961 (1925)).

J. Lindenman, 'Immunology in the 1880's: two early theories', in *The Immune System* (Basel: Karger; 1981) vol. 1, pp. 413–422.

I. Löwy, 'The impact of medical practice on biomedical research: The case of Human Leucocyte Antigen studies' *Minerva* (1987) 25(1–2): 171–200.

I. Löwy, and J.P. Gaudillière, 'Disciplining Cancer: Mice and the Practice of Genetic Purity' in *The Invisible Industrialist: Manufactures and the Production of Scientific Knowledge* Gaudillière and Löwy (eds) (London: Macmillan, 1998) pp. 209–249.

R. Maulitz, Physician versus bacteriologist: the ideology of science in clinical medicine' in M.J. Vogel, and C.E. Rosenberg, *The Therapeutic Revolution: Essays in the Social History of American Medicine* (Philadelphia: University of Pennsylvania Press, 1979) pp. 91–108.

A.M. Moulin, *Le dernier langage de la médecine: L'histoire de l'immunologie de Pasteur au Sida* (Paris: Presses Universitaires de France, 1991).

W. Muraskin, *The War Against Hepatitis B: A History of the International Task Force on Hepatitis B Immunization* (Philadelphia: University of Pennsylvania Press, 1995).

D. Pick, *Faces of Degeneration: A European Disorder, 1848–1918* (Cambridge: Cambridge University Press, 1989).

C. Richet, *L'Anaphylaxie* (Paris: Felix Alcan, 1911).

C. Rosenberg, *The Cholera Years* (Chicago and London: Chicago University Press, 1962).

S. Shapin, *The Scientific Revolution* (Chicago: Chicago University Press, 1997).

A.M. Silverstein, *The History of Immunology* (San Diego: Academic Press, 1989).

G. Weisz, 'A moment of synthesis. Medical holism in France between the wars', in Lawrence and Weisz (eds), *Greater than the Parts: Holism in Biomedicine, 1920–1950* (Oxford: Oxford University Press, 1998) pp. 68–93.

Part 1
TUBERCULOSIS

Chapter 1

MEDICINE AND THE MAKING OF BODILY INEQUALITY IN TWENTIETH-CENTURY EUROPE [1]

J. Andrew Mendelsohn

Perhaps the most memorable drama of bodily inequality before the external causes of disease was staged on 7 October 1892 in Munich by Max von Pettenkofer. On that day, the founder of scientific hygiene swallowed a pure culture of cholera bacilli. In the preceding two months, cholera had taken the lives of over 8,000 people in the city of Hamburg. Pettenkofer, however, survived his drink with nothing more than a bout of diarrhoea. Though remembered as the wrong-headed if heroic 'last stand' of anticontagionism against the bacteriological revolution of Pasteur and Koch, this simple self-experiment did create a riveting fact, a fact on whose existence everyone came to agree and puzzle over, not least Robert Koch himself (von Pettenkofer, 1892; Koch, 1893).[2] That Pettenkofer had lived with millions of cholera bacilli in his intestinal tract for over a week without suffering more than diarrhoea was a fact of immediate etiological relevance and, at the same time, of indeterminate meaning. It was a fact, moreover, of relevance to the wider structure of scientific medicine insofar as it was becoming, with the rise of bacteriology, a medicine of specific causes and insofar as it had been since the beginning of the nineteenth century a science of regular pathological reactions, a science of essentially equal bodies.

The Munich-school bacteriologist Hans Buchner, who was anathema to Koch and who publicly welcomed Pettenkofer's interpretation of the self-experiment, struggled privately with what to make of the strange result, this 'riddle,' as he called it. 'If the guinea pigs die from [the cholera] poison, why did Pettenkofer, with his massive comma [bacillus] production, remain so healthy?'[3] Even a bacteriologist not of the Koch school was surprised by the apparent failure of specific cause. Buchner prided himself on his flexible view of the host-parasite relationship. But that was apparently no help here. Nor evidently was the concept of natural immunity.

'Do you know that big news concerning cholera may be taking shape?' wrote Louis Pasteur to one of his department chiefs on 20 November 1892:

> Pettenkofer has announced in a Munich journal that he swallowed a cubic centimeter of a virulent pure culture of the comma bacillus without being inconvenienced by it. ... Here, Metchnikoff to triumph or rather to be very encouraged. For several months, actually, he has done experiment after experiment with the idea that the comma bacillus is not the determining cause of cholera; that when there is cholera, the comma bacillus must be associated with another microbe ... Let's wait for more light.[4]

In the following months, Elie Metchnikoff and his assistants at the Institut Pasteur carried out no less than 16 self-experiments with various cultures of cholera and choleralike bacilli. But more light was slow to come. Etiology still seemed about as obscure as Pettenkofer's self-experiment had suggested it to be (Metchnikoff, 1893, 1894).

In the remaining years of the century, phenomena of contamination or infection in the absence of disease came to occupy a central place in bacteriological investigation. Pettenkofer's rearguard action, the last stand of anticontagionism, was inadvertently in the vanguard. Subclinical contamination or infection of the body with the micro-organisms of diphtheria, cholera (in times of epidemic), pneumonia, typhoid, and tuberculosis was found to be more common than disease-producing infection. By 1900, tubercle infection, above all, was deemed 'ubiquitous' by many. Even the standard early-twentieth-century bacteriological handbook looked back on a decade of doubt, in which, 'the exclusive etiological significance of the pathogenic microbe seemed most threatened by the observation that it was found not only in cases of the infection it caused, but also in entirely healthy persons' (Gotschlich, 1904, p. 2).[5] Some of these observations of infected 'healthy persons' will be familiar to historians of medicine and public health as the so-called discovery of the healthy carrier, which indeed, from an epidemiological point of view, they were. All but forgotten, however, is the *etiological* problem they posed, that is, the problem for explaining disease causation (Mendelsohn, 1996).

The bacteriologists' etiological problem was, at the same time, an invitation to other experts. By their own methods, bacteriologists mapped a disjunction between infection and illness such that clinicians, social hygienists, pathologists, and soon enough eugenicists could claim equal expertise. Having once threatened to exclude consideration of heredity and environment from disease explanation, bacteriology now provided the most suggestive evidence for including them. This evidence was, as in Pettenkofer's self-experiment, differential susceptibility and, at its extreme, infection without disease. This paper is about the ways in which that disjunction, rather than the principle of necessary, specific cause, structured much of scientific medicine till the Second World War: its etiological thought, its understanding of the body, its expectations about and research on the role of heredity in common diseases, and eventually, in a new form lent by the methods and norms of biological science, its genetics. From the unravelling of deterministic bacterial etiology in the 1890s to the making of a non-deterministic, etiologically complex medical genetics in the 1930s, there stretched a continuous structure of what might be called the predisposed body, the body between danger and disease. Though the focus here will be especially on the body between infection and disease, its structure was more general. A wider lens would capture the history of the body's inequality before cancer, allergy, mental illness, and perhaps as well the intrinsic moral history of this inequality.[6]

As the object of medical knowledge, this body, this differential susceptibility or, as it was sometimes called, reactivity often went under the name of constitution or, in France, *terrain*. It displayed a continuity with the perception of individual difference that had always been central to medical practice and experience. Yet no one in 1890, not even a champion of the 'old medicine,' could have predicted it would become partly paradigmatic of the new scientific medicine. Bacteriologists called it into exact, scientific being almost despite themselves. Nor was differential reactivity among those theories and principles considered to be the scientific basis of modern medicine. It did not bear the signature of a Virchow or a Pasteur. Indeed, while individual difference had remained important to practitioners, it had been deemphasized since the eighteenth century by those academic physicians who sought a science of regular relations between symptoms, morbid anatomies or physiologies, and environments. Thus a wider purpose of the paper is to point toward an alternative periodization of scientific medicine in the nineteenth and twentieth centuries and to explore ways in which a field of knowledge may be structured by an unintended experimental and experiential effect, such as the body as a configuration of differential reactivity. The history of the medical sciences, accordingly, might be written as a succession of such body configurations.

There is certainly an irony in the argument that much of the impetus for twentieth-century constitutionalism came from the bacteriological laboratory, rather than simply from clinical medicine. The formula usually taken to sum up this period of medical history is, after all, 'physician versus bacteriologist' (Maulitz, 1979). Yet the irony arises only because bacteriology is ordinarily understood to have embodied germ theory. In fact, its legacy was plural, even contradictory. On the one hand, bacteriological methods first showed how to reduce health and disease to necessary, specific causes, be they germs, genes, vitamins, or, to use Archibald Garrod's famous phrase of 1908, 'inborn errors' of metabolism.[7] Identification of necessary rather than sufficient causes did indeed create a new and often powerful kind of diagnosis, therapy, and prevention. Above all, it made eradication thinkable (Temkin, 1977; Latour, 1988; Carter, 1991; Schlich, 1996). On the other hand, in other professional and practical contexts, bacteriological methods and results ushered in an era, indeed perhaps *the* era of constitution in European scientific medicine.[8] By 1931, Garrod could publish his rather less well-known, constitutional treatise, *The Inborn Factors in Disease*.[9] Even the deficiency diseases were soon to be understood according to more complex constitutional mechanisms (Sinding, 1991). Thanks again largely to the stimulus provided by the methods and results of bacteriology, the decades after 1890 were also, in many contexts, an era of pronounced multicausal disease explanation. Constitution and multicausality do not here refer to particular doctrines. Of these, there were many: philosophies of causation such as medical energetics, antiontologism, conditionalism, nosoparasitism (Diepgen, 1926; von Engelhardt, 1985); competing constitutional concepts such as constitution itself, predisposition, *terrain*, resistance, natural immunity, *Anlage*, diathesis, and ultimately the full-blown psychosomatic, neo-Hippocratic 'whole person' of interwar French and German holistic medicine. The topic here is what these all shared, rather than what distinguished them: that is, the more fundamental and concrete, longer-lived structure of the medical body as a thing of differential reactivity, of essential inequality.

Though stable, this body configuration was marked by a tension: differences in susceptibility could as well be acquired as inherited. Consensus on this question, the question of heredity, was remarkably unstable. Three periods may be distinguished. They form the parts of this paper. In the period 1890–1910, which saw the emergence of complex etiological thought and constitutionalism (section 1), heredity promised to explain much of the differential susceptibility that marked the behavior of infectious and other common diseases (section 2). In the second, overlapping period, 1905–1925, *acquired* constitution steadily took shape in laboratory and clinic. Acquired factors, whether immunological or physiological or biochemical, were ultimately seen to account for most phenomena previously attributed to innate difference (section 3). The third and last period began in the early 1920s. The introduction into medicine of methods and norms from biology worked a reversal of fortunes: individual organic difference in relation to external causes of disease now did indeed seem to be inherited as much as acquired (section 4).

Against this scenario of constitutionalism, it might be objected that the elements of knowledge of heredity and infection before 1940 are clear and familiar: 'germ theory,' or specific cause; August Weismann's continuity of the germplasm; persistent 'neo-Lamarckian' belief in inheritance of acquired characters; Mendel's laws and genealogy as a method for demonstrating not only the fact of inheritance, but also the dependence of traits on specific units of hereditary material, in other words, as a method of what would come to be called genetic determinism; transmission, of genes and of germs, as the unifying problem. This paper charts a rather different landscape. As Charles Rosenberg has remarked, like Darwinism, 'that other great benchmark in the history of biology, the germ theory, also played an ambiguous role in the shaping of nineteenth-century hereditarian thought. One might suppose that it would have served to reduce dependence on hereditarian models. Yet the reality was far more complex; for the vogue of hereditarian thinking—at its height between 1885 and 1920—coincided with the most enthusiastic and uncritical acceptance of the germ theory' (Rosenberg, 1974, p. 204). The story told here should revise the traditional historical understanding of germ theory shared by Rosenberg and solve the problem of its compatibility with hereditarian medicine.[10] Yet my account will I hope deepen his point about standard categories not capturing medical knowledge in this period.

Not even those hereditarian models, as we shall see in section 2, developed along the expected axes. Hereditarian medicine and hygiene of common diseases was neither neo-Lamarckian nor Mendelian in any meaningful sense. Inheritance of acquired characters was largely superfluous. Contrary to received wisdom, even French medicine and eugenics had no need of neo-Lamarckism. The mechanism of so-called pathological heredity was not acquisition and transmission of characters, but 'intoxication,' that is, physio-chemical injury to fetus, embryo, gamete, or germplasm itself. These forms of injury—sometimes clearly distinguished, sometimes not—were continuous with what biologists came to call teratogenesis and mutagenesis. And such intoxication was, as we shall see, a motif of pathology stretching far back into the nineteenth century and bound up with the origins of the theory of *dégénérescence* itself.

More generally, few academic physicians after about 1890 were content to rehearse the old common sense that 'like begets like.'[11] Instead, most hereditarian medical thought was based on current research. Yet little research on common diseases was Mendelian. Mendel's laws did not seem to hold for these.[12] And the method of pedigrees was treacherous for any disease with a recognized external cause or environmental component, as was the case with many common ailments. Thus genealogy often ceded to methods of pathology, bacteriology, clinical medicine, and epidemiology, all of which were sometimes augmented by statistics. Finally, as we shall see in sections 3 and 4, the history of heredity in scientific medicine was not dictated by the rise of eugenic and racial hygiene movements. It thus becomes possible to ask: Why heredity? and to imagine a cultural history of this aspect of medical knowledge in which all the explanatory work is not done by 'eugenics.'

FROM ETIOLOGICAL PROBLEM TO CONSTITUTIONS

German medicine witnessed an efflorescence of constitutionalism between the world wars (Klasen, 1984; Timmermann, 1996; Prüll, 1998). Yet the trend had been well underway since the 1890s. By 1922, the main German pathology review journal counted nearly 2,000 articles and books contributing to the subject of constitution and predisposition over the preceding thirty years.[13] French clinical medicine and pathology, too, were overwhelmingly constitutional. France is generally considered to have lacked constitutionalism, and the entries on constitution and diathesis in the volumes of the *Index Catalogue* of the Surgeon-General covering the early twentieth century do indeed list few French publications. But this was because French constitutionalism was organized around a term that did not index well abroad, 'le terrain,' and because it was far less theoretical than its German variant.[14]

Constitutional and hereditarian explanation of diseases such as tuberculosis would seem to have persisted simply because the medical old regime persisted beyond the bacteriological revolution of the 1870s and '80s. It is natural enough to draw a line of continuity from, say, Pasteur's opponent the clinician Hermann Pidoux at the Académie de médecine in the 1870s, with his famous phrase 'La maladie est en nous, de nous, par nous,' up to the clinician and dean of the Paris medical faculty Louis Landouzy, whose concept of inherited tubercular predisposition was rejected by the third-generation Pastorian Albert Calmette in 1910; likewise in Germany, from the clinician Friedrich Wilhelm Beneke's *Konstitution und konstitutionelles Kranksein des Menschen* (1881) to the widely read and translated constitutional textbook of the clinician Julius Bauer first issued in 1917. Twentieth-century constitutionalism would seem simply to be a chapter in the well-known story: clinic versus laboratory, physician versus bacteriologist.[15]

Yet to focus on undoubted continuities is to miss a fundamental change. For what gave the revived language of individual predisposition and constitution its power in the golden age of bacteriology and early immunology was precisely its ability to fill the new yawning gap that the bacteriologists had opened up in their own etiological explanations—the 'dark something of predisposition,' in the words of the bacteriologist who first recorded widespread asymptomatic meningococcus infection (von Lingelsheim, 1906, p. 488). The infected body, so recently lit by microscopy and bright stains, illuminated by the clean, pure-culture demonstrations of specific cause and by the decisive effects of immunological interventions (anthrax and rabies vaccines, diphtheria antiserum) was now darkened again by the bacteriologists despite themselves. The obscurity of terms like predisposition and constitution now seemed appropriate and unavoidable, rather than archaic. Moreover, the obscurity could not be dispelled by the new immunology. We may easily imagine natural immunity filling the explanatory gap. They, however, did not. The vast majority of research on natural immunity concerned the mechanisms themselves, rather than fluctuation of their function in individuals or differences among individuals, races, species. That is, relatively few studies addressed etiology, differential susceptibility. Moreover, what research there was tended only to reveal lack of fit between, say, bactericidal power of the blood (alexine content) and observed resistance to disease.[16] Thus even devotees of laboratory medicine found themselves unable to dispense with those darker, as yet irreducible constitutional concepts.

The darkness of predisposition was not only figurative. Predisposition had no ontology. It was at first almost purely an operational concept, inseparable from laboratory practice, that is, from animal experiment and from bacteriological diagnosis in clinic and epidemic. Differential predisposition, or susceptibility, named the differential outcome of experimental infections and of everyday ones insofar as they were revealed by bacteriological testing. It was the other side of insufficient bacterial causation. The bacteriological revolution in medicine was thus equally its constitutional revolution. Contemporaries sensed this when they located the origins of constitutional medicine, ironically enough, in research on infectious diseases.[17]

Constitution was less revived than remade. To be sure, both the older tradition of clinical sensitivity to individual difference and the newer 'scientific' methods of gathering information at the bedside, as well as the typologies of body and temperament derived from those observations, were not lost.[18] Eventually, for some clinicians and pathologists, these types and temperaments filled out the empty placeholder of 'predisposition.' Yet by and large, to paraphrase the Pastorian Émile Duclaux, homonyms were not synonyms thirty years apart (Duclaux, 1896). The old constitutionalism tended to be multiple and pathological: diverse constitutional anomalies and constitutional diseases; constitutions or diatheses of the lymphatic, arthritic, syphilitic, alcoholic type, themselves often being generalized pathologies, but also predisposing to further ailments. The new constitutionalism tended more toward the singular and neutral: 'very generally the soil (*le terrain* of the French) upon which at any given time under certain conditions a disease developes or does not develop' (Martius, 1914, p. 65).[19]

The difference is less subtle than it may seem. In the old medicine, constitution named a tendency for an illness to appear in individuals of a given family at a certain age, or in individuals displaying a certain body type or temperament.[20] Thus constitutional differences, in their very perceivability, were classificatory: part of a nosology or genealogy. Alternatively, and especially in the study of epidemic fevers up to the time of Edwin Chadwick, such differences could be guessed from epidemiological experience: individuals exposed to the same 'exciting cause' reacted in different ways (Hamlin, 1992). Late nineteenth-century bacteriological experiment, by contrast to both of these by-then dwindled traditions, was the first instance in which external disease causes were controlled enough that such bodily difference could be definitely known or, in effect, produced. It could now be apprehended independently of actual observation and classification of bodily difference, clinical or pathological. And it could be apprehended independently of the myriad unknowns that had once made its apprehension through epidemiological experience uncertain. Thus the new constitutional thinking, like the old epidemiologically based doctrine of predisposing causes, would take the form of experiment rather than classification. Indeed, their characteristic mode of expression was the thought-experiment. Compare two examples, one taken from the leading mid-nineteenth-century British medical text, the other from the leading twentieth-century constitutional text:

> Of a score of persons exposed to the same noxious influence—to the combined influence of wet and cold during a shipwreck, for example—one shall have catarrh, another rheumatism, a third pleurisy, a fourth ophthalmia, a fifth inflammation of the bowels, and fifteen shall escape without any illness at all. A man shall do that with impunity to day, which shall put his life in jeopardy when he repeats it next week.[21]
>
> —Sir Thomas Watson 1844

> If for example among one hundred fully healthy soldiers generally of the same age and strength, under identical external conditions, two become ill of an acute heart insufficiency, two others of a grave neurasthenia, then obviously the first two had a less robust heart, the second two a less robust central nervous system (Bauer, 1917, p. 38).
>
> —Julius Bauer 1917

Predisposition is not here established by reference to a classificatory system of types, temperaments, family patterns. Instead, it names a differential outcome. Bodily difference has hardened from Watson to Bauer: the one thought-experiment is meant to illustrate the importance of the body's changing condition, the other to illustrate intrinsic inequalities. In the same way that the relative robustness of Bauer's hundred men did not exist independently of their regimented and statistical circumstances, the body as a thing of intrinsic and exactly knowable difference, relative to external conditions of health, hardly existed before bacteriology.

There are several advantages to looking at the origins of twentieth-century constitutionalism in this way. First, the constitutional and new causal schemes of the 1890s are thereby not mistaken for sectarian, clinical, or merely 'speculative' movements (Diepgen, 1926; von Engelhardt, 1985). Instead, they become recognizable as expressions of a massive new etiological problem shared equally by bacteriological, clinical, and pathological investigators and as visible in laboratory and clinical practice as in theoretical writings. Second, this approach avoids designating constitutionalism a 'reaction' against bacteriology or a 'pendulum swing'—turns of phrase that appealed in a common-sense way to contemporaries but that fail to explain the course of events.[22] Reactions do not just happen, at least not successfully; nor does scientific opinion simply swing. Rather, a regime of knowledge must become weak in a specific way. And space for new statements must be made, else they remain irrelevant, or unthinkable. This was accomplished when bacteriology itself inadvertantly created a common problem of etiology. Thus the historical mystery is dispelled from the fact that in 1890 there was no hint in scientific medicine, whether in or out of bacteriological research, that the following decade would launch a half-century of constitutionalism to vie with the model of specific cause. The third advantage is that this approach also enables asking, and perhaps answering, the question: Why constitution, why differential reactivity at all? Other eras in the history of medicine and its sciences have witnessed other governing ideas, other dominant configurations of the body. An important part of the answer is, again, that the medical science that was then methodologically strongest, bacteriology, quite suddenly demanded predisposition in the structure of etiological explanation.

Workers in the dominant fields of medical bacteriology—bacteriological and serological diagnosis, mechanisms of specific immunity, serotherapy and vaccine development, chemistry of serological reactions—did not, in fact, need to explain etiology and pathogenesis. These were not their problems. But these were precisely the problems of pathology, clinical bacteriology, clinical medicine and its myriad subspecialties, and even of social hygiene and eugenics insofar as they dealt with the susceptibility of groups and populations to disease. It was one thing for surprised scientists to name the gap in their etiologies 'predisposition', quite another to give it an ontology and a mechanism, much less to demonstrate a hereditary component. Many of the men who did this were early contributors to bacteriological knowledge. Some were outspoken critics of bacteriology. All were deeply impressed by the disjunction of infection and disease, by the differential susceptibilities and organic reactivities that bacteriological method itself had given exact, controlled form and vast extension. Above all, their professional identities were far too diverse to permit treating their work as an episode in clinical medicine. In France, representative figures were Charles Bouchard, the leading pathologist of the late nineteenth and early twentieth centuries, and his students Albert Charrin, professor of general and comparative pathology at the Collège de France, and Georges-Henri Roger, dean of the Paris medical faculty (1917–30); Fernand Widal, discoverer of the agglutination reaction for typhoid bacilli but also a leading figure in functional diagnosis, and his student Fernand Bezançon, who held the first chair of bacteriology at the Paris faculty (1919) and moved on to hold the chairs of clinical medicine (1925) and tuberculosis (1935); Louis Landouzy, a leading French phthisiologist in the late nineteenth century and dean of the Paris faculty until his death in 1917; and various members of the 'Lyon school' of experimental pathology and bacteriology, which remained associated with Bouchard and his students throughout this period, notably Jules and Paul Courmont and Saturnin and Fernand Arloing (both *père et fils*).

Their counterparts in the German-speaking world, where the better established nineteenth-century research traditions in pathology (Virchow) and hygiene (Pettenkofer) made for a more conflicted relationship with bacteriology, were more scattered: Ottomar Rosenbach, founder of functional diagnosis and better remembered for his criticism of bacteriology; Ferdinand Hueppe, an early Koch student who became critical of the Koch school, later professor of hygiene at Graz and a leading race hygienist; Adolf Gottstein, a contributor to the bacteriological literature who went on to write the first modern German textbook of epidemiology and to become director of Prussian medical affairs after World War I; Friedrich Martius, clinician who came to be regarded as founder of constitutional pathology; Emil von Behring, who won the first Nobel Prize in medicine for his discovery of antitoxic immunity and who used his authority to establish the importance of 'ubiquitous' tubercle infection and a predispositional doctrine of tuberculosis in opposition to the teachings of his mentor Koch; Wilhelm His, Jr., son of the anatomist, holder of Ernst von Leyden's chair at the Charité after 1907, dean of the Berlin medical faculty after 1918, and rector of the university after 1928; as well as various pathology professors such as Otto Lubarsch and Hugo Ribbert.

Not Pasteur or Koch, but writers such as these articulated the principles of etiology and pathogenesis that would characterize much of scientific medicine in the first half of this century. Their motto was less that of necessary, specific causation—'no tuberculosis without the tubercle bacillus'—than that any given pathology was the result of multiple causes,[23] and that one and the same cause can produce different effects; the same effect, have different causes. The processes of disease had no more to do with the specificity of external causes than with the specific irritability or reactive capacities of organs and tissues and with intoxications produced by the body itself, the individualities of metabolism in sickness or in health. Here, there was a long-standing difference of style between French and German pathology: the one, humoral (Bernard and Bouchard); the other, solidar (Virchow). German physicians, in keeping with this solidar bias, emphasized irritability (*Reizbarkeit*, *Erregbarkeit*), which derived as much from plant and animal physiology as from cellular pathology (Virchow, 1858).[24] The mechanism of irritability entailed quantitative, non-'ontological' causation. That is, causes were not things that imparted qualities. Rather, the relation of irritant to irritated was release of potential energy. Hueppe and others derived this understanding of causation from the law of conservation of energy of Robert Mayer and Hermann von Helmholtz and from its precursor ideas in the sensory physiology of Johannes Müller (Hueppe, 1894). There was an obvious parallel to the critique of materialism and causation in the 'energetics' movement associated with the physical chemist Wilhelm Ostwald.[25] French physicians, in keeping with their humoral bias, meanwhile focused on autointoxication, which referred to injurious metabolic productions of both the well and sick organism.[26]

Finally, the other important mode of predispositional etiology was 'hypersensitivity.' Pharmacological and toxicological study had long displayed 'idiosyncrasy,' that is, exaggerated individual reactivity to doses usually without effect on the body. While developing a workable serotherapy out of his initial observation of antitoxic immunity and its passive transfer, Behring coined the term hypersensitivity in 1893 to designate the condition of those who reacted to serum injections that were harmless in others. After 1900 these kinds of observation yielded the new field of research on anaphylaxis and allergy.[27] Hypersensitivity, anaphylaxis, and allergy were not understood only as diseases or pathological states, as in 'serum-sickness' or 'hay fever.' They were also, as Behring and others made clear, modes of specific predisposition to disease (von Behring, 1914, 1914a).[28]

Bouchard and his students summed up much of this when they pronounced their 'fundamental law': 'It is the organism and not the microbe that makes the disease.'[29] To be sure, no one stopped referring to the typhoid bacillus or the tubercle bacillus as such. Yet while public health bacteriology generally dealt with notified cases and with outbreaks and epidemics, in which symptoms and postmortems were indeed more likely to conform to expectations forged in the laboratory, the majority of phenomena in pathology, clinical bacteriology, and clinical medicine did not fit the categories. The first, stunning example of this difference came in the years after 1890 when the aggregate result of bacteriological diagnoses for diphtheria fell from 95 percent to 60 percent. The reason: until that time, throat cultures had been made only from 'specially selected [*ausgesuchte*] typical diphtheria cases, by which bacteriologists and pathological anatomists wanted to satisfy themselves as to the constancy of the finding of diphtheria bacilli' (von Behring, 1901, pp. 102–13). It also soon became apparent that one and the same pathological or clinical effect could be produced by a variety of microbial agents: for example, not only cholera bacteria, but also the colon bacillus, normal inhabitant of the human intestine, could produce the characteristic 'rice water' discharges of cholera; the microbes of tuberculosis, typhoid, and pneumonia could all cause meningitis. Likewise, one and the same microbial agent could produce the most various pathologies and symptoms: the diphtheria bacilli could produce not only the typical pseudomembrane in the throat, but also paralysis or acute intoxication of the blood; pneumococci could produce typical pneumonia, but also septicemia, suppurations, meningitis.[30] Thus it is no wonder that specificity was soon given bounds, sandwiched within those more general 'two laws' that the same cause can yield different effects; the same effect, have different causes.[31]

Causation was more self-consciously and formally discussed in Germany than in France. As important as the distinction between sufficient and necessary causation was in certain contexts, German scientific medicine was most closely associated with—and seems partly to have stimulated—wider scientific movements to substitute purely quantitative for qualitative ('ontological') causes; to define causation in terms of as many factors as possible; or to give up the language of causes altogether in favor of that of conditions. Medical scientists only displayed their consonance with the physics and biology of their time when they cited the conditionalism of the physiologist Max Verworn; or the multicausal thinking of the founder of experimental embryology Wilhelm Roux; or the physicist Ernst Mach on the need to do away with that 'fetishism,' the concept of cause and effect.[32]

A first step had come already in 1887 when Hueppe proposed that pathogenic microorganisms be referred to as disease agents (*Krankheitserreger*) rather than disease causes (*Krankheitsursachen*).[33] The suggestion was widely taken up, and 'Erreger' became standard terminology in German bacteriology and medicine. A decade later Gottstein composed the first modern German textbook of epidemiology using the term *Konstitutionsreiz* ('constitutional irritation') instead of external cause.[34] Here, he also proposed an influential formula for disease causation: C/p, where C stands for constitutional strength

(*Konstitutionskraft*) and p for the parasite's degree of pathogenicity. Disease results whenever the ratio is less than one. Martius adapted this formula for all pathological processes. Both versions were frequently reproduced in early-twentieth-century German medical writings.[35] It may be wondered why. The formula is so simple as to seem self-evident. Yet read as a historical artifact, it signals a shift in etiological thought. C/p expresses a different kind of multicausality from, say, Max von Pettenkofer's much-cited model of cholera and typhoid, in which disease results only when factors X (germ), Y (condition of environment, especially ground water levels), and z (individual predisposition) come together.[36] Where XYZ denotes the combination of qualities, C/p is a purely quantitative expression of disease causation. It codified, in its very form, the body as a thing of differential reactivity rather than an object whose variations, whose individualities, were known by their place in a nosological system. For any ratio greater than one, it was a mathematical expression of infection without disease. And wherever in experience or experiment p could be assumed or made constant, C designated not only difference, but, in a newly powerful way, inequality. As Wilhelm His said in his inaugural address as rector of the University of Berlin, *Über die natürliche Ungleichheit der Menschen*, whose title summed up an age of medicine, biology, and society: 'The science of disease becomes ever more constitutional pathology: the environment does indeed affect the body, but the nature and degree of the effect depend on the innate predisposition [*Anlage*]' (His, 1928).[37]

In France, so much was the first half of this century an era of the *terrain* that when in the 1930s bacteriology came especially under criticism from clinical, pathological, and Bernardian physiological points of view, the Pastorian response was not to belittle the *terrain*, but to claim it for their own. Chief among critics of bacteriology in the name of Claude Bernard was the immunologist and blood-transfusion expert Arnault Tzanck (Tzanck, 1933).[38] In response, Pasteur Vallery-Radot, editor of his grandfather's collected works and a leading clinical investigator in his own right, presented much evidence from the classic papers of the 1870s and '80s demonstrating the essential place of the *terrain* in the thought and research of the master himself, though Pasteur never used the term in its medical sense, which seems first to have emerged strongly in the 1890s, for the reasons outlined above (Vallery-Radot, 1933). What is most interesting about the exchange between Vallery-Radot and Tzanck is not the conflict, but the closeness of the parties.[39] The two men were colleagues and agreed on the importance of the terrain.[40] Tzanck cited Vallery-Radot in his work on renal pathology and anaphylaxis. And he published his further Bernardian reflections on immunity, anaphylaxis, and the *terrain* in Vallery-Radot's *Revue d'immunologie*, a new journal which can be taken as emblematic of the common constitutional orientation of the various traditions of French academic medicine in this period (Tzanck and Cottet, 1934; Tzanck and André, 1936).[41]

These developments were not without practical import. They often proceeded apace with new developments in diagnosis and therapeutics, developments which might best be summed up as functionalism. The clinician Rosenbach, who took what he called an 'evolutionary and managerial-technical [*betriebstechnische*] perspective,' pioneered functional diagnosis in Germany, as did the bacteriologist Widal in France. Correct therapy was not necessarily etiological, in Rosenbach's words, but better 'function-regulating,' that is, regulating the abnormal, disease-producing irritability of the organism or its parts (Rosenbach, 1903, pp. iv, 3, 59).[42] When in his presidential address to the twenty-second French medical congress in 1932, Fernand Bezançon proclaimed: 'We have left behind the notion of specificity, we have left behind the etiological period: we have arrived at the physiopathological mechanism, at pathogenesis,' he described not only the intellectual but also the practical *terrain*-orientation of medicine in his country. 'The notion of cause, as important as it remains, often recedes nevertheless into the background; we can make a representation of the disease, we can even treat it scientifically, although we are ignorant of its cause ... Instead of thinking: specific disease, we think: specific reactional modalities.' Thus the future of therapeutics lay not in 'a specific medication for each disease, but varied medications against the reactional modalities presented by the patient at various stages of the illness'—not in the specific vaccinations of Pasteur, but in the 'grandes médications' such as the desensitization to shock, or peptonic desensitization, of Pasteur Vallery-Radot (Bezançon, 1932). These, note well, were not the words of a clinician, but of the first holder of a chair of bacteriology in France.

The Promise of Heredity

Nothing in constitutional thinking, multifactoral etiology, or the differentially susceptible body guaranteed a role for heredity, for inherited difference. Two older traditions of medical observation and experience, however, kept the potential etiological significance of heredity constantly, and literally, in sight: family disease and racial immunity. That certain diseases ran in families was a quotidian fact of medical practice.[43] That certain peoples were not affected by diseases that attacked others was a quotidian fact of medicine in the tropics and colonies. The former observation gained in sweep with the rise of psychiatry in the nineteenth century and with the merging of familial pathologies and wider social ones under the rubric 'degeneration.' The latter observation gained in urgency with each step of European colonial expansion in the second half of the century and with the concomitant emergence of race as a scientific category (Curtin, 1989). Both were transformed by bacteriology and by Darwin's theory of evolution by natural selection.

The expectation that the rough concept of racial immunity would soon be made an exact object of scientific experiment arose from the very first problem to beset laboratory research on human infectious diseases. This was the problem of the animal model, the problem of finding a susceptible animal species in which to reproduce the disease by inoculation. It was soon found that different animal species responded differently to a given infection—some not at all, some with partly or entirely different clinical manifestations. These facts of species susceptibility and immunity differed from those of the old racial immunities in two crucial ways. First, they were the product of controlled laboratory experiment rather than gross and usually unsystematic epidemiological experience. Second, the animal species used in bacteriological experiment—mice, rats, guinea pigs, rabbits, frogs, birds, dogs, sheep—were well-defined, uncontroversial biological, hereditary entities whereas races were held to be ambiguous products of nature, culture, and climate. Thus manifest species differences in the bacteriological laboratory did perhaps more than anything else to heighten the expectation that more subtle racial, familial, and individual inherited differences of susceptibility to infectious diseases would be identified.[44]

Outside the laboratory, at the same time, the tradition of racial immunity was transformed intellectually through the idea of natural selection. As early as the 1880s, pathologists involved in bacteriological research theorized that racial differences in susceptibility arose from the selective effects of the diseases themselves.[45] The wide-spread notion of acclimatization had served the same explanatory purpose, especially in France (Livingstone, 1987; Osborne, 1994; Harrison, 1996). Yet where acclimatization codified a continuity between physiological and hereditary modification of the organism—between acquired and natural immunity—selection kept these distinct. Selective theories represented the observed racial and regional differences in susceptibility as purely inherited ones.

Early selective theories remained plausible enough to be taken seriously in the experimental immunological work of Paul Ehrlich and others in the 1890s (Ehrlich, 1892, p. 32). Yet they proved difficult to submit to experiment or to systematic epidemiological study. They quickly reached their fullest development in the work of Adolf Gottstein between 1893 and 1903 on the epidemiology of children's diseases such as diphtheria, work which was still being cited thirty years later as exemplary (Rott, 1934). Most writers had assumed that the cycles of recurrent children's diseases such as diphtheria reflected the waxing and waning of acquired immunity in a population: diphtheria rates would decline as the number of susceptibles decreased through death or acquisition of specific immunity and would rise again as a new generation was born and reached the age of prime susceptibility. Such a model might well fit the short cycles of a disease such as measles, Gottstein argued. But it must fail to account for the far longer cycles of diphtheria. These cycles, on the order of many decades, were explainable instead by assuming that the disease declined by creating a naturally resistant population through weeding out of innately susceptible individuals over several generations. Then, as the disease itself reached ever lower levels, the conditions for its own slow resurgence reappeared: in the near absence of selection by the disease, a new population of innately susceptible individuals arose and, at a certain lag, the disease returned (Gottstein, 1903).

The two most important diseases displaying familial patterns were tuberculosis and syphilis. Though they belonged together in the nineteenth century as family diseases having something, but not everything, to do with heredity, they differed nevertheless in important ways. Where for syphilis a non-hereditary mode of transmission had always been clear—sexual contact—consumption was at most loosely associated with contagion before Koch's isolation of the bacillus in 1882. Likewise, while children of syphilitic parents were often born with unmistakable signs of the disease, children of tubercular parents were not born clinically tubercular, though some physicians, ever since Hippocrates, did claim to be able to identify a consumptive constitutional 'type.'[46]

These differences were confirmed by turn-of-the-century clinical and laboratory research on the possibility of direct transmission of infection from parent to child. The spirochaete, identified in 1905, was shown to pass from mother to child, through the placenta, or from father to mother to child.[47] Though many physicians continued to speak of hereditary or heredosyphilis, they usually meant direct transmission of infection, which they themselves acknowledged had nothing to do with true heredity.[48] Meanwhile, tubercle bacilli were found capable of passing through the placenta, but only in extremely rare cases. Thus if tuberculosis displayed familial patterns, these would clearly reduce either to actual contagion in the home or to inherited predisposition. Methodological difficulties beset the attempt to demonstrate hereditary predisposition. Tuberculosis was a chronic disease often of long latency. It was extremely widespread. And it was casually communicable, especially in the close quarters of the home. A pedigree of tuberculosis could as well reflect family contagion as family predisposition.[49]

The argument for heredity in tuberculosis would have gone nowhere but for a fortunate historical circumstance. Statistics in the nineteenth century was in large part the study of deviation, that is, 'error' and later variation. With the man who coined the word 'eugenics,' Francis Galton, the rise of statistics became coupled with the study of one particular kind of variation, namely hereditary human variety. The bearer of Galton's eugenic banner into the twentieth century, Karl Pearson, also founded mathematical statistics and strove to revolutionize medicine and biology along 'biometric' lines. Thus shortly after 1900 the clinicians, whose lack of scientific method made them the very butt of the biometricians' jokes, suddenly found on their side (the side of 'soil' over 'seed') the mathematical statisticians and keenest methodological innovators of the day. Starting in 1907, Pearson and his associates published a series of statistical studies on heredity and pulmonary tuberculosis. The main methodological innovation was to use husband-wife correlations as a control for the factor of familial contagion in parent-child correlations. The latter were found to be significantly higher than the former, thus suggesting hereditary difference in susceptibility. Yet even here, though method and mathematics might be sound, the clinical data and patient histories on which the whole analysis rested were open to question (Pearson, 1907; Pearson and Pope, 1908; Goring, 1909; Greenwood, 1909).

If epidemiological and statistical demonstration of the role of heredity in tuberculosis daunted all but a handful of mathematically armed researchers, clinical and post-mortem investigation of bodily peculiarities of the consumptive, whether anatomical or functional or chemical, remained an avenue open to practically all physicians.[50] Here, the difference of style between French and German pathology accentuated a difference in emphasis on heredity. Despite the ever more international culture of scientific medicine in the twentieth century, and despite the rise in both countries of functional over anatomical thinking, German pathology stayed remarkably true to its morphological tradition, French pathology to its chemical and physiological one. Partly because it was intuitively easier to accept that anatomical abnormalities were stable and inherited than that chemical ones were, German constitutionalism was generally more hereditarian than its French counterpart.

But how were such constitutions and predispositions produced? Did they result partly from the external causes of disease themselves, and if so, how could they be hereditary? These were the most theoretically and practically urgent questions raised by family disease patterns. They could be addressed, moreover, through a wide variety of methods—clinical, pathological, bacteriological. This is perhaps the most unexpected and interesting chapter in late-nineteenth and early-twentieth-century research on heredity and disease, because it runs against the conventional wisdom that well into the twentieth century much hereditarian medicine and social thought—and French eugenics in particular—were 'Lamarckian' or 'neo-Lamarckian' and counter to Weismann's teaching of the absolute distinction between the acquired and the innate, the soma and the germ.[51]

Contemporary medical and sociological writings do indeed seem replete with obvious examples of belief in inheritance of acquired characters. The alcoholism or the tuberculosis or the syphilis of the parent is visited on the child by an alteration of the germ cells. Inheritance is held to depend strongly on the condition of the parents at the time of conception.[52] Yet even in most French examples, there is no invocation of inheritance of acquired characters, nor of Lamarck. Indeed, the leading early-twentieth-century French medical writer on heredity, the pediatrician and eugenicist Eugène Apert, rejected the doctrine of inheritance of acquired characters in all but a highly nuanced and exceptional form, of theoretical interest only and of no relevance to alcoholism, tuberculosis, syphilis, and other problems of eugenics.[53]

A simpler and more plausible mechanism of pathological heredity was at hand: namely, the mechanism of 'intoxication.' No one needed to argue that tubercular or alcoholic pathologies, once produced by external causes, could then be passed on as characteristics to descendents. Far easier it was to postulate that those external causes in chemical form in the parental body—alcohol or microbial toxins—injured the germ cells just as they injured other cells of the body. When this mechanism has not been lumped by historians with inheritance of acquired characters, it has been considered 'closet Lamarckism,' an admission at least of external influence on the hereditary material.[54] Yet, as contemporaries were well aware, Weismann himself admitted this possibility. His leading medical advocate, the pathologist Ernst Ziegler, had suggested as early as 1886 that injury to germplasm by alcohol and other toxic substances was a likely mechanism for the production of hereditary pathologies.[55] In his opus magnum of 1902, *Vorträge über Descendenztheorie*, Weismann compared the possible injurious effects of alcohol or infection on the germplasm to those of climate and nutrition, 'environmental influences' to which he devoted an entire lecture (Weismann, 1902, ii, pp. 78–79 and ch. 31). It is later generations that have confused the absolute distinction between soma and germplasm with an absolute distinction between germplasm and environment. Contemporary medical writers attempting to reconcile external disease causes and familial pathological patterns knew well they could invoke Weismann and did not need Lamarck.[56]

Nor were those writers confused about the scope of the hereditary. By 1900, whether in France or Germany, most had a concept of true heredity yet often chose consciously to use 'hereditary' as a broad category for representing the transmission of disease through reproduction. Thus hereditarian medicine was as much a medicine of reproduction and development as of the 'true heredity' that most of its practitioners were able to distinguish. In an influential paper of 1911, for example, an emerging leader of constitutionalism in pediatrics Meinhard von Pfaundler listed three possible causes of inherited diathesis: 'true heredity' (*echte Vererbung*); *Blastophthorie*, meaning injury to the germ cell; and *Embryophthorie*, or injury to the fertilized egg.[57] Blastophthoria, a common term for injury to the 'germ' in both French and German medical writings, had been introduced by the Swiss pyschiatrist and temperance movement leader August Forel.[58] Others, such as Louis Landouzy and the early leaders of French puericulture and eugenics Adolphe Pinard and Eugène Apert, worte simply of heredo-intoxication. Heredo-intoxication, though often discussed with heredo-contagion, was recognized as a mechanism of constitutional pathology, a *terrain* idea. Already in 1891 Landouzy, who was to be vice-president of the French Eugenics Society at its creation in 1912, postulated that heredo-intoxication was the mechanism by which a tubercular *terrain* was produced. Koch having revealed his tubercle culture extract 'tuberculin' in 1890–91, France's leading phtisiologist as well as experimentalists of the Bouchard and Lyon schools, such as Charrin and Courmont, then investigated the possible effects of metabolic products of the tubercle bacillus, released in the infected maternal body, on gamete, embryo, and fetus.[59] In Germany, the ambiguity of such research and of terms like heredo-intoxication and blastophthoria, which included both heritable and non-heritable injury to the germ cell, was eventually dispatched. The bible of German eugenics, *Grundriß der menschlichen Erblichkeitslehre und Rassenhygiene* published by Erwin Baur, Eugen Fischer, and Fritz Lenz in 1921 ('Baur-Fischer-Lenz'), treated environmental influence on hereditary material only, for which Lenz had coined the term 'Idiokinese'.[60] The 'control of idiokinetic noxious substances,' or 'genetic poisons' as they came to be called, held a place second only to selective measures in German race hygiene.[61] In sum, whether injury to the reproductive process was construed narrowly or broadly, there was no question of inheritance of acquired characters. Some experimenters, such as Albert Charrin and his students, were indeed interested in that possibility. Others aimed doctrinal polemics at Weismann in particular and biologists in general. Yet such work was controversial, such polemics marginal to the robust and ubiquitous notion of intoxication of the germplasm.[62]

In the end, the clinically based belief of generations of physicians and psychiatrists that alcohol could produce 'variations'[63] in the germ cells, as much as it rings of medical Lamarckism, was to be considered experimentally confirmed after World War I by Agnes Bluhm at the Kaiser Wilhelm Institute for Biology. Upon nomination by the geneticist Carl Correns, co-rediscoverer of Mendel and director of the Institute, Bluhm received the Silver Leibniz Medal of the Prussian Academy of Sciences for her demonstration of 'injury to the germplasm, more exactly, to the sex chromosomes of the male' through exposure to alcohol (Bluhm, 1930).[64] The seemingly neo-Lamarckian, French doctrine of heredo-intoxication found its scientific apotheosis in German classical genetics. Bluhm's demonstrations followed many biologists' attempts at mutagenesis using various chemicals and forms of radiation. The first widely acknowledged success had come in 1927 when H.J. Muller used x-rays to induce mutation in drosophila (Dunn, 1965). Thus the medical and eugenic idea of heredo-intoxication, far from being a rut of old, physicians' assumptions, was part and parcel of that aspect of the post-Weismannian conception of heredity that held open the possibility of environmental influence on heredity or, in the biologists' language that Muller used in the title of his famous paper, artificial mutation of the gene. In any case, neither neo-Lamarckism nor the mutation theory of Hugo de Vries, indeed no biologist's theory of inheritance at all, sustained and organized these medical expectations. They arose, rather, from the need to explain the appearance in time of hereditary pathologies. And they belonged to the pervasive medical topos of intoxication.

It is worth adding that French hereditarian medicine, usually thought of as the last Lamarckian and anti-Mendelian backwater, reserved in fact an important place for Mendelism.[65] This was for precisely the same reason that led French *biologists* to resist Mendelism. In France, a distinction was often drawn between normal heredity and exceptional, Mendelian heredity, that is, between hereditary mechanisms that had to account for the development of the complex traits that made up the organism and exceptional, simple characters that were seen to follow Mendel's laws.[66] Prominent among these exceptional, Mendelian characters were certain familial abnormalities and diseases. Thus the same understanding of heredity that prevented many French biologists from embracing Mendelism made Mendelism the special province of the French physician, at least insofar as he was concerned with those rare familial diseases and abnormalities. If these physicians elaborated their own principles for understanding more complex 'morbid heredities,' such as hereditary predispositions, this was not out of ignorance of or resistance to Mendelism. The point is thus not only that heredo-intoxication had little to do with neo-Lamarckism, but more generally that important axes of hereditarian medical thought and controversy were not defined according to Lamarck and Mendel.

If thought and research on heredo-intoxication, at its height between 1890 and 1920, was more developed in France than elsewhere, this was due, in the first instance, to the pronounced physiological and humoral orientation of French experimental medicine, by contrast to the morphological and cellular tradition of German pathology. Intoxication was, of course, a chemical phenomenon and, as we have seen, was the motif of French pathology in the era of Bouchard. The well-known French national anxiety of depopulation and the special development of puericulture and pediatrics in France also reinforced the medical perception of mother and embryo, mother and child, as a single unit of health or sickness.[67] Thus not heredity, but reproduction and 'prenatal pathology,' encompassing morbid heredity, germinal pathology, and pathology of the blastula, embryo, and fetus, was the overarching conceptual field. A notable contributor to this field, itself perhaps reflecting the national trauma of *denatalité*, was the peculiar development and prestige of experimental teratology in France in the last quarter of the nineteenth century. For physicians concerned with 'heredity' and disease, the teratologist Camille Dareste was a far more important source than was Jean Baptiste de Lamarck.[68] Neither the Chevalier nor his namesake doctrine is referred to in writings of Pinard and Apert before 1920.[69] Later invocations of Lamarck, by Apert himself and other French eugenicists in the 1930s, were couched in the telling neologisms 'lamarckisme social' and 'somato-lamarckien' by physicians who had never needed the doctrine of inheritance of acquired characters but who now needed to differentiate 'positive' French eugenics from its 'negative' German and Anglo-American forms and who thus rechristened old doctrines of germ-cell intoxication such as Forel's blastophthoria according to the name of the great French naturalist.[70] 'Lamarckism' could be, in short, many things.

The most interesting source of late-nineteenth and early-twentieth-century medical preoccupation with what Landouzy called ancestral intoxication lay in the very nature of the topos of *dégénérescence* itself. 'Degeneration,' Western civilization's self-diagnosis in the latter half of the nineteenth century, is remembered primarily as the opposite of development. It was the form that collective anxiety was bound to take in the great age of evolution and progress marked by the publication of the *Origin of Species* in 1859 and in France by the subsequent revival of Lamarck (Nye, 1984; Chamberlin and Gilman, 1985; Pick, 1989). Yet the foundational text of degeneration, the *Traité des dégénérescences* published by Bénédict-Augustin Morel in 1857, though sometimes read as evolutionary and arch-hereditarian, was nothing of the kind. The pious *aliéniste*'s treatise, which would do much to shape several generations of psychiatry, anthropology, medicine, literature, and perhaps politics, began with a meditation on the manifest harmony between the *Book of Genesis* and the science of natural history. Morel's 'type primitif' is not the ape-man, but Adam. If one individual, one race was inferior to another, this was not through descent or reversion. Rather, according to the old climatological understanding of human diversity, such inequality was the product of environment. The natural history of Buffon, Gall, Cuvier, Serres, and Flourens taught Morel to distinguish 'between *natural varieties and morbid varieties* in the human species' (Morel, 1857, p. xiii), The morbid ones were his special subject, what he called the *dégénérés*. Thus the *Traité des dégénérescences* is not organized around time, heredity, evolution, and decay, but around human variety in the milieux—injurious or salubrious—which Morel understood to make it. And since the external causes of morbid variety were, above all, poisons, whose exact effects on the animal organism Morel was pleased to discover in the experimental physiology of his friend Claude Bernard (Morel, 1857 p. xiii), the *Traité* is a book about intoxication: chapter 1 on alcohol; chapter 2 on vegetable and mineral poisons such as opium, tobacco, lead; chapters 3 and 6 on intoxication by alteration or insufficiency of alimentary substances; chapter 4 comparing diverse modes of intoxication among various races and peoples; chapter 5 on curative effects of racial crossing; the seventh and last chapter on the telluric intoxications cretinism and paludism (malarial fevers), including those produced in man-made environments such as the intoxications of factory and mine and the 'malaria of big cities.'[71]

To be sure, the *Traité* is not only a book about intoxication. What distinguished it from countless contemporary hygienic and medical treatises on the chemical dangers of environment, and thus what made the *Traité* historically so important a book, was that Morel took as the object of these miasma and toxic agents not individuals and social groups (Chadwick's laboring population, Villermé's rich and poor), but natural-historical human varieties. A biological rather than medical or social pathology had become possible. The new word *dégénérescence* named this possibility. Inherited dangers and environmental dangers became one problem. From Morel in 1857 to Agnes Bluhm in 1930, there then stretches a whole history of the body in chemical milieux, its modes of generation and degeneration and their instabilities, a history intertwined with the better-known social history of intoxication and having more to do with pathology, psychiatry, toxicology, pharmacology, cytology, embryology, teratology, and pediatrics than with Lamarck, Darwin, Weismann, and Mendel.

The intoxication tradition was robust, probably because of the continuity of these fields and their practices. Overshadowed after 1910, as we shall see in the next section, it surfaced again in the early 1930s, in France notably through a flurry of medical interest in a 'mutant' strain of rabbits apparently resulting from syphilis infection. In an article pronouncing belated last rites over the notion of inheritance of acquired characters, a Mendelian and emerging figure in French eugenics Raymond Turpin wrote optimistically that experimental reproduction of the rabbit strain by infection with the syphilis Treponema would demonstrate 'the possible infectious etiology of mutations' (Turpin, 1932, p. 689). A more speculative spirit, at meetings of the Society of Dermatology and Syphilography in 1933, imagined such mutations as syphilitic 'lesions reduced to fine cytological alternations' provoked by an 'intragametic, intrachromosomal' viruslike form of the microorganism 'involving only a limited number of "genes"' (Tourraine, 1933, p. 845).[72] Postulating an intrachromosomal form in a supposed 'life cycle' of the Treponema was not quite as exotic as it may seem. Interwar bacteriology in many countries was marked by variation research, theories of cyclogeny (bacterial life cycles involving change of form) and of endogeneous formation of viruses, and interest in bacteriophage (Amsterdamska, 1987; Summers, 1991; Amsterdamska, 1991; van Helvoort, 1994). In France there was an outpouring of bacteriological research, led by Calmette himself, on a supposed 'ultra-virus' form of the tubercle bacillus (Calmette *et al.*, 1925; Arloing and Dufourt, 1932). Despite the new elements of cyclogeny, ultra-viruses, and genetic mutation theory, all the characteristics of the earlier topos were there: the irrelevance of inheritance of acquired characters; the confident application of Mendel's laws even by French physicians to apparent patterns of dominant and recessive inheritance; the overarching interest not only in transmission, but also in the etiology of the altered hereditary *terrain*; intoxication, or by now 'intra-gametic, intra-chromosomic' infection, as the mechanism of constitutional change; the expectation that infection and heredity were somehow intertwined.

ACQUIRED INEQUALITIES AND THE MUTABLE TERRAIN

Here, then, to return to 1900, were all the preconditions for the establishment of differences in inborn susceptibility to infectious disease as an object of scientific knowledge: an etiology cognizant of multiple factors as well as necessary causes; the disjunction between infection and disease produced by bacteriological method itself, authorizing predisposition as a problem capable of exact definition, if not yet necessarily of exact solution; the precedent of species differences as revealed in the bacteriological laboratory; the establishment of a Darwinian selectionist theory of natural immunity and its long-term fluctuations in populations; a wide constitutional movement in clinical medicine and pathology, beginning to yield a vast if inconclusive medical literature on the physical, chemical, and functional diversity of the human body in its relation to disease and on family patterns of such diversity; a mechanism to explain how infectious processes themselves might produce specific or nonspecific predispositions in descendents, which soon promised to link with research on genetic mutation; a synthesis of eugenics with wider movements against the social diseases tuberculosis, syphilis, alcoholism, and infant mortality, supported by an emerging alliance of bacteriology and social hygiene in Germany and of medicine and eugenics in France. And yet the promise of heredity was to be disappointed. The two decades after about 1905 would see less the blending than the separation of heredity and infection in etiological explanation. It was indeed to be an era of the *terrain*, but above all of the acquired *terrain*.

Predisposition named the gap between infection and disease. What predisposition was would depend on what served, in experiment and perhaps also in epidemiological observation, to open and close that gap. There were essentially three variables in any experiment on infectious disease: infection (including dosage, virulence, and mode), organism (including hereditary difference), and environment. Experimenters and relatively little trouble varying environment and infection; varying heredity, so as to produce regular effects, was far more difficult.

Methods for demonstrating the environmental dependence of susceptibility were simple. Individuals of naturally resistant animal species were inoculated and then run to exhaustion on a treadmill, or fed inadequate diets, or exposed to abnormal temperatures. The classic such experiment was by Pasteur in 1878. He and his associates were able to render naturally resistant fowl susceptible to anthrax simply by standing them in cold water and thus lowering their body temperature. By the 1890s, such experiments formed a substantial genre.[73] These methods were as powerful as they were simple. They were able dramatically to *separate* predisposition from heredity. In the absence of other experiments producing equally dramatic effects according to other variables, they made differential individual susceptibility a function of environment, an acquired rather than innate property of the organism.

No one doubted that innate individual differences would be much subtler and harder to demonstrate than species differences. Though these subtleties might help explain disease patterns on the large scale of a population, they were likely to be drowned out by irregularities of environment and, above all, of dose and virulence of infection. Thus, next to the sizeable literature on individual predisposition as a function of environment, there simply was no genre of experiment showing inherited differences. To be sure, bacteriologists had experience with animals of different stock. They had constantly to procure guinea pigs, mice, rabbits, and so on from breeders. And there was a tradition of informal commentary on differences among breeders' stocks. Thus Behring wrote in 1899 of the existence in Bologna of a race of rabbits far more sensitive to tetanus toxin than those observed in Germany; of races of pigeons varying greatly in sensitivity to tetanus toxin; of the considerable inherited diphtheria immunity of guinea-pigs in England; of diphtheria-immune guinea-pigs obtained by Ehrlich from a particular dealer.[74] Yet almost no attempt to control and study these differences was made. A short look at one of the few exceptions will suggest why not.

In 1905 the American comparative pathologist and bacteriologist Theobald Smith published a paper titled, 'Degrees of Susceptibility to Diphtheria Toxin among Guinea-Pigs: Transmission from Parents to Offspring' (Smith, 1905). This and a following paper, though they led the heredity question nowhere at the time, would be cited in the 1930s as researchers, under a more widely shared ethos of biological research and using other methods, took up where Smith had left off. Smith had observed that some female guinea-pigs gave birth to young with more than average resistance to diphtheria toxin. Though he eventually found that such difference was due to transmission of passive immunity (antibodies formed in the mother) through the placenta and perhaps through nursing, he also concluded that difference in the degree to which the mother reacted to toxin and thus also in degree of passive immunity transmitted was an inherited factor. In short, 'the capacity for producing antibodies ... varies much from family to family' (Smith, 1907, p. 378). Smith's papers make clear that such observations required unusually large numbers of animals bred for long periods under controlled conditions and marked as to descendence, submitted to precisely known doses of infection or toxin.

Where were such prerequisites likely to be met? Nowhere but in exactly the kind of state serum laboratory of which Smith was then director, the Antitoxin and Vaccine Laboratory of the Massachusetts State Board of Health. Here, as in several other such institutions around the world, all modeled ultimately on Paul Ehrlich's original state serum institute in Berlin-Steglitz and his methods, large numbers of animals were routinely submitted to precisely measured, standardized doses of diphtheria toxin and antitoxin. Thus it is no accident that Smith's papers issued from such a laboratory and that the scientist with whom he corresponded about this work was Enrlich. Yet such laboratories were not conceived, built, and funded to explore the etiology of infectious disease and the possible role of heredity in it. They had been built to produce, improve, and maintain bacteriological standards and products. The correspondence between Ehrlich and Smith illustrates the problem perfectly. Even as Ehrlich repeatedly expressed keen interest in Smith's work and wrote that it would be 'due to your efforts that we will obtain a sure footing on knowledge of variations in resistance,' he could not pull himself away from the toxicity measurement and standardization tasks at hand. Criticizing the physical chemist Svante Arrhenius's tendency to report precise toxicity values based on experiments with only three or four guinea pigs, Ehrlich wrote to Smith:

> If one has many animal experiments and can use a relatively uniform [animal] material, one can exclude an occasional insensitive or hypersensitive animal to arrive at a reasonably correct result statistically. Should the animal material be non-uniform, I immediately stop all experiments, because I think it is then impossible to obtain reliable numbers [i.e., values of toxin activity] from such material. I prefer to wait until the middle of the summer when from experience [I know that] more uniform material is available.[75]

It would be difficult better to express the paradoxical way in which studying variable resistance was incompatible with the very serological standardization work on which such study in fact depended.

Even when a bacteriologist exploited those facilities to other ends, as Smith did, the differences of susceptibility he was able to observe were small compared to those an experimenter could produce by varying environment or dosage. The experimental phenomenology of inherited predisposition was weak. The most powerful experimental effects were not in fact those of variable nonspecific resistance under varied environmental conditions, but rather those of acquired specific immunity and hypersensitivity. Ironically, exactly the wide-spread subclinical infections that had first opened the door to predisposition and heredity in the 1890s increasingly seemed themselves to account for observed differential susceptibility. Infection itself in atypical or subclinical forms and its specific immunological effects was apparently going to fill the observed gap between infection and disease.

The paradigmatic study was August Wassermann's 'On Individual Predisposition and Prophylaxis against Diphtheria' of 1895. Wassermann, then at the Koch institute in Berlin, set out to account for the phenomenon of widespread asymptomatic diphtheria infection. He tested for diptheria antitoxin in the blood of 17 healthy children and 34 adults, none known to have had the disease. The serum of 11 of the children and 28 of the adults was antitoxic. From the fact that the serum of naturally immune species, such as the white rat, provided no protection when injected into susceptible animals, and from the fact that the percentage of individuals with antitoxin in their blood increased with age, Wassermann concluded that the property was not inborn. The source of the antitoxin was, instead, unrecognized prior exposure. Individual predisposition to diphtheria, then, was simply a matter of never having acquired specific antitoxic immunity through subclinical infection (Wassermann, 1895). The question of why those infections remained subclinical in the act of stimulating antitoxin production in a given individual remained unasked, but could presumably be explained by haphazard differences of dosage and environment. Wassermann's explanation of individual predisposition to diphtheria overshadowed Gottstein's selectionist, hereditary theory. It remained unchallenged, as we shall see, until 1924 when the Polish bacteriologists Hanna and Ludwik Hirszfeld inaugurated 'constitutional' serology by claiming to show that the antitoxins in normal blood were not acquired, but inherited in a Mendelian pattern correlated with blood group.

In the intervening 30 years, phenomena previously classed as natural immunity were reinterpreted as acquired immunity.[76] Though first exemplified by Wassermann's diphtheria study, this reinterpretation gained momentum when it was applied to differential resistance among races rather than individuals. Here, the paradigmatic work was Robert Koch's studies on malaria in 1897–1900. Physicians in the tropics and colonies had long noted that native populations seemed resistant to the disease that so threatened European travelers and colonists. Two findings put Koch on a different track. He observed during his expeditions to Africa, India, and southest Asia that cases of tropical malaria could be found among all 'races.' More important, he found that while the disease might be rare or even non-existent among the adults of a native population, it was common among their children. Those who survived illness in childhood acquired immunity, which could, in some cases, coexist for long periods with 'latent' infection (Koch, 1898, 1900). The apparent inherited 'racial' immunity was an illusion. The reinterpretation of other racial immunities followed suit (Anderson, 1996).

First, however, came the reinterpretation of differential resistance to domestic diseases, especially tuberculosis. Around 1900, physicians and public alike were jolted by the revelation that nearly 100 percent of the adult population of Europe was infected with tubercle bacilli. Only a fraction, of course, suffered from tuberculosis. Most physicians initially understood this radical disjunction between infection and disease as proof of the determining role of environmental and social factors or of differences in inborn susceptibility. But two bacteriologists who would give tuberculosis research its most distinctive direction in the first third of this century, Emil von Behring and Albert Calmette, began to shift the focus to acquired immunological factors—to a complex interplay of specific immunity, latency, and hypersensitivity.[77]

Becoming consumptive, Behring taught, was not a matter of inhaling the bacilli from dust or droplets of consumptives' sputum, but an awakening of latent infection contracted in infancy, usually from milk of tuberculous cows. This primary, latent infection, while potentially dangerous, did prevent later exposure from producing the acute and rapidly fatal forms of the disease, such as miliary tuberculosis, which were encountered in 'virgin' populations. Consumption, rather than acute tuberculoses, was the common form of the disease in the industrialized world because the nearly ubiquitous primary infection 'partially' immunized the population (von Behring, 1903, p. 224). Still, the line between latency and immunity could be drawn only with difficulty. Tuberculosis research in the first decades of this century circled around this problem. The Pastorian Albert Calmette offered an influential answer. After 1906 he took up and extended Behring's doctrine of partial immunity, as well as the Nobel laureate's controversial teaching that the intestinal mode of infection was more important than the respiratory one. What determined whether a first subclinical infection would confer, on the one hand, immunity or, on the other hand, increased susceptibility to disease was, according to Calmette, whether the lesions it produced had a chance to heal and thus encase the bacilli in a kind of internal scar formation. Immunity was a function of the continued presence of these still living tubercle bacilli in the healed, closed lesion (Calmette and Guérin, 1905; Calmette, 1920). The work of Calmette and his collaborators and of Behring and his student Paul Römer fueled arguments against the role of heredity in susceptibility to tuberculosis.[78]

The climax came at the Ninth International Tuberculosis Conference, in Brussels in 1910, for which Louis Landouzy had organized a plenary session on the question of hereditary predisposition. It proved to be Landouzy's last stand. The Paris dean adduced a considerable clinical and experimental literature to his argument that tubercular mothers could bequeath anatomical and functional dystrophies to their babies as well as an immunologically specific heredo-predisposition (hypersensitivity) to tubercle infection (Landouzy, 1910). None of a dozen international speakers gave much support.[79] More important, Calmette presented a paper entitled 'L'hérédo-prédisposition tuberculeuse et le terrain tuberculisable,' which established a counter-position that was to achieve wide assent. Calmette acknowledged the antiquity of medical observation of a constitution predisposing to tuberculosis: the shoulder blades winglike, the thorax cylindrical with projection of the sternum, the skin fine and transparent or dull and swarthy. Modern clinicians had added other characteristics such as anomalies in development of the heart and arteries, exaggerated respiration, 'demineralization' of the body, and a host of dystrophies supposedly resulting from what Landouzy called 'ancestral tuberculinous toxemia.' Calmette argued, however, that these anomalies of form and function were the product of the infection itself, acquired at a young age: a 'precocious familial contagion' rather than an 'intoxication of descendents.' His evidence was correlation between occurrence of such heredo-dystrophies and positive reaction to tuberculin tests. Clinicians, he claimed, had yet to find a case of such constitutional anomalies in an individual who did not react to the various tuberculin tests (Calmette, 1911).[80] This view carried the day, not only at the conference but thereafter. Landouzy closed the session by summing up policy implications. 'La lutte contre le bacille' was paramount, as it had always been, since even the dystrophies and predispositions were its products. This *lutte* was not one-sided germ hunting, but part and parcel of puericulture.[81] Exactly why 'campaign against the bacillus' could be a unifying banner in France and the point of division in Germany remains unclear.[82] The problem demands a more differentiated understanding of bacillus-centered public health policy. It took many forms, some including more attention to the *terrain*, some less.

Either way, after about 1910 heredo-intoxication was separated from heredo-predisposition. It ceased to be an etiological mechanism, even in the view of those who emphasized its importance for public health. In Landouzy, Courmont, and Charrin, heredo-intoxication had been most importantly the process by which a heredo-predisposition, a *terrain* tuberculisable, was created. In the influential eugenical writings of the pediatrician Eugène Apert, it could create only a nonspecific 'para-tuberculous' dystrophy. The new prefix 'para' is crucial. Children of tubercular parents were no more predisposed to the disease than were children in tuberculosis-free families.[83] Thus, whatever its continuing importance for the overall health of the population (there were no statistics on the frequency of such dystrophies), heredo-intoxication had given up the ground of etiology. This was not only a theoretical weakness. Practically speaking, it was the vast morbidity and mortality of pulmonary tuberculosis itself, not certain unusual, congenital physiological and anatomical dystrophies, that commanded the attention of the social hygienist, the public, and the state.

Meanwhile, Behring and Calmette's teaching that the etiology and epidemiology of tuberculosis depended on the vicissitudes of acquired immunological and infection factors—immunity, partial immunity, latency in both protective and endangering forms, 'allergy' or hypersensitivity–became the most widely accepted doctrine of the *terrain*. The term coined by Fernand Bezançon, *le terrain bacteriologique*, summed up the irony of the situation. The receptivity of the soil depended on hidden interplay with the seed itself.[84] Leading French clinicians and pathologists such as Albert Robin and Émile Sergent found themselves in a minority position in continuing to promulgate a nonspecific humoral doctrine of the *terrain* as they had done since 1900. They produced clinical and experimental evidence for the existence of various nonspecific 'biochemical predispositions,' involving demineralization, exaggerated respiration, hypoacidity, and so on. They assumed that these metabolic predispositions could be inherited as well as acquired: 'We often resemble our parents physically, why not then chemically?' But they neither had evidence for heredity, nor tried to produce it. Their argument was for the *terrain*, for clinical medicine, not for heredity (Sergent, 1914, p. 301).

By the 1920s, in a new edition of the great turn-of-the-century *Traité de médecine* that had been created by Bouchard and Jean-Martin Charcot, one read that 'heredity plays essentially no role' in tuberculosis.[85] Likewise in a book devoted to the importance of the *terrain* in infectious and other diseases, Jules Héricourt, early collaborator of Charles Richet, could write that, 'The influence of the heredity of the disease is no longer discussed ... Tuberculosis includes no heredity of the *terrain*' (Hericourt, 1927, pp. 91, 109).[86] Even the more research-oriented, far larger community of German constitutionalists, though still affirming an etiological role for certain predisposing morphological and functional anomalies such as the thin-chested 'Habitus asthenicus,' could make little headway.[87] Ever since Behring and others had revealed 'the complex interplay between infection and immunization, between hypersensitivity and increase of resistance through [natural] immunizing processes,' the Viennese clinician Julius Bauer wrote in 1917, 'sharp delimitation of constitutional factors in the predisposition to tuberculosis has become frequently impossible' (Bauer, 1917, p. 52).

From the viewpoint of practice as well as theory, the central place of tuberculosis in the discourse of eugenics and hereditarian hygiene, not to mention its longer history as a 'social disease' to be controlled by social means, can be misleading. The interwar French tuberculosis movement was characterized as much by the development, promotion, and implementation of a vaccine, a Pastorian vaccine at that (Bacille Calmette-Guérin, or B.C.G.), as by legal and social regulation of reproduction to prevent propagation of hereditary predispositions or 'paratubercular' degenerative dystrophies. Nor was it the case that Calmette and the Institute Pasteur had to face a recalcitrant medical and eugenics community committed to the *terrain.* On the contrary, clinical trials of B.C.G. were carried out under the auspices of the Paris medical faculty's École de Puériculture. Most of the papers reporting these trials, moreover, were co-authored by Raymond Turpin, who emerged as an important figure in French eugenics in the 1930s and went on to establish medical genetics after the Second World War.[88] Apparently emboldened by his own campaign on behalf of B.C.G., Calmette went so far as to undertake an experimental refutation of the doctrine of *terrain* in tuberculosis. In guinea pigs of the same age and weight, kept under identical conditions and diet, the severity of symptoms after inoculation with tubercle bacilli was directly proportional to the number and virulence of the bacilli, he reported to the Académie des sciences in 1923. 'Individual variations in resistance,' indeed 'le facteur terrain,' Calmette argued rather cavalierly from these results, played no role in the etiology and clinical course of the disease (Calmette, Boquet and Nègre, 1923).

In Germany, medical discussion of mechanisms of inheritance and relations of body, environment, and germplasm remained speculative and a source of frustration to participants. The outlines of debate over genotype versus phenotype as the proper object of constitutional medicine stayed essentially unchanged for decades, a stagnancy of which participants were keenly aware (Martius, 1914; Bauer, 1929). The method of genealogy, for which great expectations had been built up at the turn of the century, had by 1920 produced what practitioners themselves described as a disappointingly small and ambiguous list of inherited pathologies.[89] A book like Julius Bauer's popular *Die konstitutionelle Disposition zu inneren Krankheiten* of 1917, which was reissued several times and translated into English, reads a bit like Pliny. One set of cases follows another in a menagerie of ailments and various sorts of habitus, status, and body type. Observations suggesting a relation between, for example, the status thymico-lymphaticus and certain infectious diseases were doubtless of clinical interest but did not add up to general statements about etiology, much less heredity.[90] Even as some leading constitutionalists began to grasp for the methods and status of 'general biology,' both the relevance of a past made of 'clinical impressions' and the way into the future of 'exact scientific research' remained obscure.[91]

In sum, the first decades of this century were indeed an era of constitutionalism, of the *terrain*, in European scientific medicine. Yet, as far as infectious diseases were concerned, it turned out to be the era of acquired rather than innate predisposition, and above all of the *terrain bacteriologigue*. On the model of Wassermann on diphtheria and Koch on malaria, apparent natural and racial immunities were reinterpreted as acquired immunity. German constitutional pathologists had little to say about most infectious diseases, and, when it came to tuberculosis, acknowledged the weakness of their research in the face of what the subclinical interplay of immunological factors could explain. Demonstration of hereditary differences through animal experiment was apparently undertaken by no one after Theobald Smith left off in 1907. Instead, a whole genre of experiment displayed the dramatic effects of varying environment, dosage, and mode of infection. Heredo-intoxication ceased to be a mechanism of predisposition and was thus relegated to the sidelines. Finally, the use of genealogy and statistics to demonstrate the role of heredity in infectious and other common diseases was fraught with difficulty. It was nearly impossible to control for contagion and myriad, interrelated environmental and social factors.[92] Ingenious solutions by a handful of statisticians, such as Pearson and the physician and mathematical Mendelian Wilhelm Weinberg, tended to remain ingenious—relatively inaccessible and ultimately unconvincing to the majority of medical scientists.[93] Weinberg's investigations, in fact, led him to stronger conclusions about the etiological importance of social conditions and contact with clinical tuberculosis within the home than about the inherited 'constitutional factor.'[94] In short, while hereditarian medicine and hygiene were on the rise in the first decades of the century and reserved a central place for those social diseases and 'racial poisons' tuberculosis and syphilis, knowledge of heredity in infectious disease went into near eclipse.

FROM CONSTITUTIONAL MEDICINE TO MEDICAL GENETICS

A passing observer of Western medicine around 1920 could safely have concluded from a wide range of writings that individuals differed little in their innate susceptibility to infectious diseases. Had this same traveller returned ten years later, he would have been astonished at the reversal of fortunes. Such inherited differences, it could now be claimed, dominated the etiology and epidemiology of infectious and other common diseases. Moreover, the scientific bases of such knowledge were changed, as were the very norms of the scientific in medicine. Constitutional medicine as an ultimately unsuccessful science of the inequality of bodies, under the purview of clinicians and pathologists, was giving way to what came to be called in France after World War II *génétique médical* and in Germany, *Erbpathologie*, later *Humangenetik*, which derived from work by biologists and by a few bacteriologists and physicians who, schooling themselves in biology and statistics, acted like biologists.

Three methods instantiated innate differential susceptibility after the First World War and especially after about 1925. These were constitutional serology, plant and animal breeding research, and twin research. The first of these was pioneered by Ludwik and Hanna Hirszfeld and taken up largely by German-speaking bacteriologists as on offshoot of the wider field of blood-group genetics. Families and populations were tested for antibody to an infectious disease, notably diphtheria, and for blood group, which was known to be inherited along Mendelian lines. Antibody to a given microorganism or toxin, it was found, seemed to run in families roughly according to Mendelian expectations. And its presence or absence seemed to correlate with blood group. If prior exposure to infection could be ruled out (a point of continuing difficulty and controversy), then Mendelian inheritance of specific disease resistance was held to be demonstrated. The Hirszfelds' early results were greeted with great interest by a wide variety of medical scientists. But over the following years, the whole program of 'constitutional' serology all but foundered on the problem of ruling out prior exposure.[95]

In the second method, individual animals or plants of species known to be susceptible to a given infectious disease were inoculated and observed for differences in length of clinical course or survival. Some individuals proved more resistant, some less so. Resistant and susceptible individuals, or their offspring, were then bred in various combinations and the differences in reaction to inoculation observed over one or more generations. Alternatively, such research was carried out on previously inbred or crossbred lines. Despite problems of statistical significance and of controlling immunological and environmental variables, this work provided the first sustained experimental evidence of individual differences in innate susceptibility to infectious disease. An elaboration of the method, which was developed mainly by the experimental epidemiologist Leslie T. Webster at the Rockefeller Institute for Medical Research, produced still more dramatic effects. The presumed natural selective action of the diseases themselves was imitated by mating survivors and testing the susceptibility of their offspring by further inoculation. As few as three generations of such selective breeding could produce highly resistant animals. Results did, however, remain open to question because of variable mortality of control groups and, above all, difficulties in screening survivors for latent infection and specific antibody in order to rule out passage of acquired immunity from parent to offspring. All of this breeding research on infectious disease was almost exclusively Anglo-American. German geneticists certainly bred animals and plants but do not seem to have addressed problems of infectious disease.[96]

Meanwhile, especially in Germany, anthropologists, physicians, and psychologists had by 1933 studied some 5,000 twin pairs in the hope of being able to distinguish inherited from acquired characters by finding differences between mono- and dizygotic pairs.[97] This take-off was fueled by the same emerging ethos of scientificity that encouraged plant and animal breeding research. Twin research was not experimental. But unlike the construction of pedigrees, it was an attempt at controlled, quantitative study, as if nature itself had carried out the experiment by providing monozygotes and dizygotes.

As if to announce the coming biological transformation of heredity as an object of medical knowledge, the first animal breeding work on innate susceptibility to infection was carried out by the co-founder of population genetics Sewall Wright, who was then at the Bureau of Animal Industry of the U.S. Department of Agriculture, in collaboration with Paul A. Lewis of the Henry Phipps Institute for tuberculosis at the University of Pennsylvania. Their 1921 paper, 'Factors in the Resistance of Guinea Pigs to Tuberculosis, with Especial Regard to Inbreeding and

Heredity,' appeared not in one of the many medical journals, but in the *American Naturalist*. They found marked differences in resistance among various inbred lines. In crosses between the most resistant line and more susceptible ones, they were able to attribute 30 percent of variation in longevity after inoculation in crossbreeds to heredity, that is, to 'amount of blood' from the resistant line. Their collaboration was emblematic of methodological and technical novelty: such research could not yet be done at one kind of institution and by one kind of scientist. Wright had to provide the controlled animal stocks, products of years of breeding work; the rationale and methods of research on inbreeding, crossbreeding, sex linkage, and so on; experience in analyzing diverse variables of environment and heredity and the mathematical statistics for evaluating the relative roles of these variables. From Lewis had to come precise and controlled bacterial dosage and virulence, without which no conclusions about differential susceptibility were possible; methods of animal inoculation; experience in observation of clinical course and in running animal experiments without unwanted infections and immunizations (Wright and Lewis, 1921).

Where pathological and bacteriological experiments had once involved perhaps a dozen animals, they now might involve hundreds or thousands. Wright and Lewis's conclusions rested on the inoculation of several hundred guinea-pigs and the controlled breeding of 30,000 of them. The research on hereditary effects of alcoholism conducted by Agnes Bluhm at the Kaiser Wilhelm Institute for Biology involved 32,000 mice. Constitutionalists and eugenicists could back up their statements about inherited predisposition to cancer by citing Maud Slye's breeding research on 40,000 mice.[98] Hand in hand with these large numbers of animals of known ancestry, bred under controlled conditions, went quantitative analysis. Medical knowledge of predisposition would now be statistical. Thus parallel to the emergence of standardized laboratory animals, a whole new material world of experimental heredity came into being without which the statements of the 1930s would not simply have been uncertain or more ambiguous, but impossible to imagine in their very form. What remains unclear is exactly where the demand for this kind of knowledge came from and what were the modes of its articulation, both in and out of the medical scientific community.[99] This, rather than the application of Mendel's laws as such, was what instantiated innate differential reactivity and accomplished the synthesis of pathology and classical genetics.

The effect on medical thought of the three lines of research—animal and plant breeding, twin research, and constitutional serology—was not merely additive. They created a field of mutually reinforcing relevance and plausibility for each other, for earlier statistical work, and for some of the earlier clinical research on constitution.[100] The tuberculosis studies of 1907–09 by Pearson and his associates, who had fought for the importance of 'soil' over 'seed' in the etiology of tuberculosis, had never been integrated into the French and German medical literature on heredity and the *terrain*.[101] Now, in the 1930s, they found their place (Turpin, 1936).

In France the changes were most visible. Compare the leading synopses of medical doctrines of heredity before and after this period: Apert's *Hérédité morbide* of 1919 and Raymond Turpin's *Hérédité des prédispositions morbides*, which, thought it appeared in 1951, opened with chapters on infectious disease and tuberculosis taken almost directly from a review article Turpin had published in 1936. The subject has not changed. Both books concern less the rare Mendelian disorders than the major social diseases: infectious disease, tuberculosis, cancer, and so on. Yet the content and, above all, the knowledge base could not be more different. In Apert, as we saw in the previous section, most 'morbid heredity' does not include what he himself regarded as 'real' or 'true' heredity, nor even does it include predisposition. In Turpin, the focus on predisposition has entirely replaced heredo-intoxication and heredo-contagion, a change also exemplified by the absence of any reference to syphilis, which had deserved a whole chapter in Apert. Moreover, Apert's book is based almost entirely on French clinical and pathological research; Turpin's book, on American, British, German (and some French) biological research, notably breeding and statistical studies. Turpin's mainly Anglo-American, experimental, biological bibliography, already in 1936, signalled the onset of a rapid shift in the very nature of the scientific in French medicine (Turpin, 1936, 1951).[102] As late as 1927, in a paper on '*terrain* factors' delivered at the fourth French tuberculosis congress, Turpin and his 'maître,' the clinician and phtisiologist Émile Sergent, had recapitulated the traditional humoral doctrine of 'terrains physiologiques électifs.' Legitimating the existence of non-immunological *terrain* factors was more important than establishing their inheritance. Wright and Lewis's guinea-pig study was indeed mentioned, but with no sense of the disjunction of scientific worlds that was so soon to become inescapable. 'The most demonstrative arguments,' Sergent and Turpin concluded in 1927, 'rest on clinical facts more than on experimental ones.'[103]

Parallel changes can be seen in Germany. Turpin's counterpart was the internist Otmar Freiherr von Verschuer, whose career from 1927 to 1945 was closely associated with the Kaiser Wilhelm Institute for Anthropology, Human Genetics, and Eugenics, of which he became director in 1942, and who is better remembered for his assistant the Auschwitz camp doctor Josef Mengele.[104] In the 1930s, Verschuer and his peers created what he called 'Erbpathologie,' which by 1940 deserved its own volume of Baur-Fischer-Lenz (von Verschuer, 1934; Lange *et al.*, 1940). It was based above all on twin research and contrasts nearly as much with earlier German constitutional pathology as Turpin's writings do with Apert's. In 1928 Verschuer had teamed with a tuberculosis expert and later Nazi Party member Karl Diehl to begin an unprecedented study of tuberculosis in monozygotic and dizygotic twin pairs. The transition from constitutional pathology to twin research was a transition from casuistic to statistics and, in effect, from descriptive to experimental science. The control of variables, which was supposed to approach experimental conditions, was provided by the difference between mono- and dyzygotic twins. In constitutional pathology, there had simply been no such attempt at controlled knowledge of variables, even when genetic claims were made. Julius Bauer's book was not an argument, but rather, as Garrod put it, a store-house of clinical observation.[105] Its 500 pages covered the whole alphabet of ailments; Diehl and Verschuer's 500 pages were devoted to 127 twin pairs from tubercular milieux. Their copious, standardized case histories, lung x-rays, serological test results, tables of numbers, and statistical correlations are recognizable to the modern eye as 'data' and analysis. The habitus asthenicus of the constitutionalists was not gone, but it was reduced to a small, causally secondary role.[106]

This return of heredity in German and French medicine of common diseases was not simply dictated by experimental results. In the same year, 1923, that Albert Calmette at the Pasteur Institute claimed to have ruled out the role of the innate *terrain* in the etiology of tuberculosis, Leslie Webster at the Rockefeller Institute inaugurated a series of papers claiming to rule out *all but* the innate *terrain* in the epidemiology of mouse typhoid—Calmette, by holding environmental and bodily factors equal and varying only the quantity of bacilli; Webster, by holding all other variables equal and varying the descendence of his mice (Calmette, Boquet, and Nègre 1923; Webster, 1923).[107] The capacity of experiment to construct the dominance of any given causal factor lies perhaps in the very nature of etiological research. Yet it may still be asked why a given experimenter chose to emphasize a given factor; and when and why verisimilitude of an experiment to 'nature' was accepted or rejected or at all made visible as a problem. Some contemporaries were keenly aware of the difference between showing that inherited differences existed and determining whether they were important in the common situations of infection. Some even argued that there was no 'nature' by which to judge the relevance of experiment; there were only different real-world situations. Thus Turpin judged Pearson's contention that hereditary constitution was five times more important than environment in the etiology of tuberculosis to be 'an opinion applicable only to the community that he studied. In a collectivity in which the risks of tubercle infection are the same for everyone, the hereditary differences emerge; if, on the contrary, the risks of contamination are very unequal, the role of contagion is more apparent' (Turpin, 1951, p. 138). Nor was the return of heredity in French and German medicine of common diseases in the interwar period dictated by eugenics or race ideologies. As we saw in the previous section, the rise of eugenics and hereditarian social hygiene in the first two decades of the twentieth century emphatically had not proceeded apace with the development of a research basis and a consensus on innate differential susceptibility to infectious disease. On the contrary, this view lost ground. Now, in the 1930s, the ground was regained. Yet if French eugenics had always been a movement of puericulture and natalism, and if what Apert called 'real' or 'true' heredity rather than heredo-contagion and heredo- intoxication had anyways never been essential to it,[108] then it remains unclear why hard heredity should suddenly have become important to a eugenicist such as Turpin in the 1930s.

The German case poses less a problem of discontinuity than of diversity. Many German-speaking medical scientists who had nothing to do with eugenics or even social hygiene received the animal breeding, twin, and serological studies as reason to change their minds about the relative importance of acquired and inherited predisposition. The reversal of views and its independence from eugenics is perhaps most strikingly visible at the Robert Koch Institute. Its director, Fred Neufeld, who had previously paid little or no attention to innate difference in susceptibility, emerged after 1925 as its leading proponent in German laboratory medicine. He and others proclaimed the end of the Behring-Römer era, in which resistance to tuberculosis had been attributed largely to acquired immunological factors. Despite their new interest in the innate inequality of bodies, Neufeld and his associates stayed clear of eugenics (Neufeld, 1925, 1935; Lange, 1936).[109] Neufeld's fellow co-editor of the *Zeitschrift für Hygiene und Infektionskrankheiten* (the journal founded by Koch) the leading virus researcher and Basel professor of hygiene Robert Doerr took interest in the new heredity research but rejected its eugenic and racial hygienic implications in favor of its potential importance for health care of individuals.[110] In short, had the biological transformation of differential susceptibility been carried out in Germany solely by Verschuer and the eugenics establishment or by a Max Gruber or a Hans Reiter—that is, by bacteriologists who became leaders of racial hygiene and, in Reiter's case, of Nazi medicine—then these larger social and political forces could indeed explain much. But the enthusiasm of a Neufeld and the interest of a Doerr show that the question: why heredity? will need be addressed in other ways. They embraced heredity in its new biological form not because of eugenics, but despite it.

Even the *Erbpathologie* of Verschuer, who enthusiastically greeted the new 'Führer' and promoted a model of the physician as 'Erbarzt' in the new, National Socialist Germany, is perhaps best understood according to a longer term devotion to the making of a world of hereditary inequality by twin research: Verschuer did his twin research under social democracy in the Weimar Republic; he did it after 1933 under Nazism; he did it during the war, now experimentally as well as clinically, using body parts and data sent by his former student Mengele from Auschwitz; and he did it after the war as a professor of human genetics in the new Federal Republic.[111] No particular arrangement of society and technology appears to have been necessary to the making of the innate inequality of bodies, in relation to external causes of disease, as an object of biomedical knowledge.

Why heredity achieved such importance in the understanding of common diseases after about 1925, having gone into near eclipse in the first decades of the century, might instead be attributed to the emergence of new methods and norms of biological knowledge and to the social and intellectual condition of European medicine in the troubled interwar period. The new, biometric, experimental scientificity seemed especially important to elite physicians at a time when medicine was widely perceived to be in intellectual and institutional crisis.[112] Outside of medicine, statisticians and geneticists had made heredity into the preeminent quantitatively and biologically knowable parameter of difference among bodies. In this form, heredity could now be the anchor of a scientific medicine that was just then besieged by a veritable neo-

Hippocratic, holistic health movement, both popular and academic and in both Germany and France, which itself advanced under the banners of *Konstitution* and *terrain*. An older pioneer of constitutional medicine such as Julius Bauer found himself having to defend his genotypic definition of the field not only against fellow academic pathologists of a phenotypic bent, but also against dilution by 'this strange 'renaissance of humoral pathology",' by doctrines of the physical and psychic unity of the 'person', and by wider neo-Paracelsian, homeopathic, 'theosophic, anthroposophic, magnetopathic' movements all laying claim to constitution.[113] The boundaries of such newly established scientific fields as endocrinology and immunology were increasingly difficult to maintain. Medical journals blared forth attributions of one pathology or another to problems of the 'inner secretions.' In 1933 the immunologist and self-appointed Bernardian champion of the *terrain* Arnault Tzanck counted as many as ten current meanings of 'allergy,' as expansive as 'L'allergie, c'est la vie' (Tzanck and Oumansky, 1933). In other words, it may be no coincidence that at the very moment when constitution achieved its widest currency and its most inclusive and predominantly non-hereditarian meanings in European medicine, a relatively few physicians undertook to transform its hereditary aspect into an object of exclusive and exact biological, statistical, experimental knowledge.

To bring this story to a close, there remains one step. Even as hereditarian thought and research were remade according to a new kind of experimental, biological, and statistical evidence, the object of knowledge remained the body as a thing of differential reactivity. This older deep structure, so to speak, was preserved in translation. This meant that the genetics or eugenics of common diseases would be organized not around the search for 'genes' as single causes, but around the *terrain*, the body as the site of complex interplay of heredity and environment.

Two careers are exemplary of both the transformation and the continuity, namely those of Turpin in France and Verschuer in Germany. The parallel is uncanny because Verschuer's name is associated with Nazi science and his student Mengele; Turpin's, with postwar genetics and the discovery of trisomy in Down's syndrome patients. Both came out of their country's respective constitutional traditions. Both led the biological transformation of constitutional medicine into medical genetics and, in so doing, emphasized the complexity of common diseases over the Mendelian patterns of rare ones. Verschuer did his important research on tuberculosis, and infectious diseases played a central role in Turpin's radical biologization of the French medical tradition of the *terrain* and morbid heredity.

In one of the first uses of the concept of 'genetic' predisposition, Verschuer and his twin-research collaborator Diehl argued that predisposition to tuberculosis was probably not the product of 'a large number of genes' but also recognized their inability to determine 'whether the genotype 'tuberculosis predisposition' is a dominant or recessive gene and whether this gene has multiple alleles, or whether it is made up of several gene pairs (polymerism).'[114] Either way, they emphasized not the gene as cause, but rather the complex interplay of genetic and environmental ('peristatic') causes, or better, 'forces.' Likewise, Turpin and his associates understood the extra chromosome (trisomy-21) they discovered in 1959 in patients with Down's syndrome less as a first cause than as a 'passage point' between constitution of the mother, conditions of pregnancy, and abnormal child (Gaudillière, 1997).

The medical tradition of constitution, differential reactivity, and complex etiology was not the only source for this model of hereditary pathology. The other source was the developmental and physiological orientation of genetics at large in both France and Germany, an orientation seen to contrast with the transmission genetics that prevailed in America and Britain (Burian, Gayon, and Zallen, 1988; Harwood, 1993). Verschuer cast himself as the herald of a third and ultimate stage in the 'history of human hereditary science,' a stage that followed the pre-Mendelian and Mendelian phases and that was characterized by 'developmental physiology.' Here he was following unmistakably in the footsteps of German geneticists led by Richard Goldschmidt, whose *Physiologische Theorie der Vererbung* (1927) had appeared just as Verschuer was beginning his career. Mendelian analysis, Verschuer argued in line with other critics, tended to presuppose relatively simple relations between genotype and phenotype, and encouraged the 'static view' that the latter was a kind of 'direct projection' or 'enlargement' of the former. The new hereditary pathology, by contrast, would comprehend genes as 'developmental forces' and 'the development of the organism as a dynamical system, whose parts stood in continuous functional relationship.' Both the phenotypic manifestation of predisposition and the clinical manifestation of disease—that is, both ontogeny and etiology—were to be understood as dynamical systems.[115]

Like all aspects of the body in constitutional medicine, the gene was not an object in direct causal correspondence with a symptom, but rather an object of differential reactivity. By studying the reactive possibilities (*Reaktionsmöglichkeiten*) of each gene and which of these possibilities was realized under which environmental influences, Verschuer believed, the goal of all pathology could finally be realized, which was to determine 'the reaction of man to particular environmental factors' and especially 'individual reaction differences.' Thus the problematic of *Erbpathologie* was a continuation of constitutional medicine.[116] The successive names of the journal Verschuer edited after 1954 testify to this continuity: *Zeitschrift für angewandte Anatomie und Konstitutionslehre* (1914–1920), *Zeitschrift für Konstitutionslehre* (1921–1934), *Zeitschrift für menschliche Vererbungs- und Konstitutionslehre* (1936–1963), *Humangenetik* (1964–).[117] France had no such specialized constitutional journals. Nor did its humoral constitutional medicine find a place in medical genetics as was the case in Germany. Instead, a relation to clinically and constitutionally sensitive laboratory medicine was displayed when Turpin chose to publish the first installment of what would become his book *L'hérédité des prédispositions morbides* in the *Revue d'immunologie*.

The interwar conjunction of constitutional medicine and physiological traditions of genetics thus completes this story of continuity. The blow to deterministic bacteriological etiology in the last decade of the nineteenth century continued to reverberate 30 years later in the making of a non-deterministic medical genetics. This was not inevitable. The prewar medical Mendelism of pedigrees and of 'inborn errors of metabolism,' to use Archibald Garrod's famous phrase, had certainly not become defunct for the rarer diseases. In its 1940 edition, Baur-Fischer-Lenz was still full of pedigrees of albinism, diabetes, dwarfism, hemophilia, muscular dystrophy, schizophrenia, as well as a number of more common, less well-defined conditions (Lange *et al.*, 1940). Though I hope to have supplied part of an answer through the story of constitution and etiology after 1890, exactly why that Mendelism made room for another model of heredity in medicine remains unclear. It could as well have remained the procrustean bed into which the genetic component in the etiology and transmission of the more common diseases would have to be pressed. Instead, Garrod himself now wrote his less well-known book *The Inborn Factors in Disease*, which amply displayed the continuity with constitutionalism and its renewal through the new biological methods. He firmly cast 'inborn errors' as factors in what he and others considered to be complex etiologies. The hypothetical inborn metabolic error leading to gout,

Garrod cautioned, was not to be attributed to a 'rare mutation,' but rather to 'an alternative and slightly divergent path of metabolism met with in a large part of the total population' and producing disease under certain environmental circumstances.[118]

The contrast to the genealogies, typologies, and mental tests by which rarer diseases, race, criminality, feeblemindedness, and so on were given a genetic essence is remarkable.[119] It speaks for the specificity of the story told here of a competing model based on constitutionalism and the medicine of infectious and other common diseases. The end of belief in inheritance of acquired characters, the acceptance of Mendelian particulate inheritance, and the eugenic marriage and sterilization laws have always seemed to mark out the discourse and practice of genetic determinism, the hard nature/nurture distinction. Yet the assumption that eugenics always went hand in hand with genetic determinism (and that neo-Lamarckism was the only alternative) evidently belongs to the world after 1945—a world that saw a divide between nature and culture and usually declared itself for culture because of the evils that had been done in the name of nature. Before 1945, by contrast, as long as a hereditary factor in, say, tuberculosis could be made plausible, no matter how conditional the ontogeny and multifactoral the etiology, an urgent genetic danger could still be recognized for the following reason, which may be taken as characteristic of the social and biological imagination in the century after 1850: the vision of population as poised on a precipice of exponential change through differential rates of reproduction.[120] That is how one could be subtle on nature/nurture and yet advocate laws to prevent marriages of individuals with family histories of tuberculosis and even their elective sterilization. The eugenicists and race hygienists were, in short, not necessarily genetic determinists. That was in part because of the physiological and developmental orientation of French and German genetics at large.[121] But it was also the ultimate legacy of the crisis of bacterial etiology at the end of the nineteenth century. With the advent of antibiotics and of molecular biology after 1940, that crisis and its legacy were finally lost to view.

To conclude, the twentieth-century medical body that spanned from the bacteriological and Mendelian revolutions up to molecular biology was no doubt made partly of genes and germs. But it was also, to the most varying degree, predisposed, susceptible, reactive, irritable, (auto) intoxicated, sensitized, with the emphasis on the variability. If consensus on the nature of this inequality was remarkably unstable, the fact of it seemed undeniable. This had by no means always been the case. The general tendency of medicine since the Renaissance, as Henry Sigerist once emphasized, had been to favor a typology of diseases over the Hippocratic and Galenic typology of people. Difference among disease entities, not difference among bodies, was the proper subject of the science of medicine (Sigerist, 1929). In a further refinement, emphasis moved from the disease entities to their anatomical loci. With the birth of the clinic around 1800, as Michel Foucault described, 'the medicine of diseases has come to an end; there now begins a medicine of pathological reactions, a structure of experience that dominated the nineteenth century, and, to a certain extent, the twentieth, since the medicine of pathogenic agents was to be contained within it, though not without certain methodological modifications' (Foucault, 1963, p. 191). The medicine of pathological reactions subsumed the perennial debate over disease 'ontologies' and took several forms: the pathological anatomy of the French after Bichat and Broussais, the cellular pathology of Virchow, the fever medicine of the hygienists and epidemiologists after Chadwick, the bacteriology of the last third of the century. In all these forms, in this structure of experience, equality—the equality of bodies—was the chief virtue, the very condition of a science of pathology. The differences among bodies, which physicians knew better than anyone, were to be ignored in the interest of establishing regular, lawlike relations between symptoms, morbid anatomies or physiologies, and environments.

Around 1900, however, equality ceased to be the chief virtue. The medicine of pathological reactions gave way to the medicine of differential pathological reactions, a would-be exact science of difference among species, races, individuals, and states of an individual. The body as an entity displaying regular reactions and thus causal correspondence between symptom, lesion, and environment yielded pride of place to the differentially reactive body. With the advent of molecular biology, this configuration, in turn, was to be displaced by the body as an ensemble of mechanisms of recognition. The intervening, somewhat overshadowed half century of scientific medicine saw the medical making of bodily inequality, acquired or innate. Medicine, in the structure of its experience and explanations, was thus in step with other sciences of man, social and natural, in the century after Darwin.

NOTES

1. I am grateful to Jean-Paul Gaudillière and George Weisz for critical readings of drafts and to Hans-Jörg Rheinberger for discussion and encouragement.
2. Descriptions of the famous self-experiment are numerous; see esp. Evans (1987), pp. 490–507.
3. Hans Buchner to Max Rubner, 31 Dec. 1892; Nachlaß Rubner, Archiv der Max-Planck-Gesellschaft, Berlin-Dahlem. On Buchner and the Munich school: Mazumdar (1995), pt. 1.
4. Pasteur to Joseph Grancher, 20 Nov. 1892; Vallery-Radot (1951), iv: 343–44.
5. For the contemporary medical reader's sake, I include the term contamination here, since the meaning of infection has been narrowed to penetration of the skin or mucus membranes.
6. On susceptibility and resistance to cancer: Löwy (1989); on medicine and social inequality, see esp. Coleman (1982).
7. The Croonian Lectures of 1908, reprinted as: Archibald E. Garrod, *The Inborn Errors of Metabolism* (London, 1909).
8. Constitutionalism also drew strength from early immunology, endocrinology, and nutrition research insofar as the determinants of health and disease were identified with humors throughout the body rather than localized in particular cells or tissues. This aspect lies outside the scope of the present study; but see Shryock (1947) , pp. 304–314.
9. A recent re-edition casts Garrod as a neglected precursor in a reductionist era: Scriver and Childs (1989). In fact, contemporaries took little notice of the book because they already shared its views. The older medical historiography recognized twentieth-century constitutionalism: Faber (1930), ch. 6; Shryock (1947); Diepgen (1955), ii, pp. 140–144. More recently: Weindling (1989), ch. 3. Misleadingly focused on the fortunes of a single term: Ackerknecht (1982). On the decline of constitution and predisposition in American medicine after 1900: Ludmerer (1972), pp. 63–73; Rosenberg (1976), p. 203; Rushton (1994); on their persistence: Tracy (1992).
10. On rethinking the reception of the germ theory of disease, see special issue of the *Journal of the History of Medicine* (1997), no. 1.
11. By contrast to the previous period: Rosenberg (1974).
12. See, for example, 'Discussion on the Influence of Heredity on Disease, with Special Reference to Tuberculosis, Cancer, and Diseases of the Nervous System' *Proceedings of the Royal Society of Medicine* General Reports (1908–09), 2: 9–142.
13. The leading German-speaking pathologists and clinicians of the period are well represented, including many who contributed fundamentally to the bacteriological literature, such as Baumgarten, Escherich, Kißkalt (a hygienist), Lubarsch, Weichselbaum; authors also included race hygienists and the occasional strictly bacteriological investigator such as Emil von Behring; Hart (1922).
14. One important experimentalist's perception of the growing importance of the *terrain* in pathology: Charrin and Riche (1897), p. 355; the school of Charles Bouchard, the so-called Lyon school, and the clinician Louis Landouzy are discussed below; two other leading *terrain*-oriented academic physicians were the pathologist Albert Robin and the clinician Émile Sergent.
15. Pasteur against Pidoux's motto: Pasteur and Joubert (1877).
16. Hahn (1904), pp. 267–69, 303–07.
17. Martius (1898), pt. 1, pp. 90–110; Martius (1914), pp. 17–19; Bauer (1917), pp. 1–3, 46; Tzanck (1933), p. 1452.
18. On the new clinical science in late-nineteenth-century Germany: Faber (1930); on persistence of older clinical traditions in Britain: Lawrence (1985).
19. See also Martius (1909), ch. 4 and pp. 431–32; more ambiguous: von Pfaundler (1911), pp. 48–49. Rejecting the category of constitutional *diseases*: Martius (1900), p. 433; Bauer (1917), p. 42.
20. On the long history of constitution: Sigerist (1929); a brief survey of predispositional concepts, especially in mental medicine: Genil-Perrin (1913), pp. 26–38; see also Bulloch and Greenwood (1911); López-Beltrán (1994).
21. Watson, *Lectures on the Principles and Practice of Physic*, as quoted by Hamlin (1992), p. 51.

22. For example, 'reaction': Faber (1930), p. 186; 'pendulum': Garrod (1931), repr. in Scriver and Childs (1989), p. 19 of Garrod's text; on pendular swings in the history of biology: Canguilhem (1952).
23. See various entries in Bouchard and Roger (1912).
24. For the importance of irritability in German physiology and medicine in the 1890s: Posner (1893).
25. See Hiebert (1971); medical energetics was not necessarily derivative of Ostwald's movement: Rosenbach (1904).
26. The classic text on autointoxication is Charles Bouchard, *Leçons sur les autointoxications dans les maladies* (Paris, 1887); by far the longest section of Bouchard and Roger's 1912 treatise on general pathology was Georges-Henri Roger, 'Les intoxications et les auto-intoxications', in Bouchard and Roger (1912), ii: 1–486; though Roger identified all infection with intoxication, he emphasized that it was 'very complex': besides the microbial toxins, 'il faut faire une large place aux poisons formés par l'organisme malade' (p. 317). For favorable reception and deployment of Bouchard's teachings at the Institut Pasteur, see: Duclaux (1887); Metchnikoff (1908). The subject was not ignored in Koch's journal: Pfeiffer (1906). Gottstein identified autointoxication as an important cause of weakening of resistance: Gottstein (1893). For the view that fatigue modifies the terrain through autointoxication just as microbes modify the terrain through intoxication: Héricourt (1927), p. 172.
27. See generally Silverstein (1989), ch. 9.
28. These papers became part of the constitutional literature: Hart (1922); Garrod, (1931), repr. in Scriver and Childs (1989), pp. 17–18 of Garrod's text; see also von Pfaundler (1911), pp. 73–77 on 'Sensibilisierung'.
29. Georges-Henri Roger, 'Introduction à l'étude de la pathologie générale' in Bouchard and Roger (1912), i: 1–118, quoting Bouchard on p. 71.
30. Hueppe (1894), p. 152.
31. Georges-Henri Roger, 'Notions générales sur les infections' in Roger, Widal and Teissier (1929), i: 1–116, p. 6 ('two laws'); other examples: Gottstein (1897), pp. 44–45, quoting Max Verworn; Charrin (1902); Georges-Henri Roger, 'Introduction à l'étude de la pathologie générale' in Bouchard and Roger (1912), i: 1–118, p. 71.
32. This large literature was reviewed many times, notably in Lubarsch (1919); bibliography: von Engelhardt (1985); for favorable reception in Britain, see 'The Harben Lectures [editorial]' *Lancet* (1903), ii: 1244–45; Hamer (1909); quotation of Mach in Martius (1898), p. 92.
33. Diepgen (1926), p. 319.
34. Gottstein (1897), passim, esp. ch. 6.
35. Martius (1898), p. 105; His (1911), p. 25; von Pfaundler (1911), p. 61.
36. Pettenkofer first proposed such a formula in 1869; this final version: von Pettenkofer (1892), p. 807.
37. As quoted in Garrod (1931), repr. in Scriver and Childs (1989), p. 146 of Garrod's text.
38. See also Kopaczwekski (1936); on Tzanck and transfusion: Moulin (1991), pp. 159–61.
39. Grmek (1991) draws attention to the conflict.
40. Tzanck readily acknowledged the historical errors pointed out by his 'friend' Vallery-Radot: Tzanck (1934).
41. Founded in 1935, the *Revue* was co-edited by Vallery-Radot, the bacteriologist Robert Debré, and the Pastorian Gaston Ramon.
42. On the move toward function in France, see Roger, Widal, and Teissier (1929), i: vi–vii; on Rosenbach, Widal, and functional diagnosis see Faber (1930), ch. 5.
43. On family disease, see Patrice Pinell in this volume; on its special status in France see Hildreth (1994).
44. See, for example, Hueppe (1894), p. 151; Martius (1898), p. 103; Lämmerhirt (1903); Garrod (1931), repr. in Scriver and Childs (1989), p. 126 of Garrod's text.
45. For example: Ziegler (1886).
46. See Bulloch and Greenwood (1911), esp. p. 151.
47. As late as the 1920s, a few investigators argued for the possibility of paternal transmission; Harrison (1931), p. 198.

48. Even the bacteriologist and inventor of serodiagnosis for syphilis August Wassermann could refer to 'hereditary' syphilis, meaning vertically transmissible: Wasserman (1910).
49. Controversy centered on Riffel (1890); see Martius (1901); Cornet and Meyer (1904), pp. 156–58.
50. German work reviewed in Bauer (1917); Hart (1922). Examples of French work: Robin and Binet (1901); Robin (1910); Sergent (1914).
51. On France: Schneider (1982); Clark (1984), pp. 156–57; Nye (1984); Schneider (1990), esp. pp. 70–77; Hildreth (1994), p. 198; on Britain and the United States: Nye (1985), pp. 64–67; Mazumdar (1992), pp. 67, 79, 92, 146–48; Rosenberg (1974), pp. 224–25; Rosenberg (1976), p. 203; on neo-Lamarckism in various eugenics movements: Adams (1990).
52. Examples may be found in the citations in the note above.
53. Apert held that an acquired character might exceptionally continue to appear in later generations but only so long as the organism remained under the environmental influence that had first produced the character; Apert (1919), ch. 18.
54. The phrase is from Nye (1985), p. 66.
55. Churchill (1976), p. 146.
56. Martius (1909), ch. 6, esp. pp. 369–71; 'Discussion on the Influence of Heredity on Disease, with Special Reference to Tuberculosis, Cancer, and Diseases of the Nervous System' *Proceedings of the Royal Society of Medicine* General Reports (1908–09), 2:12; Bauer (1917), pp. 8–9.
57. Von Pfaundler (1911), p. 55; for a parallel French example, see my discussion of Apert below.
58. Other examples of usage of Forel's term in contradistinction to inheritance of acquired characters: Martius (1909), p. 371; Bauer (1917), p. 9; Hart (1922), p. 197. Forel influenced the early German race hygienists; Weindling (1989), pp. 71–72, 84–85.
59. Landouzy (1891), p. 428; Charrin and Riche (1897). For continuing relevance: Le Gendre (1912), esp. p. 378; Achard (1912), esp. p. 549; Courmont (1912), esp. pp. 579–80. Even the Koch school's strict-contagionist tuberculosis spokesman, Georg Cornet, though he rejected a role for true heredity, warned of the possibility that infected mothers could bequeath predisposition to their children in the form of tuberculin-induced hypersensitivity; Cornet and Meyer (1904), p. 157.
60. Baur, Fischer and Lenz (1921), i: 252–66.
61. Baur, Fischer and Lenz (1921), ii: 113–18; Proctor (1988), pp. 237–41.
62. Charrin and Gley (1896), esp. pp. 231–33. Polemic: Adami (1901), esp. p. 1322. See also 'Medical Doctrines of Heredity [correspondence]' *Lancet* (1903), ii: 56–57, 264–65. On the marginality of Charrin's views: Vignes (1924), pp. 196–99.
63. For example, Virchow's student and successor Orth (1909), pp. 21–23. By this time, Orth had accepted Weismann's teachings and given up his earlier belief in inheritance of acquired characters, on which see Churchill (1976).
64. I thank Annette Vogt, Max-Planck-Institut für Wissenschaftsgeschichte, for sharing with me her unpublished manuscript 'Agnes Bluhm—eine Gastwissenschaftlerin an einem Kaiser Wilhelm Institut,' from which comes the quotation of Correns' nomination.
65. I refer to the generation before that of Raymond Turpin, who in the 1930s began to resituate French eugenics in an obviously Mendelian frame. Apert's treatise opened with two chapters on familial diseases that followed Mendel's laws: Apert (1919); on Apert's Mendelism, see also Mazumdar (1996), p. 618. Other examples are Henri Vignes and Albert Touraine (see below). Even the prolific neo-Lamarckian and anti-Mendelian polemicist Felix Le Dantec accepted the *fact* of discontinuous Mendelian characters even as he called them 'microbes' and distinguished their pathological behavior from normal, continuous heredity: Le Dantec (1904).
66. Burian, Gayon and Zallen (1988), pp. 379–80.
67. My thanks to Jean-Paul Gaudillière for his thoughts on this issue. Germany too, on the other hand, had a strong movement against infant mortality; Weindling (1989).
68. The whole tradition, beginning in the 1890s, was reviewed in a series of lectures to the Ecole d'Anthropologie in 1924: Vignes (1924). Vignes does admit the possibility of inheritance of acquired characters as a first cause of inherited pathologies of 'degeneration,' pp. 145–46; Dareste is discussed in the third lecture; the teratological research of the psychiatrist Charles

Féré, in the first and second; for Dareste, Féré, and Isidore Geoffroy Saint-Hilaire, see also Apert (1919), p. 263.

69. On the lack of reference to Lamarck or Lamarckism in Pinard, see Schneider (1990), p. 73; Apert (1919) contains no reference to Lamarck or Lamarckism.
70. Lefaucheur (1992), pp. 430–32.
71. Commentators on Morel have recognized diversity of causes rather than a common mechanism of intoxication: Genil-Perrin (1913), pp. 56–57; Burgener (1964), pp. 40–41; Ackerknecht (1985), pp. 53–58; Rosenberg (1974), pp. 217–18; Weingart, Kroll and Bayertz, (1988), pp. 47–49.
72. See also Touraine (1933a).
73. This literature is reviewed in Ribbert (1893); Hahn (1904), pp. 303–07.
74. As quoted in Smith (1907), p. 377n.
75. Ehrlich to Smith, 7 Feb. and 29 July 1905; Paul Ehrlich Papers, Box 24, Kopirbuch XVI, pp. 6–8 and XVII, pp. 468–70, Rockefeller Archive Center, North Tarrytown, New York. I am grateful to the donor for permission to consult the Ehrlich papers.
76. Hahn (1904), pp. 269–72. Differences in exposure and hygienic habits also played a role.
77. Mendelsohn (1996), ch. 8.
78. For example: 'Discussion on the Influence of Heredity on Disease, with Special Reference to Tuberculosis, Cancer, and Diseases of the Nervous System' *Proceedings of the Royal Society of Medicine* General Reports (1908–09), 2: 9–142, p. 52.
79. *Neunte Internationale Tuberkulose-Konferenz, Brüssel, 6.–8. Oktober 1910: Bericht* (Berlin: Internationale Vereinigung gegen die Tuberkulose, 1911), pp. 19–116.
80. Also printed in *Annales de l'Institut Pasteur*, (1910), 24: 771–77.
81. *Neunte Internationale Tuberkulose-Konferenz, Brüssel, 6.–8. Oktober 1910: Bericht* (Berlin: Internationale Vereinigung gegen die Tuberkulose, 1911), pp. 33, 113.
82. For the earlier period, see Barnes (1995).
83. Apert (1919), p. 128.
84. In France, the important figures here were Calmette, Bezançon, Roger, Léon Bernard, and Robert Debré; see Bezançon (1922) on terrain doctrines; Georges-Henri Roger, 'La tuberculose' in Roger, Widal, and Teissier (1929), iv: 153–270, esp. p. 197. On the hegemony of the 'Behring-Römerschen Lehre' in German pathology, clinical medicine, and epidemiology: Lange (1936), p. 805.
85. Maurice Letulle and Paul Halbron, 'La tuberculose pulmonaire,' in Roger, Widal, and Teissier (1929), xii: 1–379, p. 17.
86. Extremist defense of the idea of non-contagiousness of tuberculosis, notably by Auguste Lumière, nevertheless persisted into the 1930s.
87. Debate as to whether the habitus asthenicus was cause or consequence of tuberculosis continued for decades; Probst (1982).
88. The first papers in the series were Calmette *et al.* (1924); Weill-Hallé and Turpin (1925). A later paper was Weill-Hallé, Turpin and Maas (1932).
89. Hart (1922), pp. 214–15; von Verschuer (1934), p. 8.
90. Bauer (1917), pp. 47–56 on infectious diseases.
91. Lubarsch (1921), quotations on pp. 816–17, 819. The first major synthesis of constitutional medicine and Mendelian genetics was Siemens (1921), which however was limited to rare conditions such as albinism and hemophilia.
92. See, for example, the rejection of a role for heredity in tuberculosis by Britain's leading veterinary scientist, John McFadyean (1909).
93. Pearl (1921) testifies to the weakness of genealogical and statistical study up to 1920.
94. Weinberg (1913), chs. 14–15 and pp. 156–57; his earlier research had produced better evidence for inherited predisposition: Mazumdar (1992), p. 94.
95. The first paper was Hirszfeld, Hirszfeld, and Brokman (1924); Keating (1998); Moulin (1991), pp. 169–77. On Hirszfeld and blood-group serology: Mazumdar (1995), ch. 14; Schneider (1983).
96. The best review and critique of this literature is Hill (1934); analogous research on plants had already begun before the First World War and is reviewed in Hayes (1930); on Webster, see

Olga Amsterdamska in this volume; the only German contributions before 1940 known to me are Neufeld and Etinger-Tulczynska (1933) and Küster and Kröning (1938); the only French contribution is Lesné and Dreyfus-Sée (1928); no such work appears in Baur's bibliography: Elisabeth Schiemann, 'Erwin Baur' *Berichte der deutschen Botanischen Gesellschaft* (1934), 52: 51–114.

97. Diehl and von Verschuer (1933), p. 60.
98. Baur, Fischer and Lenz (1927), i: 325.
99. On the problems of quantification in medicine: Matthews (1995).
100. The first edition of Baur, Fischer and Lenz (1921) included no discussion of infectious diseases except the role of the (presumably inherited) asthenic constitution in tuberculosis (pp. 213–14); the third edition (1927) included a whole section on susceptibility to infectious diseases, based on the new literature (pp. 311–22).
101. In the plenary session on predisposition at the International Tuberculosis Conference of 1910, only a delegate from London cited Pearson's work: *Neunte Internationale Tuberkulose-Konferenz, Brüssel, 6.–8. Oktober 1910: Bericht* (Berlin: Internationale Vereinigung gegen die Tuberkulose, 1911), p. 104. Pearson and his associates are not cited in Hart (1922); nor in Apert (1919); nor in Baur, Fischer, and Lenz (1921).
102. The chapter on cancer includes more French work; that on rheumatism, almost only German work.
103. Sergent and Turpin (1927), pp. 200–21, quotations on pp. 214, 219; 'mon maître': Turpin (1951), p. 110.
104. See Weindling (1985); Weingart, Kroll, and Bayertz (1988); Weindling (1989); Bergmann, Czarnowski, and Ehmann (1989); Deichmann (1992).
105. Garrod (1931), repr. in Scriver and Childs (1989).
106. Diehl and von Verschuer (1933). A 1936 sequel brought the total to 239 pairs.
107. See Olga Amsterdamska in this volume.
108. Apert (1919), pp. 9, 13, passim.
109. The liberal and internationalist atmosphere of the Koch Institute under Neufeld contrasted with the nationalist conservative Imperial Health Office laboratories and with the many university hygiene institutes where bacteriology was combined with hereditarian social hygiene. With the Nazi seizure of power in 1933, Neufeld stepped down from the directorship. See Hubenstorf (1994); Mendelsohn (1998).
110. Doerr (1937), pp. 645–46, 653.
111. On Verschuer after 1945: Weingart, Kroll, and Bayertz (1988), pp. 572–81; Ash (1995), pp. 915–18.
112. On the interwar medical crises and holistic health movements: Klasen (1984); Weisz (1998); Weindling (1985), p. 304, suggests that Weimar eugenics and human genetics should be understood partly as a program for biological reconstruction in a time of social and economic crisis.
113. Bauer (1929), pp. 149–50.
114. Diehl and von Verschuer (1933), pp. 457, 462.
115. von Verschuer (1934), pp. 7, 37–38; Diehl and von Verschuer (1933), pp. 90–91, 463.
116. von Verschuer (1934), pp. 8–9; Diehl and von Verschuer (1933), p. 90.
117. Though Verschuer was relieved of the editorship in 1964 by a younger generation anxious to distance itself from the Nazi past (Weingart, Kroll, and Bayertz, 1988, p. 580), he had titled his own postwar textbook *Genetik des Menschen: Lehrbuch der Humangenetik* (Munich and Berlin: Urban & Schwarzenberg, 1959). The older constitutionalism also retained a place in Verschuer's 1934 *Erbpathologie*; in the quasi-official publication of the Nazi eugenic and racial laws with medical and legal commentary: Gütt, Linden and Maßfeller (1936); and in the new, *Erbpathologie* volume of the 1940 edition of Bauer-Fischer-Lenz (Lange *et al.*, 1940).
118. Garrod (1931), repr. in Scriver and Childs (1989), p. 108 of Garrod's text.
119. On race and psychology: Mazumdar (1996), pp. 639–57; Paul (1995), chs. 4, 6; Haller (1963), chs. 7–10.
120. For example: von Verschuer (1934), p. 3.

121. Even Richard Goldschmidt, leader of physiological genetics in Weimar Germany, advocated sterilization laws; Weindling (1985) pp. 304, 315.

REFERENCES

C. Achard, 'Immunités et prédispositions morbides' in Bouchard and Roger (eds.) (1912) i: 484–556.

E.H. Ackerknecht, 'Diathesis: The Word and the Concept in Medical History' *Bulletin of the History of Medicine* (1982) 56: 317–25.

E.H. Ackerknecht, *Kurze Geschichte der Psychiatrie* 3rd ed. (Stuttgart: Ferdinand Enke, 1985).

J.G. Adami, 'An Address on Theories of Inheritance, with Special Reference to the Inheritance of Acquired Characters in Man' *British Medical Journal* (1901) i: 1317–23.

M.B. Adams (ed.), *The Wellborn Science: Eugenics in Germany, France, Brazil, and Russia* (New York and Oxford: Oxford University Press, 1990).

O. Amsterdamska, 'Medical and Biological Constraints: Early Research on Variation in Bacteriology' *Social Studies of Science* (1987) 17: 657–87.

O. Amsterdamska, 'Stabilizing Instability: The Controversy over Cyclogenic Theories of Bacterial Variation during the Interwar Period' *Journal of the History of Biology* (1991) 24: 191–22.

W. Anderson, 'Immunities of Empire: Race, Disease, and the New Tropical Medicine, 1900–1920' *Bulletin of the History of Medicine* (1996) 70: 94–118.

E. Apert, *L'hérédité morbide* (Paris: Flammarion, 1919).

F. Arloing, and A. Dufourt, 'Réflexions sur le cycle de l'infection tuberculeuse humaine' *Presse médicale* (1932) 40: 1877–80.

M.G. Ash, 'Verordnete Umbrüche—Konstruierte Kontinuitäten: Zur Entnazifizierung von Wissenschaftlern und Wissenschaften nach 1945' *Zeitschrift für Geschichtswissenschaft* (1995) 43: 903–23.

D.S. Barnes, *The Making of a Social Disease: Tuberculosis in Nineteenth-Century France* (Berkeley: University of California Press, 1995).

J. Bauer, *Die konstitutionelle Disposition zu inneren Krankheiten* (Berlin: Julius Springer, 1917).

J. Bauer, 'Wandlungen des Konstitutionsproblems' *Klinische Wochenschrift* (1929) 8: 145–50.

E. Baur, E. Fischer, and F. Lenz, *Grundriß der menschlichen Erblichkeitslehre und Rassenhygiene* 2 vols., (Munich: J.F. Lehmann, 1921).

E. Baur, E. Fischer, and F. Lenz, *Menschliche Erblichkeitslehre und Rassenhygiene* 3rd edn., 2 vols. (Munich: J.F. Lehmann, 1927).

E. Baur, E. Fischer, and F. Lenz (eds.), *Menschliche Erblichkeitslehre und Rassenhygiene* 5th edn., 2 vols. (Munich: J.F. Lehmann, 1940).

E. von Behring, *Diphtherie (Begriffsbestimmung, Zustandekommen, Erkennung und Verhütung)* (Berlin: Hirschwald, 1901).

E. von Behring, 'Disposition und Diathese' *Hamburgische medizinische Überseehefte* (1914) 1: 2–7.

E. von Behring, 'Über Idiosynkrasie, Anaphylaxie, Toxin-Überempfindlichkeit, Disposition und Diathese' (1914a) in *Gesammelte Abhandlungen: Neue Folge, 1915* (Bonn: Marcus & Weber, 1915) pp. 175–97.

E. von Behring, 'Tuberkulosebekämpfung' (1903) in *Gesammelte Abhandlungen: Neue Folge, 1915* (Bonn: Marcus & Weber, 1915) pp. 209–33.

A. Bergmann, G. Czarnowski, and A. Ehmann, 'Menschen als Objekte humangenetischer Forschung und Politik im 20. Jahrhundert: Zur Geschichte des Kaiser Wilhelm-Instituts für Anthropologie, menschliche Erblehre und Eugenik in Berlin-Dahlem (1927–1945)' in C. Pross and G. Aly (eds.), *Der Wert des Menschen: Medizin in Deutschland, 1918–1945* (Berlin: Edition Hentrich, 1989).

F. Bezançon, *Les bases actuelles de la problème de la tuberculose* (Paris: Gauthier-Villars, 1922).

F. Bezançon, 'Les tendances nosographiques actuelles et la notion de spécificité' *Presse médicale* (1932) 40: 1549–51.

A. Bluhm, *Zum Problem 'Alkohol und Nachkommenschaft': Eine experimentelle Studie* (Munich: J.F. Lehmann, 1930).

C. Bouchard, and G-H. Roger (eds.), *Nouveau traité de pathologie générale* 2 vols. (Paris: Masson, 1912).

W. Bulloch, and M. Greenwood, 'The Problem of Pulmonary Tuberculosis Considered from the Standpoint of Disposition' *Proceedings of the Royal Society of Medicine* (1910–11) Epid. Sect., 4: 147–84.

R.M. Burian, J. Gayon, and D. Zallen, 'The Singular Fate of Genetics in the History of French Biology, 1900–1940' *Journal of the History of Biology* (1988) 21: 357–402.

P. Burgener, *Die Einflüsse des zeitgenössischen Denkens in Morels Begriff der 'dégénérescence'* Med. Diss. (Zurich: Juris-Verlag, 1964).

A. Calmette, *L'infection bacillaire et la tuberculose chez l'homme et chez les animaux. Processus d'infection et de défense: Étude biologique et expérimentale* (Paris: Masson, 1920).

A. Calmette, and C. Guérin, 'Origine intestinale de la tuberculose pulmonaire' *Annales de l'Institut Pasteur* (1905) 19.

A. Calmette, 'L'hérédo-prédisposition tuberculeuse et le terrain tuberculisable' in *Neunte Internationale Tuberkulose-Konferenz, Brüssel, 6.–8. Oktober 1910: Bericht* (Berlin: Internationale Vereinigung gegen die Tuberkulose, 1911) pp. 49–57.

A. Calmette, A. Boquet, and L. Nègre, 'Rôle du facteur *terrain* dans l'évolution de la tuberculose expérimentale chez le lapin et chez le cobaye' *Comptes rendus des séances de l'Académie des sciences* (1923) 176: 1197–99.

A. Calmette *et al.*, 'Essais d'immunisation contre l'infection tuberculeuse' *Bulletin de l'Académie de médecine* (1924) 91.

A. Calmette *et al.*, 'Infection expérimentale transplacentaire par les éléments filtrables du bacille tuberculeux' *Comptes rendus de l'Académie des sciences* (1925).

G. Canguilhem, 'Aspects du vitalisme' (1952) in *La connaissance de la vie* (Paris: Vrin, 1989) pp. 83–100.

K.C. Carter, 'The Development of Pasteur's Concept of Disease Causation and the Emergence of Specific Causes in Nineteenth-Century Medicine' *Bulletin of the History of Medicine* (1991) 65: 528–48.

J.E. Chamberlin, and S.L. Gilman (eds.), *Degeneration: The Dark Side of Progress* (New York: Columbia University Press, 1985).

A. Charrin, 'Ueber die Multiplizität der krankheitserzeugenden Sekrete ein und derselben Bakterie' *Deutsche medizinische Wochenschrift* (1902) 28: 281–83.

A. Charrin and E. Gley, 'Sur l'action héréditaire et l'influence tératogène des produits microbiens' *Archives de physiologie normale et pathologique* 5th ser. (1896) 8: 225–37.

A. Charrin and A. Riche, 'Hérédité et tuberculose: Modifications héréditaires de l'organisme' *Comptes rendus de la Société de biologie* (1897) 49: 355–357.

F.B. Churchill, 'Rudolf Virchow and the Pathologist's Criteria for the Inheritance of Acquired Characteristics' *Journal of the History of Medicine* (1976) 31: 117–48.

L.L. Clark, *Social Darwinism in France* (Birmingham: University of Alabama Press, 1984).

W. Coleman, *Death Is a Social Disease: Public Health and Political Economy in Early Industrial France* (Madison: University of Wisconsin Press, 1982).

G. Cornet, and A. Meyer, 'Tuberkulose' in Kolle and Wasserman, (1903–04) ii: 78–177.

P. Courmont, 'De l'anaphylaxie' in Bouchard and Roger, (1912) i: 557–654.

P.D. Curtin, *Death by Migration: Europe's Encounter with the Tropical World in the Nineteenth Century* (Cambridge: Cambridge University Press, 1989).

F. Le Dantec, 'L'hérédité des diathèses ou hérédité mendélienne' *Revue scientifique* 5th ser. (1904) 1: 513–17.

U. Deichmann, *Biologen unter Hitler: Vertreibung, Karrieren, Forschung* (Frankfurt a.M. and New York: Campus, 1992).

K. Diehl, and O. von Verschuer, *Zwillingstuberkulose: Zwillingsforschung und erbliche Tuberkulosedisposition* (Jena: Gustav Fischer, 1933).

K. Diehl, and O. von Verschuer, *Der Erbeinfluß bei der Tuberkulose (Zwillingstuberkulose II)* (Jena: Gustav Fischer, 1936).

P. Diepgen, 'Krankheitswesen und Krankheitsursache in der spekulativen Pathologie des 19. Jahrhunderts' *Sudhoffs Archiv* (1926) 18: 302–27.

P. Diepgen, *Geschichte der Medizin* 2 vols. (Berlin: Walter de Gruyter, 1955).

R. Doerr, 'Die erblichen Grundlagen der Disposition für Infektionen und Infektionskrankheiten' *Zeitschrift für Hygiene und Infektionskrankheiten* (1937) 119: 635–59.

É. Duclaux, 'Sur les phénomènes généraux de la vie des microbes' *Annales de l'Institut Pasteur* (1887) 1: 145–52.

É. Duclaux, *Pasteur: The History of a Mind* trans. E.F. Smith and F. Hedges (orig. 1896, Philadelphia: W.B. Saunders, 1920).

L.C. Dunn, *A Short History of Genetics: The Development of Some of the Main Lines of Thought: 1864–1939* (New York: McGraw-Hill, 1965).

P. Ehrlich, 'Über Immunität durch Vererbung und Säugung' (1892) in F. Himmelweit, M. Marquardt, and Sir H. Dale (eds.), *The Collected Papers of Paul Ehrlich* 3 vols. (London: Pergamon Press, 1957) ii: 31–44.

D. von Engelhardt, 'Kausalität und Konditionalität in der modernen Medizin' in H. Schipperges (ed.), *Pathogenese: Grundzüge und Perspektiven einer Theoretischen Pathologie* (Berlin: Springer, 1985) pp. 32–58.

R.J. Evans, *Death in Hamburg: Society and Politics in the Cholera Years, 1830–1910* (Oxford: Clarendon, 1987).

K. Faber, *Nosography: The Evolution of Clinical Medicine in Modern Times* 2nd ed. (New York: Hoeber, 1930).

M. Foucault, *The Birth of the Clinic: An Archaeology of Medical Perception* (orig. 1963; New York: Vintage, 1975).

J-P. Gaudillière, 'Whose Work Shall We Trust? Genetics, Pediatrics and Hereditary Diseases in Postwar France' in P.R. Sloan (ed.), *Controlling Our Destinies: Philosophical, Ethical, Social and Historical Studies of the Human Genome Project* Notre Dame: University of Notre Dame Press, 1997) pp. 63-93.

P. Le Gendre, 'L'hérédité et la pathologie générale' in Bouchard and Roger (1912) i: 335–483.

G. Genil-Perrin, *Histoire des origines et de l'évolution de l'idée de dégénérescence en médecine mentale* (Paris: Alfred Leclerc, 1913).

C. Goring, *On the Inheritance of the Diatheses of Phthisis and Insanity: A Statistical Study Based on the Family History of 1,500 Criminals* (London, 1909).

E. Gotschlich, 'Allgemeine Prophylaxe der Infektionskrankheiten' in Kolle and Wassermann (1903–04) iv: 1–65.

A. Gottstein, 'Die Contagiosität der Diphtherie' *Berliner klinische Wochenschrift* (1893) 30: 594–98, p. 598.

A. Gottstein, *Allgemeine Epidemiologie* (Leipzig: Georg H. Wigand, 1897).

A. Gottstein, *Die Periodizität der Diphtherie und ihre Ursachen: Epidemiologische Untersuchung* (Berlin: August Hirschwald, 1903).

M. Greenwood, 'The Problem of Marital Infection in Pulmonary Tuberculosis' *Proceedings of the Royal Society of Medicine* (1908–09) Epid. Sect., 2: 259–82.

M. Grmek, 'Louis Pasteur, Claude Bernard et la méthode expérimentale' in M. Morange (ed), *L'Institut Pasteur: Contributions à son histoire* (Paris: Éditions la Découverte, 1991) pp. 21–44.

A. Gütt, H. Linden, and F. Maßfeller, *Blutschutz- und Ehegesundheitsgesetz* (Munich: J.F. Lehmann, 1936).

M. Hahn, Natürliche Immunität (Resistenz)' in Kolle and Wasserman (1903–04) iv: 266–331.

M.H. Haller, *Eugenics: Hereditarian Attitudes in American Thought* (New Brunswick: Rutgers University Press, 1963).

W.H. Hamer, 'Some Bacteriological Problems Considered from an Epidemiological Point of View' *Proceedings of the Royal Society of Medicine* (1908–09) Epid. Sect., 2: 89–132.

C. Hamlin, 'Predisposing Causes and Public Health in Early Nineteenth-Century Medical Thought' *Social History of Medicine* (1992) 5: 43–70.

L.W. Harrison, 'Syphilis' in Medical Research Council (ed.), *A System of Bacteriology in Relation to Medicine* 9 vols. (London: H.M.S.O., 1929–31) viii: 185–285.

M. Harrison, ' "The Tender Frame of Man": Disease, Climate, and Racial Difference in India and the West Indies, 1760–1860' *Bulletin of the History of Medicine* (1996) 70: 68–93.

C. Hart, 'Konstitution und Disposition' *Ergebnisse der allgemeinen Pathologie und pathologischen Anatomie des Menschen und der Tiere* (1922) I. Abt., 20: 1–435.

J. Harwood, *Styles of Scientific Thought: The German Genetics Community, 1900–1933* (Chicago: University of Chicago Press, 1993).

H.K. Hayes, 'Inheritance of Disease Resistance in Plants' *American Naturalist* (1930) 64: 15–36.

T. van Helvoort, 'History of Virus Research in the Twentieth Century: The Problem of Conceptual Continuity' *History of Science* (1994) 32: 185–235.

J. Héricourt, *Le terrain dans les maladies* (Paris, 1927).

E.N. Hiebert, 'The Energetics Controversy and the New Thermodynamics' in D. H. D. Roller (ed.), *Perspectives in the History of Science and Technology* (Norman, Oklahoma, 1971) pp. 67–86.

M. L. Hildreth, 'Doctors and Families in France, 1880–1930: The Cultural Reconstruction of Medicine' in A. La Berge and M. Feingold (eds.), *French Medical Culture in the Nineteenth Century* (Amsterdam and Atlanta: Rodopi, 1994) Clio Medica 25: 189–209.

A.B. Hill, *The Inheritance of Resistance to Bacterial Infection in Animal Species: A Review of the Published Experimental Data* Medical Research Council, Special Report Series, No. 196 (London: H.M.S.O., 1934).

H. Hirszfeld, L. Hirszfeld, and H. Brokman, 'Untersuchungen über Vererbung der Disposition bei Infektionskrankheiten, speziell bei Diphtherie' *Klinische Wochenschrift* (1924) 3: 1308–11.

W. His, 'Geschichtliches und Diathesen in der inneren Medizin' *Verhandlungen des deutschen Kongresses für innere Medizin* (1911) 28: 15–35.

W. His, *Über die natürliche Ungleichheit der Menschen* (Berlin: R. von Decker, 1928).

M. Hubenstorf, '"Aber es kommt mir doch so vor, als ob Sie dabei nichts verloren hätten": Public Health in Exile und der wissenschaftliche Unterbau des nationalsozialistischen "Volksgesundheitsdienstes": Zum Exodus von Wissenschaftlern aus den staatlichen Forschungsinstituten Berlins im Bereich des öffentlichen Gesundheitswesens' in W. Fischer *et al.* (eds.), *Exodus von Wissenschaften aus Berlin: Fragestellungen—Ergebnisse—Desiderate—Entwicklungen vor und nach 1933* (Berlin and New York: Walter de Gruyter, 1994) pp. 355-460.

F. Hueppe, 'Ueber die Ursachen der Gährungen und Infectionskrankheiten und deren Beziehungen zum Causalproblem und zur Energetik' *Verhandlungen der Gesellschaft deutscher Naturforscher und Ärzte: 65. Versammlung zu Nürnberg, 11.–15–September 1893* (1894) pt. 2.1: 134–58.

P. Keating, 'Holistic Bacteriology: Ludwik Hirszfeld's Doctrine of Serogenesis between the Two World Wars' in Lawrence and Weisz (1998).

E-M. Klasen, *Die Diskussion über eine 'Krise' der Medizin in Deutschland zwischen 1925 und 1935* (Med. Diss., Johannes-Gutenberg-Universität Mainz, 1984).

R. Koch, 'Die Cholera in Deutschland während des Winters 1892 bis 1893' (1893), in J. Schwalbe, G. Gaffky, and E. Pfuhl (eds.), *Gesammelte Werke von Robert Koch* 2 vols. (Leipzig: Georg Thieme, 1912) ii: 207–61.

R. Koch, 'Ärztliche Beobachtungen in den Tropen' (1898) and 'Ergebnisse der vom Deutschen Reich ausgesandten Malariaexpedition' (1900) in J. Schwalbe, G. Gaffky, and E. Pfuhl (eds.), *Gesammelte Werke von Robert Koch* 2 vols. (Leipzig: Georg Thieme, 1912) ii: 326–43, 435–47.

W. Kolle, and A. Wassermann (eds.), *Handbuch der pathogenen Mikroorganismen* 4 vols. (Jena: Gustav Fischer, 1903–04).

W. Kopaczwekski, 'Le terrain, le microbe et l'état infectieux: Claude Bernard ou Pasteur?' *Revue scientifique* (1936) 74: 417–25.

E. Küster, and F. Kröning, 'Der Einfluß des Genotyps und der Einfluß äußerer Faktoren auf die Tuberkuloseresistenz beim Meerschweinchen' *Arbeiten aus dem Staatsinstitut für experimentelle Therapie und dem Georg Speyer-Hause zu Frankfurt a.M.* (1938) 35: 38–68.

F. Lämmerhirt, 'Erblichkeit und familiärer Faktor bei den tuberkulösen Erkrankungen' *Politisch-Anthropologische Revue* (1902–03) 1: 789–98.

L. Landouzy, 'Hérédité tuberculeuse,' *Revue de médecine* (1891) pp. 410–31.

L. Landouzy, 'Voies conceptionnelle et transplacentaire de pénétration de la tuberculose (tuberculose congénitale). Hérédo-tuberculose: Hérédité de graine (hérédité bacillaire); Hérédité de terrain (hérédité dystrophiante). Les prédispositions à la tuberculose du fait de terrains viciés, innés ou acquis' *Revue de la tuberculose* 2nd ser. (1910) 7: 337–89, 417–70.

B. Lange, 'Äußere und innere Ursachen der Infektionskrankheiten, dargestellt am Beispiel der Tuberkulose' *Die Naturwissenschaften* (1936) 24: 802–09.

J. Lange *et al.*, *Erbpathologie* Baur, Fischer and Lenz (1940) vol. 1.2.

B. Latour, *The Pasteurization of France* trans. Alan Sheridan and John Law (Cambridge, Mass.: Harvard University Press, 1988).

C. Lawrence, 'Incommunicable Knowledge: Science, Technology, and the Clinical Art in Britain, 1850–1914' *Journal of Contemporary History* (1985) 20: 503–20.

C. Lawrence, and G. Weisz (eds.), *Greater than the Parts: Holism in Biomedicine, 1920–1950* (New York: Oxford University Press, 1998).

N. Lefaucheur, 'La puériculture d'Adolphe Pinard' in P. Tort (ed.), *Darwinisme et société* (Paris: Presses Universitaires de France, 1992) pp. 413–35.

E. Lesné, and G. Dreyfus-Sée, 'Séléction d'espèces animales à caractères immunitaires fixes: Transmission de ces caractères selon les lois mendéliennes et modifications durables obtenues par des vaccinations répétées' *Comptes rendus de la Société de biologie* (1928) 98: 922–24.

W. von Lingelsheim, 'Die bakteriologische Arbeiten der Kgl. Hygienischen Station zu Beuthen O.-Schl. während der Genickstarreepidemie in Oberschlesien im Winter 1904/05' *Klinisches Jahrbuch* (1906) 15: 373–488.

D.N. Livingstone, 'Human Acclimatization: Perspectives on a Contested Field of Inquiry in Science, Medicine and Geography' *History of Science* (1987) 25: 359–94.

C. López-Beltrán, 'Forging Heredity: From Metaphor to Cause, a Reification Story' *Studies in History and Philosophy of Science* (1994) 25: 211–35.

I. Löwy, 'Biomedical Research and the Constraints of Medical Practice: James Bumgardner Murphy and the Early Eiscovery of the Role of Lymphocytes in Immune Reactions' *Bulletin of the History of Medicine* (1989) 63: 356–91.

O. Lubarsch, 'Ursachenforschung, Ursachenbegriff und Bedingungslehre' *Deutsche medizinische Wochenschrift* (1919) 45: 1–4, 33–36.

O. Lubarsch, 'Zur Konstitutions- und Dispositionslehre' *Die Naturwissenschaften* (1921) 9: 812–19, (special issue on 'Pathology as Biological Science').

K.M. Ludmerer, *Genetics and American Society: A Historical Appraisal* (Baltimore and London: Johns Hopkins University Press, 1972).

F. Martius, 'Krankheitsursachen und Krankheitsanlage' *Verhandlungen der Gesellschaft deutscher Naturforscher und Ärzte* (1898) pt. 1, pp. 90–110.

F. Martius, 'Pathogenetische Grundanschauungen, Saecular-Artikel' *Berliner klinische Wochenschrift* (1900) 37: 429–34.

F. Martius, 'Die Vererbbarkeit des constitutionellen Faktors der Tuberculose' *Berliner klinische Wochenschrift* (1901) 38: 1125–30.

F. Martius, *Pathogenese innerer Krankheiten nach Vorlesungen für Studierende und Aerzte* (Leipzig and Vienna: Franz Deuticke, 1899–1909).

F. Martius, *Konstitution und Vererbung in ihren Beziehungen zur Pathologie* (Berlin: Julius Springer, 1914).

J.R. Matthews, *Quantification and the Quest for Medical Certainty* (Princeton: Princeton University Press, 1995).

R.C. Maulitz, '"Physician versus Bacteriologist": The Ideology of Science in Clinical Medicine' in M. J. Vogel and C. E. Rosenberg (eds.), *The Therapeutic Revolution: Essays in the Social History of American Medicine* (Philadelphia: University of Pennsylvania Press, 1979) pp. 97–107.

P.M.H. Mazumdar, *Eugenics, Human Genetics and Human Failings: The Eugenics Society, its Sources and its Critics in Britain* (London and New York: Routledge, 1992).

P.M.H. Mazumdar, *Species and Specificity: An Interpretation of the History of Immunology* (Cambridge: Cambridge University Press, 1995).

P.M.H. Mazumdar, 'Two Models for Human Genetics: Blood Grouping and Psychiatry in Germany between the World Wars' *Bulletin of the History of Medicine* (1996) 70: 609–57.

J. McFadyean, in 'Discussion on the Influence of Heredity on Disease, with Special Reference to Tuberculosis, Cancer, and Diseases of the Nervous System' *Proceedings of the Royal Society of Medicine* General Reports (1908–09) 2: 80–86.

J.A. Mendelsohn, *Cultures of Bacteriology: Formation and Transformation of a Science in France and Germany, 1870–1914* (Ph.D. diss., Princeton University, 1996).

J.A. Mendelsohn, 'From Eradication to Equilibrium: How Epidemics Became Complex after World War I' in Lawrence and Weisz (1998).

E. Metchnikoff, 'Recherches sur le choléra et les vibrions: Deuxième mémoire, Sur la propriété pathogène des vibrions' *Annales de l'Institut Pasteur* (1893) 7: 562–87.

E. Metchnikoff, 'Recherches sur le choléra et les vibrions: Quatrième mémoire, Sur l'immunité et la réceptivité vis-à-vis du choléra intestinal' *Annales de l'Institut Pasteur* (1894) 8: 529–89.

E. Metchnikoff, 'Études sur la flore intestinale' *Annales de l'Institut Pasteur* (1908) 22: 929–55.

B-A. Morel, *Traité des dégénérescences physiques, intellectuelles et morales de l'espèce humaine et des causes qui produisent ces variétés maladives* (Paris: J.B. Baillière, 1857).

A.M. Moulin, *Le dernier langage de la médecine: Histoire de l'immunologie de Pasteur au Sida* (Paris: Presses Universitaires de France, 1991).

F. Neufeld, 'Seuchenprobleme, II: Neue Ergebnisse der experimentellen Forschung' *Deutsche medizinische Wochenschrift* (1925) 51: 341–44.

F. Neufeld, 'Die Entwicklung der epidemiologischen Forschung seit Robert Koch' *Klinische Wochenschrift* (1935) 14: 737–41.

F. Neufeld, and R. Etinger-Tulczynska, 'Experimentelle Untersuchungen über die zeitlichen Schwankungen der natürlichen Empfänglichkeit für Infektionen' *Zeitschrift für Hygiene und Infektionskrankheiten (1933) 115: 573–93.*

R.A. Nye, *Crime, Madness, and Politics in Modern France: The Medical Concept of National Decline* (Princeton: Princeton University Press, 1984).

R.A. Nye, 'Sociology and Degeneration: The Irony of Progress' in E. Chamberlin and S. L. Gilman (eds.), *Degeneration: The Dark Side of Progress* (New York: Columbia University Press, 1985) pp. 49–71.

J. Orth, *Über die Bedeutung der Vererbung für Gesundheit und Krankheit* (Munich and Berlin: R. Oldenbourg, 1909).

M.A. Osborne, *Nature, the Exotic, and the Science of French Colonialism* (Bloomington: Indiana University Press, 1994).

L. Pasteur, and J. Joubert, 'Étude sur la maladie charbonneuse' and 'Charbon et septicémie' (1877) in P. Vallery-Radot (ed.), *Oeuvres de Pasteur* 7 vols. (Paris: Masson, 1922–39) vi: 164–71, 172–88.

D.B. Paul, *Controlling Human Heredity, 1865 to the Present* (Atlantic Highlands, N.J.: Humanities Press, 1995).

R. Pearl, 'The Relative Influence of the Constitutional Factor in the Etiology of Tuberculosis' *American Review of Tuberculosis* (1920–21) 4: 688–712.

K. Pearson, *A First Study of the Statistics of Pulmonary Tuberculosis* (London, 1907).

K. Pearson, and Pope, *A Second Study of the Statistics of Pulmonary Tuberculosis: Marital Infection* (London, 1908).

M. von Pettenkofer, 'Ueber Cholera mit Berücksichtigung der jüngsten Cholera-Epidemie in Hamburg' *Münchener medizinische Wochenschrift* (1892) 39: 807–17.

M. von Pfaundler, 'Diathesen in der Kinderheilkunde' *Verhandlungen des deutschen Kongresses für innere Medizin* (1911) 28: 36–85.

H. Pfeiffer, 'Experimentelle Studien zur Lehre von den Autointoxikationen' *Zeitschrift für Hygiene und Infektionskrankheiten* (1906) 54: 419–90.

D. Pick, *Faces of Degeneration: A European Disorder, c.* 1848–c. 1918 (Cambridge: Cambridge University Press, 1989).

C. Posner, '65. Versammlung der Gesellschaft Deutscher Naturforscher und Aerzte' *Berliner klinische Wochenschrift* (1893) 30: 955–56.

J. Probst, *Zur Entwicklung der Konstitutionslehre zwischen 1911 und 1980* (Med. Diss., Freiburg i. Br., 1982).

R.N. Proctor, *Racial Hygiene: Medicine under the Nazis* (Cambridge, Mass.: Harvard University Press, 1988).

C-R. Prüll, 'Holism and German Pathology (1914–1933)' in Lawrence and Weisz (1998).

H. Ribbert, 'Neuere Beobachtungen über die Disposition' *Deutsche medizinische Wochenschrift* (1893) 19: 12–15.

Riffel, *Die Erblichkeit der Schwindsucht und der tuberculösen Processe* (1890).

A. Robin, 'Les principes de la reminéralisation organique—Essai d'antisepsie pulmonaire directe—L'orientation du traitement de la phtisie pulmonaire' *Bulletin de l'Académie de médecine* 3rd ser. (1910) 63: 62–75.

A. Robin, and M. Binet, 'Les conditions de la diagnostic du terrain de la tuberculose' *Bulletin de l'Académie de médecine* 3rd ser. (1901) 45: 351–63.

G-H. Roger, F. Widal, and P.J. Teissier (eds.), *Nouveau traité de médecine* 22 vols. (Paris: Masson, 1925–29).

O. Rosenbach, *Das Problem der Syphilis und die Legende von der specifischen Wirkung des Quecksilbers und Jods* (Berlin: August Hirschwald, 1903).

O. Rosenbach, *Energetik und Medizin (Die Organisation als Transformator und Betrieb)* 2nd ed. (Berlin: August Hirschwald, 1904).

C. Rosenberg, 'The Bitter Fruit: Heredity, Disease, and Social Thought in Nineteenth-Century America' *Perspectives in American History* (1974) 8: 189–235.

C.E. Rosenberg, *No Other Gods: On Science and American Social Thought* (Baltimore: Johns Hopkins University Press, 1976).

F. Rott, 'Konstitution, Morbidität und Mortalität im frühen Kindesalter' in W. Jaensch (ed.), *Konstitutions- und Erbbiologie in der Praxis der Medizin* (Leipzig: Johann Ambrosius Barth, 1934) pp. 310–20.

A.R. Rushton, *Genetics and Medicine in the United States, 1800–1922* (Baltimore and London: Johns Hopkins University Press, 1994).

T. Schlich, 'Die Konstruktion der notwendigen Krankheitsursache: Wie die Medizin Krankheit beherrschen will' in C. Borck (ed.), *Anatomien medizinischen Wissens* (Frankfurt a.M.: Fischer, 1996) pp. 201–29.

W.H. Schneider, 'Towards the Improvement of the Human Race: The History of Eugenics in France' *Journal of Modern History* (1982) 54: 268–91.

W.H. Schneider, 'Chance and Social Setting in the Application of the Discovery of Blood Groups' *Bulletin of the History of Medicine* (1983) 57: 545–62.

W.H. Schneider, *Quality and Quantity: The Quest for Biological Regeneration in Twentieth-Century France* (Cambridge: Cambridge University Press, 1990).

C.R. Scriver, and B. Childs (eds.), *Garrod's Inborn Factors in Disease, Including an Annotated Facsimile Reprint of 'The Inborn Factors in Disease' by Archibald E. Garrod (1931)* foreword by Joshua Lederberg (Oxford: Oxford University Press, 1989).

É. Sergent, 'Le rôle du terrain dans la tuberculose' *Bulletin médical (Paris)* (1914) 28: 298–303.

É. Sergent, and R. Turpin, 'Les facteurs de terrain, autres que l'allergie, dans l'infection tuberculeuse' *Revue de la tuberculose* ser. 3, 8 (1927).

R.H. Shryock, *The Development of Modern Medicine: An Interpretation of the Social and Scientific Factors Involved* (New York: Knopf, 1947).

H.W. Siemens, *Einführung in die allgemeine Konstitutions- und Vererbungspathologie* (Berlin: Julius Springer, 1921).

H.E. Sigerist, 'Wandlungen des Konstitutionsbegriffs' *Karlsbader ärztliche Vorträge* (1929) 10: 97–108.

A.M. Silverstein, *A History of Immunology* (San Diego: Academic Press, 1989).

C. Sinding, *Le clinicien et le chercheur: Des grandes maladies de carence à la médecine moléculaire (1880–1980)* (Paris: Presses Universitaires de France, 1991).

T. Smith, 'Degrees of Susceptibility to Diphtheria Toxin among Guinea-Pigs: Transmission from Parents to Offspring' *Journal of Medical Research* (1905) 13: 341–48.

T. Smith, 'The Degree and Duration of Passive Immunity to Diphtheria Toxin Transmitted by Immunized Female Guinea-Pigs to their Immediate Offspring' *Journal of Medical Research* (1907) 16: 359–79.

W.C. Summers, 'From Culture as Organism to Organism as Cell: Historical Origins of Bacterial Genetics' *Journal of the History of Biology* (1991) 24: 171–90.

O. Temkin, *The Double Face of Janus and Other Essays in the History of Medicine* (Baltimore: Johns Hopkins University Press, 1977).

C. Timmermann, *Concepts of the Human Constitution in Weimar Medicine, 1918–1933* (M.A. thesis, University of Manchester, 1996).

A. Touraine, 'Hérédité mendélienne récessive du lapin "rex": Hérédo-dystrophie tréponémosique' *Bulletin de la Société francaise de dermatologie et de syphilographie* (1933) 40: 840–46.

A. Touraine, 'Tréponémose hérédo-dystrophique et virulente chez une lapine "castorrex"' *Bulletin de la Société francaise de dermatologie et de syphilographie* (1933a) 40: 1284–87.

S.W. Tracy, 'George Draper and American Constitutional Medicine, 1916–46: Re-inventing the Sick Man' *Bulletin of the History of Medicine* (1992) 66: 53–87.
R. Turpin, 'L'avenir des caractères acquis' *Progrès médical* (1932) i: 682–89.
R. Turpin, 'De l'influence des qualités héréditaires sur la sensibilité des animaux à l'égard des maladies infectieuses' *Revue d'immunologie* (1936) 2: 54–95.
R. Turpin, *Hérédité des prédispositions morbides* (Paris: Gallimard, 1951).
A. Tzanck, 'Les doctrines médicales: Introduction à l'étude de l'immunologie moderne' *Presse médicale* (1933) 41: 1449–53.
A. Tzanck, 'La notion du terrain: son importance doctrinal' *Presse médicale* (1934) 42: 315–17.
A. Tzanck, and V. Oumansky, 'Allergie (dix sens différents pour un même terme)' *Presse médicale* (1933) 41: 690–93.
A. Tzanck, and J. Cottet, 'Les intolérances rénales' *Presse médicale* (1934) 42: 415–19.
A. Tzanck, and R. André, 'L'immunité: le mot, le fait, l'idée' *Revue d'immunologie* (1936) 2: 273–87.
P. Vallery-Radot, 'Quelques considérations sur l'oeuvre de Pasteur' *Presse médicale* (1933) 41: 1877–80.
P. Vallery-Radot (ed.), *Louis Pasteur, Correspondance*, 4 vols. (Paris: Flammarion, 1940–1951).
O. von Verschuer, *Erbpathologie: Ein Lehrbuch für Ärzte* (Dresden and Leipzig: Theodor Steinkopff, 1934).
H. Vignes, 'Pathologie anténatale et hérédité morbide', 'La pathologie germinale', 'La pathologie anténatale de la fécondation à la maturité foetale', 'Les échéances de la pathologie anténatale' *Progrès médical* (1924) pp. 129–31, 142–46, 196–99, 231–33.
R. Virchow, 'Reizung und Reizbarkeit' *Virchows Archiv* (1858) 14: 1–63.
A. Wassermann, 'Ueber die persönliche Disposition und die Prophylaxe gegenüber Diphtherie' *Zeitschrift für Hygiene und Infektionskrankheiten* (1895) 19: 408–26.
A. Wassermann, 'La syphilis héréditaire' *Association française pour l'avancement des sciences, Compte rendu de la 38e Session, Lille, 1909: Notes et mémoires* (1910) pp. 1023–26.
L.T. Webster, 'Microbic Virulence and Host Susceptibility in Paratyphoid-Enteritidis Infection of White Mice, II' *Journal of Experimental Medicine* (1923) 38: 45–54.
B. Weill-Hallé, and R. Turpin, 'Premiers essais de vaccination antituberculeuse de l'enfant par le bacille Calmette-Guérin (BCG)' *Bulletin et mémoires de la Société médicale des hôpitaux de Paris* (1925) 49: no. 39.
B. Weill-Hallé, R. Turpin, and A. Maas, 'Étude clinique des réactions à l'infection tuberculeuse des nourrissons vaccinés par ingestion de BCG' *Presse médicale* (1932) 40: 1605–07.
W. Weinberg, *Die Kinder der Tuberkulösen* (Leipzig: S. Hirzel, 1913).
P. Weindling, 'Weimar Eugenics: The Kaiser Wilhelm Institute for Anthropology, Human Heredity and Eugenics in Social Context' *Annals of Science* (1985) 42: 303–18.
P. Weindling, *Health, Race and German Politics between National Unification and Nazism, 1870–1945* (Cambridge: Cambridge University Press, 1989).
P. Weingart, J. Kroll, and K. Bayertz, *Rasse, Blut und Gene: Geschichte der Eugenik und Rassenhygiene in Deutschland* (Frankfurt a.M.: Suhrkamp, 1988).
A. Weismann, *Vorträge über Descendenztheorie* 2 vols. in 1 (Jena: Gustav Fischer, 1902).
G. Weisz, 'A Moment of Synthesis: Medical Holism in France between the Wars' in Lawrence and Weisz (1998).
S. Wright, and P.A. Lewis, 'Factors in the Resistance of Guinea-Pigs to Tuberculosis, with Especial Regard to Inbreeding and Heredity' *American Naturalist* (1921) 55: 20–50.
E. Ziegler, 'Können erworbene pathologische Eigenschaften vererbt werden und wie entstehen erbliche Krankheiten und Missbildungen?' *Beiträge zur pathologischen Anatomie und Physiologie* (1886) 1: 361–406.

Chapter 2

FROM HEREDITY TO INFECTION: TUBERCULOSIS, 1870–1890

Michael Worboys

Many writers have claimed that the medical understanding of tuberculosis was transformed following Koch's identification of the tubercle bacillus as the essential cause of the disease in 1882.[1] Their argument is that tuberculosis, especially its pulmonary form then known as phthisis or consumption, ceased to be regarded as an inherited, constitutional disease and instead became a specific, acquired infection; put another way, its origin moved from generation-to-generation to person-to-person transmission. In this paper I challenge this received account by discussing changing treatments and ideas about the disease over the period 1860–1890 in Britain, focusing on the two groups most interested in the disease at this time—clinicians (mostly those in specialist consumption hospitals) and those who speculated on the pathology of the disease.[2]

There are three parts to my argument. Firstly, I consider what it meant to manage consumption was an inherited, constitutional disease c. 1860 and warn against equating the Victorian notion of constitutional proclivities with post Mendel ideas of inheritance.[3] Secondly, I suggest that the major change in the pathological understanding of tuberculosis occurred before the 1880s when the dominant model of tubercle formation in tissues changed from one of *degeneration* to *inflammation*. I show how this change influenced clinical practice and was associated with growing optimism that the disease could be treated and prevented. Thirdly, I discuss the reception of Koch's claims for the tubercle bacillus in the early 1880s. Here I contrast the ready acceptance of Koch's work by those versed in the new pathology with the resistance of rank and file clinicians. The former tended to see the bacillus as the irritant responsible for *inflammation*, while the latter were not resistant to the bacillus as such, but to the idea that consumption was ordinarily contagious. I conclude that an accommodation between the views of clinicians and 'pathologists' quickly developed around the metaphor of 'seed and soil', where the irritant bacillus (the seed) had to find constitutionally vulnerable tissue (the soil) to produce tubercular disease. The elaboration of this position led to the view that tuberculosis had limited infectivity and was contingently contagious, spreading more easily within and into weak bodies, especially in overcrowded and insanitary conditions.

CLINICIANS AND CONSUMPTION, 1860–70

There was a clear consensus amongst clinicians after mid-century that consumption ran in families and that inheritance was the major factor in its development. That said, aetiology was not a particular concern or interest of hospital doctors and general practitioners. They were usually confronted by those already sick and hence they valued the skills of diagnosis, prognosis and treatment, and the knowledge that underpinned them. Reflections on past or remote causes, which they could not verify nor alter, had little practical value. The causes of disease were only of major interest in public health medicine and although consumption was the single largest cause of death, being neither epidemic nor zymotic, it was not a 'public health' disease. Attempts were made in the 1860s and later to recategorise tubercular diseases as zymotic infections, but little came of these and it was not until the late 1890s that the consumption was the subject of initiatives and legislation in preventive medicine (Bryder, 1988).

The distinction common in public health medicine between predisposing and exciting causes was applied to many diseases in the third quarter of the century, including consumption (Hamlin, 1992).[4] The predominant predisposing cause was heredity, which included everything from feeble parents to having developed weak tissues. The commonly cited exciting causes were: cold and chilly winds, injuries to the bronchi and lungs, occupation, and exposure to bad air. The essence of the disease, as the name consumption indicates, was that these processes produced degeneration and a wasting of tissues. Predisposition worked in a number of ways, at one extreme being a passive 'proneness' to exciting causes, while at the other extreme, an active tendency for the spontaneous development of tubercles.[5] A person's constitution was not fixed at conception or birth, it could be altered by other diseases, diet, behaviour and environmental factors, hence a tubercular constitution could be acquired by unhealthy living, or countered by healthy living (Hogg, 1860). In many cases clinicians reported that it was impossible to separate predisposing and exciting causes, especially when tissues were weakened by other diseases.

While some clinicians did speculate explicitly on the aetiology and pathology of consumption, more was said about the management of the disease. These writings and reports suggest that after mid-century medical opinion became decidedly more optimistic about preventing, or at least arresting the disease by countering degeneration, and by strengthening weakened or devitalised tissues. These treatments did not aim to remove causes, rather they sought to modify morbid processes by aiding the healing powers of nature. Regimens were prescribed to 'raise' patients by 'improving faulty nutrition' and, if a phthisical constitution could not be remade, then at least it might be compensated for.[6] John Henry Bennett summed up the key elements of the restorative treatment as a good diet, cod-liver oil, exercise, a pure atmosphere, and bathing. In other words, stimulant, phlogistic, or 'sthenic treatments, aided by climate hygiene and medicine'.[7] Cod-liver oil became somewhat of a panacea, being described by E.A. Parkes in 1873 as 'an article of commerce on an enormous scale' and was said by some physicians to have brought a therapeutic revolution (Parkes, 1873).[8] Drugs were used largely to manage symptoms, especially linctuses to control coughs and promote expectoration. The variability of the disease, the patient's background, their temperament, their surroundings, and their income meant there was no formula for successful therapy, everything depended on the fine judgement of the doctor. The rich were advised to spend time in the warm Mediterranean and it was sometimes suggested to the poor that they visit relatives in the country to escape bad air and insanitary conditions. However, not all doctors were optimistic about treating the disease as many believed that consumption was not a disease at all, rather 'a mere mode of dying' so that all clinicians could do was to 'prolong a life naturally drawing to its close'.[9]

The idea that consumption was contagious—i.e. a zymotic disease that could be communicated person to person by 'specific germs contained in tuberculous matter'—was promulgated by William Budd in 1867 (Budd, 1867). He argued that there were many similarities between consumption and a disease like typhoid fever: both were specific, involved the multiplication of tissue in the body, led to the casting off *materies morbi*, and were infectious, especially in families. Coming from such a well-known figure these claims provoked a reaction. John Simon, the government's senior medical officer, was quite receptive to the idea and he speculated that both tuberculosis and cancer might be zymotic diseases. This led Lionel Beale to write to John Burdon Sanderson, two leading metropolitan experts on pathology, that Simon must have been 'Some wicked little microzyme ... pirouetting in ... the cortical portion of his cerebral convolutions'.[10] However, the most vigorous response to Budd's ideas came from clinicians. The attack was led by Richard Payne Cotton, Physician at the Brompton Hospital for Consumption in London, and Samuel Wilks (1824–1911), the senior physician at Guy's Hospital (Cotton, 1867; Wilks, 1867). Cotton offered detailed evidence that over his long career only a handful of Brompton staff had ever suffered from consumption and that contagion explained few, if any, of these cases. He reiterated that clinical experience had shown repeatedly that a particular constitution, inherited or acquired, was needed for consumption to manifest itself in an individual. Cotton had his own pathological model, namely, that tubercles were the product of consumption just as sugar was of diabetes. Wilks argued that consumption was quite unlike any zymotic disease. He pointed out that such fevers were specific and ran a predictable course, while consumption was one of the most variable of all diseases, liable to wax and wane erratically. In addition, epidemiological data showed that mortality for consumption was constant from month to month, which suggested that whatever was producing the disease was neither seasonal nor environmental (Hardy, 1993).

TUBERCULAR PATHOLOGY: DEGENERATION TO INFLAMMATION, 1870–80

The change from degeneration to inflammation in the pathogenesis of tuberculosis followed many decades of fraught debate. Until the 1860s, the dominant model in Britain was that deriving from the work and ideas of Laennec, whose assumption had been that tubercles were 'new formations', the primary consequence of a specific degeneration. Hughes Bennett's restorative treatment was based on combating such a process. However, the alternative view of Broussais, that tubercles were neither specific nor primary, but were the secondary products of inflammation, had some supporters in Britain and an increasing number in Germany. From the 1850s the work of Reinhardt, Neimeyer and especially Virchow, offered new evidence that supported Broussais's ideas. First, they distinguished true grey tubercular lesions from other processes producing caseous nodules, but then further histological studies led to the increasingly fine differentiation of lesions and to calls for the term 'tubercle' to be dropped altogether—another casualty of the change from gross morbid anatomy to cellular pathology.[11] Those following Laennec's ideas became known as the 'French School', while those following Broussais, or rather Virchow, with his modern ideas of inflammation and non-specificity, became known as the 'German School'. Both degenerative and inflammatory models were compatible with the hereditary character of the disease, in the former tubercles emerged *de novo* as 'new formations' as in cancers, while in the latter injuries prompted the emergence of the latent tubercular potential of cells.

A good insight into pathological thinking in Britain in the early 1870s can be gleaned from the writings and work of T.H. Green, whose textbook was the most popular pathology text with medical students from its first edition in 1871 until its last revised appearance in the 1900s (Green, 1871).[12] Green was a graduate of University College, London, who had studied in Berlin before becoming Assistant Physician, Lecturer on Pathology, and Superintendent of Post Mortem Examinations at Charing Cross Hospital, London. He was also Assistant Physician at the Brompton Hospital. The structure of Green's textbook remained unchanged through the 1870s and 1880s, being organised around the assumption that there were broadly two kinds of change in diseased tissues and cells—*degenerative* or *inflammatory*. In fact, this gave three processes: (i) degeneration, (ii) degenerative change combined with inflammation, (e.g. tumours); (iii) inflammation. When it came to causes, the many types of degeneration were classified as being either spontaneous (an expression of physiological weakness, poor cellular nutrition, or constitutional proclivities) or prompted by injury or irritation. Most inflammations were ascribed to injuries.

In the first edition of Green's book, pulmonary tuberculosis was discussed in the chapter on 'New Formations' along with cancer, both being examples of degeneration associated with inflammation. In the second edition, published two years later in 1873, tuberculosis was placed with 'Inflammations', Green had seemingly changed from the French to the German School.[13] As elsewhere in his initially slim volume, Green concentrated on the different processes and results of tubercular diseases, not their causes. However, in his account of the causes of inflammation four types of injury were recognised: traumatic (mechanical, chemical or temperature), infective, idiopathic, and specific (as in smallpox). Tuberculosis was said to be predominantly infective or idiopathic, terms which then had quite different meanings to now. Some pathologists, like Sanderson, proposed a restricted use of the term 'infective', applying it only to morbid materials transmitted by the blood vessels and lymphatic system *within* the body, as with metastases in cancer (Sanderson, 1877, 1878). Infectious or contagious diseases should have been confined to examples of the transmission of morbid matter or poisons between bodies, however, 'infective' was used as an adjective for both infect and infectious, and the restricted meaning never caught on. Idiopathic did not then carry its modern connotation of 'unknown cause', instead it implied to a primary morbid state, 'Arising by itself', and reminds us how the *de novo* development of disease was a very common nineteenth century aetiological assumption (Onions, 1993).[14]

New support for the specificity and unity of tubercular diseases had been given in the late 1860s by Villemin's experimental work, which claimed to show that tuberculous lesions could be produced in laboratory animals by the inoculation of tubercular material from a consumptive. Followers of Virchow in Germany and those opposed to Laennec's doctrines in Britain, argued that Villemin had only produced 'artificial tuberculosis' and that a laboratory procedure offered no guide to the natural development of the disease. The experiments seemed to have few implications for aetiology as they were so contrived, how could tubercular matter find its way so quickly and precisely from deep inside one body to the interior of another in ordinary life? However, one reading of inoculation studies, made by those in public health medicine, was that it warned of the dangers of ingesting tuberculous meat and milk. Such possibilities fed increasing anxieties in the 1870s about the transmission of diseases from animals to the humans, for example: glanders, anthrax, rabies and now tuberculosis.[15] The Sale of Food and Drugs Act, 1875, only regarded tubercular matter as a contaminant of food not infectious matter, however, the Public Health Act, 1875, contained clauses that enabled sanitary authorities to seize unsound tubercular meat as a nuisance. Medical Officers of Health (MOHs) soon began to use these powers. For example, in April 1877, the MOH for Rotherham, Yorkshire, seized a carcass because it contained tubercles, which he claimed to be 'the active agent which produces disease in humans'.[16] Veterinary journals gave most attention and support to the notion that tubercular diseases were contagious and some veterinarians, like George Fleming, were disappointed that the Contagious Diseases (Animals) Act, 1878, did not schedule tuberculosis in cattle for stamping out.[17] Such views spilled over into the columns of medical journals in the late 1870s, with discussions of reports of infectious cases, in the experimental studies of Villemin, Tappeiner and others, plus an increasing awareness that both medical and popular opinion in southern Europe regarded tubercular diseases as 'catching'.[18]

Historians have discussed Villemin's work in detail because of its seeming anticipation of modern views of the aetiology of tuberculosis;

indeed, a recent assessment says that he was the first to reveal the truth about the disease and bemoans the fact that 'British doctors continued to dismiss his results for the next two decades' (Smith, 1988). Yet, Villemin's experiments were not dismissed, they were repeated in Britain by Wilson Fox and James Clark, and in state-funded research by Sanderson.[19] Their results were published and debated, but crucially the British researchers obtained different results and interpreted these as having more to say about pathology than aetiology. Sanderson was not content just to inoculate tubercular matter, he also injected pus and left cotton setons in wounds as controls. He found that tubercles could form around the site of any inoculation or injury, which he took to mean that tubercles were produced by irritation and were the secondary products of non-specific inflammations. Thus, his work addressed the key pathological question of the 1870s, were tubercular diseases specific or not? Sanderson's alignment with the German School was clear to all and his influence may have been behind the change in Green's ideas as both men held part-time positions at the Brompton Hospital. One advantage of new irritation-inflammation model was that it removed one of the contradictions between clinical experience and Virchow's ideas, namely, how an affliction that presented clinically as a wasting disease could involve inflammation and the formation of new tissue.[20] Sanderson suggested tubercular diseases might usefully be seen as analogous to septic poisonings, as in both cases tissues contiguous with inflammations lost their vitality and degenerated, perhaps due to the diffusion of chemical poisons. In principle, at least, the new pathological ideas made tuberculosis a treatable disease, but only if clinicians could rid themselves of the belief that tubercles had some 'specific malignity' and approach lesions as the 'unabsorbed residue of common inflammatory processes' (Sanderson, 1869a).

Changing views of the disease were reflected in Green's writing on the pathology of tuberculosis. By 1874 he was siding with Virchow in calling for the abandonment altogether of the term 'tubercle'. However, he still sought to interpret tubercular pathology in clinical terms, differentiating the following types of disease: acute, scrofulous, tuberculo-pneumonic, catarrhal, fibroid, heamorrhagic, laryngeal and chronic (Green, 1878). Some spoke of this type of classification as the product of the 'British School' of tubercular pathology, which was based on solid anatomical work and clinical experience, without speculative excesses.[21] This approach allowed for the mutability of diseases, for example, catarrh and pneumonia worsening and developing into pulmonary tuberculosis. What mattered, perhaps for all lung diseases including the non-tubercular, was the strength and duration of inflammation. Thus, miliary tuberculosis would often follow from croupous pneumonia, a disease characterised by a brief but intense irritation that went on to affect the blood. Consumption on the other hand resulted from chronic inflammations that remained localised in the lungs. The key variable was the 'intensity' of the inflammation, which was determined by two factors: 'severity of injury, and susceptibility of tissue injured'.[22]

Green's views on the aetiology of consumption were expressed only briefly. Unsurprisingly, he favoured an inclusive, multi-factorial model. The first component was an 'inherent weakness of the lungs ... in most cases an inherited one', which could be unique to the lungs, or due to 'a general weakness of the tissues'.[23] Openness to consumption might also develop after a respiratory or debilitating illness, for example, bronchitis, pneumonia or pleurisy. His second aetiological component was injury, of which he listed three kinds: i) indirect injuries, like chills on the surface of the body; ii) direct injuries to the bronchi, such as catarrh and pneumonic products spreading into the lungs; and iii) infection by tuberculous matter from nodules in other parts of the body. The role of 'bad air' was explained as either an indirect injury, as re-breathed or devitiated air starved the body of oxygen, or a direct injury, as its high carboniferous (carbon dioxide) content irritated the alveoli, leading to some of this 'carbon' being encrusted in tubercular deposits. While he acknowledged an important role for 'internal' infection, Green made no mention of the disease being contagious from person to person despite the recent suggestions of Budd and others (Thompson, 1880).

The reshaping of consumption as an inflammatory affliction brought the addition of anti-inflammatory and antiseptic treatments to clinicians' range of therapies. These principally involved inhalations (with the vapours of carbolic acid, creosote, pine oil and hydrogen sulphide) and the removal of sufferers to cold, mountainous climates (Yeo, 1877).[24] Antiseptics were first used as deodorants because the breath of consumptives was offensive, though clinicians hoped that their acrid character would stimulate expectoration, reduce irritation and inflammation, and deepen breathing. However, some practitioners did see this approach as antiseptic in the Listerian sense, especially following Sanderson's speculations about parallels between tuberculosis and septic poisoning. Sanderson hoped that antiseptics would halt the absorption of poisons into body and generally act as an anti-putrefactant. Writing in Quain's *Dictionary of Medicine* in 1882, C.T. Williams, then Senior Physician at the Brompton, recommended a three part therapeutic strategy based on Green's pathological ideas: 1) medicinal—to improve nutrition, reduce inflammation and relieve symptoms; 2) dietetic and hygienic—to build up the patient; and 3) climatic—to help breathing and stimulate the system.[25] None of these offered a 'cure' as such, rather he believed they might help arrest the affliction by aiding the patient's ability to counter a disease whose hereditary character was 'too well known to be required to be more than stated'.

BACILLARY THEORY, 1880–90

The question of the contagiousness of tubercular diseases re-emerged from time to time in the 1870s, but at the end of the decade claims that the actual 'virus' of the disease had been isolated were made independently by several investigators across Europe.[26] Toussaint had found micro-organisms in diseased meat and reported producing the disease in pigs by feeding and inoculation (Cameron, 1881). In 1879 Klebs reported finding a micrococcus as the causative and contagious agent, which Cohnheim also found and called the *Monas tuberculosum*.[27] This work was controversial in Germany for its pathological rather than aetiological implications, as it suggested that tuberculosis was a unitary and specific disease. This work was widely publicised and discussed, being typical of the claims about germ aetiologies made for many other diseases at this time (Harley, 1881). In Britain, however, the status of bacteria and independent pathogenic organisms was less secure than elsewhere in Europe (Kern, 1972). For example, despite the success and acclaim that Lister received, as late as 1883 he was still not insisting that surgeons accept that germs were the actual cause of septic infections.

> '[W]e do not require any scientific theory to enable us to believe in antiseptic treatment. You need not believe in the truth of the germ theory of putrefaction and of septic agencies generally, no matter whatsoever with reference to antiseptic practice.'[28]

Other germ theories were still current, for example, Beale's bioplasm (defective protoplasm), as was evident in Charles Creighton's description of tubercular matter in 1881.

> The juices and particles of the primarily diseased body acquired a kind of spermatic virtue which gave them the power to communicate the specific disease . . . to another body in which they happened to lodge. But it is hardly possible to think of a neutral living organism being charged with the power of conveying so complex details of form and structure from one and structure from one body to another (Creighton, 1881, pp. 103–04).

Another possibility was that the cells of those with a tubercular constitution had within them, 'transmitted in semen, ova or placenta' and lying dormant, tubercular 'germs' whose development could be triggered by an injury or debility.[29] In this context, the *de novo* origin of tubercular germs had not been ruled out completely and it is significant that Charlton Bastian, the champion of spontaneous generation, wrote the entries on 'Bacteria' and 'Germs of Disease' in the Quain's *Dictionary of Medicine* in 1882.[30]

What a micrococcal or bacterial aetiology might mean was spelt out in August 1881 by Emmanuel Klein, soon to become 'the father of British bacteriology' (Bulloch, 1921). He suggested provocatively that the only aspect of inheritance that would remain was a 'delicacy and vulnerability of the tissues', that was 'neither more nor less than a secondary or accidental condition'.[31] Such views were resisted by clinicians who argued that it was the presence of the bacillus that was accidental as the disease as 'expression of a subtle organic dyscrasia operating in the individual . . . not so much a factor in the downfall of the physiological status of the individual, as it is proof of a previously operating decay in the organism' (Smith, 1881).

The last extended discussion of tuberculosis in Britain before Koch's now famous lecture in March 1882, took place in Glasgow in February and March 1881 (Coats, 1881). Joseph Coats and David Hamilton, the two leading teachers of pathology in Scotland had both sided with the German School that the disease was 'always the result of irritation'.[32] However, Coats had moved with Cohnheim to adopt the notion that the irritant was a 'self-propagating virus . . . introduced from without or *formed within the body*' (emphasis added). He qualified this by saying that 'all persons and all tissues are not equally susceptible to the virus, just as all people are not equally susceptible to the viri of typhus or any infectious disease'. Hamilton had remained faithful to the views of Sanderson and Virchow that the irritant was non-specific and that the capacity for tubercle formation lay in certain types of cell or a more general predisposition.[33]

At the end of the debate Coats summed up as follows, 'that pathology undoubtedly points to a virus as the cause of consumption, and clinical facts point to the state of the system as at the bottom of it'.[34]

Koch's now famous paper 'On the aetiology of tubercular disease' was read to the Physiological Society in Berlin on 24 March 1882 and published in Germany on 10 April. First notices appeared in Britain on 22 April in an article in the *Times* by John Tyndall and editorials in the *Lancet* and *Medical Times and Gazette*.[35] Tyndall emphasised: (i) Koch's claim that tubercular diseases were communicable, and (ii) how his work was a triumph of experimental research. The timing of the announcement of Koch's work was opportune for several reasons. Tyndall, along with leaders of the medical profession, senior scientists and key public figures, had recently founded the Association for the Advancement of Medicine by Research (A.A.M.R.) to promote experimental researches in medicine and counter the activities of anti-vivisectionists.[36] The great Darwin had died only three days before and his subsequent burial in Westminster Abbey was made an opportunity to celebrate British scientific achievement (Desmond and Moore, 1991). But most important was that consumption was the largest single cause of death in Britain, so any suggestion of new methods of prevention or cure were bound to be newsworthy. Most of the diseases linked to pathogenic bacteria before 1882 were somewhat obscure, for example anthrax, relapsing fever, and the septicaemia in mice, but consumption and then the controversy of cholera in 1883–84 gave bacteriology public and professional visibility. The bacillus quickly made numerous public appearances at prestigious medical and scientific meetings, such as at Lister's department at King's College, London, and at the Annual Meetings of the British Medical Association and the British Association for the Advancement of Science.[37] William Watson Cheyne, E.M. Nelson, G.A. Heron, Heneage Gibbes, and Alexander Ogston amongst others, showed tubercle bacilli to audiences at metropolitan and provincial medical meetings.[38] However, they did not just show this single bacillus, the germs of other diseases were demonstrated, along with the methods of staining, culturing, and observing bacteria. In short, bacterial germ theory itself was promoted through the tubercle bacillus. The reverse tack was taken by James Pollock in his Croonian Lectures on Consumption in March 1883, he gave a separate lecture on 'the whole theory of the induction of diseases from germs introduced from without' before discussing the tubercle bacillus and consumption in detail.[39]

Medical reactions to Koch's claims were mixed. At one extreme it was observed that the medicine had rarely experienced 'so sudden and complete casting aside of tradition' (Roberts, 1882). It was even said that tuberculosis had become a disease for the administrator and public health medicine rather than clinicians.[40] In 1883 Julius Dreschfeld, the German born Professor of Pathology at Owen's College Manchester, summed up the shibboleths Koch had challenged:

> 1. The fact that tubercle is hereditary; 2. That phthisis is not easily communicable from man to man; 3. That inoculation experiments on animals do not allow conclusions to be drawn as to their causal agency in man (Dreschfeld, 1883).

However, at the other extreme it was said that Koch had changed very little as clinicians rapidly harmonised the bacillus with 'the fact of heredity' by suggesting that 'physico-chemical changes must precede botanical aggression' (Neale, 1885). In other words, for phthisis to develop degenerative changes arising from a tubercular constitution or poor general health, inherited or acquired, had to have made the body 'open' to the bacillus (Thorowgood, 1885). This was, of course, a variant of the prevailing irritation-inflammation model, where the bacillus now became a specific irritant and tubercles its inflammatory product. Green expressed the accommodation as follows: 'two conditions ... are necessary in order to produce the disease: the presence of the tubercle bacillus, and some abnormal state of the pulmonary tissue'.[41] C.T. Williams, Senior Physician at the Brompton Hospital was also comfortable with such ideas.

> If we are to accept the bacillary theory at all, we must suppose that the various and well known predisposing causes of phthisis, such as dampness of soil, bas ventilation, bad confinements, and other debilitating conditions, must act by preparing a fit soil for the bacillus either by bringing about some low inflammatory condition ... or by weakening the resisting powers of the constitution (Williams, 1883, p. 176).

However, Williams believed that bacilli spread secondary inflammations rather than caused the disease and hence contended that non-bacillary tuberculosis was possible.

Amongst 'pathologists', many of whom were developing bacteriological interests and skills, the tubercle bacillus was readily accepted.[42] Given Koch's reputation, it was entirely plausible that he had isolated the 'virus' that had eluded Klebs and Cohnheim.[43] Clifford Allbutt objected to talk of the 'bacillus theory' saying 'it was the bacillus fact'.[44] Green accepted that Koch's bacillus was the 'injurious influence' that either caused the inflammation that led to tubercle formation, or produced tubercular lesions directly (Green, 1882). Joseph Coats, in his *Manual of Pathology* published in 1883, incorporated 'the recent discoveries of Koch, already confirmed by several others' as proven (Coats, 1883). A review of his book attacked Coats for his immediate acceptance of the tubercle bacillus and assured readers this agent would not be 'the whole story'.[45] In a rejoinder, Coats said that he had drafted the section on tuberculosis in February 1881, over a year before Koch's work appeared and yet, 'Koch's facts could thus be inserted without dislocating anything'.[46] However, not all pathologists took the bacillus in their stride. A particular problem was that Koch's work came on top of two decades of fraught debate in which tubercular pathology was said to have been swayed by 'every wind of doctrine' and where no idea lasted more than four or five years (Coupland, 1882).[47] Why should 'microbe pathology' be any different? David Hamilton, for example, remained loyal to the ideas of Virchow that tubercles formed when certain cells were altered by irritants (Hamilton, 1883).

Discussions amongst clinicians in 1882 and 1883, focused not on the pathology of the disease but on the question of its communicability, in what looked likely to become a classic conflict between experimental, laboratory studies and clinical experience (Yeo, 1882; Williams, 1882a; Pirrie, 1882; Smith, 1883; Skerrett, 1883). In January 1883 opinion on the contagiousness of consumption was explored in two surveys under the auspices of a drive to distill the wisdom of clinical opinion in 'collective investigations'. One of the aims of the movement was to bolster the authority of clinicians and counter the growing standing of laboratory and experimental investigators (Gull, 1884; Maclagan, 1884). The first survey, taken amongst members of the Cambridge Medical Society, revealed that 34 out of 38 doctors had never seen any instances of communicability.[48] A larger national survey was organised by the National Collective Investigations Committee, with the support of the British Medical Association asking 'Have you observed any case or cases in which Pulmonary Phthisis appeared to be communicated from one person to another'.[49] By the end of the year 1078 replies had been received, of which 62% said 'no' and 24% 'yes', leaving 14% undecided or equivocal.[50] The question was about communicability by any means, not person-to-person contagion as such. Indeed, respondents were happiest talking about indirect transmission via the inhalation of contaminated dust or air, the ingestion of contaminated food, or transmission from husband to wife by sexual intercourse. The idea that tubercular diseases were contagious had little direct support, unless the concept of contagion was widened. Most clinicians had at best become contingent contagionists, but few accepted that the bacillus was the single, necessary cause.[51]

The tubercle bacillus was announced in the wake of Pasteur's studies on the attenuation of germs, which reopened questions about the fixity of bacteria species and the possibilities of their attenuation. Two practical points were taken from this work in Britain; first, its potential to produce protective immunity, and second, its value in explaining the variability of tubercular diseases.[52] The former was discussed in an Editorial in the *Times*, no doubt informed by Tyndall, which accompanied the first notice of the bacillus in Britain.

> It is characteristic of many of the disease-producing bacilli, and probably all of them, that they can be so altered by cultivation as to produce a mild disease instead of a severe on, and that the designed communication of the former will afford protection against the latter. PASTEUR has lately shown how completely this may be accomplished in the case of the bacillus which causes splenic fever in cattle; and vaccination itself is now regarded merely as inoculation with the smallpox bacillus, after it has been modified in its character by being cultivated in the bodies of the bovine race. The experiments of Dr KOCH ... seem as yet to have been carried no further than to the repeated cultivation of the tubercle bacillus in it original virulence; but they will speedily be followed, as a matter of course, by attempts at cultivation in diminished intensity ... At this point, therefore, we come into manifest contact with the high probability that the thousands of human lives which are now sacrificed every year to the disease produced by bacilli may at no distant period be protected against these formidable enemies. (Capitals in original)[53]

It is significant how the tubercle bacillus was associated so quickly and powerfully with the potential for the control and perhaps, the eventual elimination of tuberculosis. The other attractive feature of Pasteur's work was that micro-organisms were not inevitably associated with specific disease processes. Indeed, Koch's experiments had shown that the susceptibility of different species to the bacilli varied, a fact that could be extended to intraspecies variability to explain differential individual, sexual and racial susceptibilities. Lister himself speculated that it might be possible to alter the blood using medicines to make humans or any other species for that matter non-susceptible to tubercular or any other bacterial infection.[54]

The major impact of the bacillary theory on the treatment of consumptive patients was the renewed impetus it gave to antiseptic treatments (Yeo, 1882a; Handford, 1885). These had been tried for many years, but new efforts were now made to deliver various antiseptics directly to lesions in the lungs, by inhalation and injection (Hill, 1885). New rationales for hygienic and climatic treatments were invented in terms of the antiseptic properties of Alpine air, sunlight or low temperatures.[55] Climate was talked about less with regard to temperature and place, and more about the properties of the air, especially the density of floating matter and temperature (Weber, 1885). Hermann Weber, the acknowledged expert on climatic therapies, began to speak of 'the hardening open-air treatment', which sought to toughened up lung tissue and the constitution (Worboys, 1992). The influence of the bacillary theory was also evident amongst surgeons treating tubercular joints and scrofula. Two new approaches were established: one line of treatment, which saw the bacillus as the enemy, aimed to cut away all infective tissue, while the other aimed to work from 'within' using rest and medicines to build up bodily resistance.

CONCLUSION

In January 1885, an editorial in the Medical Times offered the following assessment of the bacillus two years on:

> [T]he discovery and even the full acceptance of the etiologic doctrine associated with Koch's tubercle bacillus leaves the subject of tubercular phthisis as regards its diagnosis, prognosis and treatment, exactly where it was before.[56]

I would support this view and add that its pathology had also changed little. Certainly, the disease came to be regarded as specific, but the role of the bacillus was mostly understood to be that the irritant that sparked 'the hereditary tendency to phthisis' programmed into the tissues of those with tubercular constitutions (Heron, 1883, 1890). Otherwise, clinicians' work was little altered. The routine use of bacterial cultures or microscopy to decide the diagnosis of tuberculosis developed only slowly over the next two decades. The enthusiasts for bacteriology anticipated that bacterial diagnosis would become as easy and common as the use of the stethoscope and thermometer, and all the early textbooks assumed that clinicians would all become their own bacteriologist. This did not happen as bacteriological techniques proved difficult and expensive to master, and their regular use had to await the establishment of laboratory services in hospitals and public health departments. Without bacterial information, prognosis remained governed by clinical assessments of the patient's physiological and psychological strength. This was expressed in terms of constitution proclivities until the late 1880s when the language of resistance and immunity came to the fore, and with this peculiar 'predispositions' changed to a general weakness or vulnerability. Existing treatments were given new rationales and anti-irritant measures were pursued more vigorously, but few new therapies emerged, other than the open-air or sanatorium treatment, the initial aims of which were to produce constitutional progress in patients by improving breathing and lung capacity, weight gain and hygienic discipline. If anything, the main benefits promised by bacteriological understanding of the disease was in prevention (Jaccoud, 1885). For most of the 1880s this was only applied to the individual not to the public at large. Sufferers and potential sufferers consulting clinicians were encouraged to improve their bodily condition and to change their lifestyle to help counter any 'proneness' to the bacillus.[57] There were few injunctions, even amongst clinicians, to avoid exposure to the bacillus. They were confident that if the body and lungs were healthy and the effects of other lung 'irritants' minimised, the body would remain 'closed'—strong human soil ought to be able to counter the growth and effects of tubercular seeds. My argument that the impact of Koch's identification of the tubercle bacillus in Britain was quite limited would seem to be confirmed by the fact that public health approaches to control the transmission of the disease in Britain only began in earnest in the late 1890s. Initially, the bacillus was shaped by those with pathological interests to support prevailing views more than to challenge them, while clinicians debated contagion, grudgingly accepting a version of contingent contagionism where 'seed' and 'soil' were necessary and equal causes. Over the next decade the bacterial aetiology was re-interpreted within medicine and in the growing welfare movement to support social measures to prevent tuberculosis as a contagious disease. However, these programmes also sustained clinical approaches, notably treatment in sanatoria and hygienic measures to strengthen constitutions.[58]

NOTES

1. Grange and Bishop (1982); Cummins (1949); Dubos and Dubos (1952); Spink (1978); Brock (1988), pp. 129–39.
2. It would be anachronistic to call this group 'pathologists' as a distinctive group or specialism, with a particular identity and occupational role had yet to emerge. Many of those who taught pathology and wrote on the subject in the period 1860–90 earned their living as clinicians.

3. See Rosenberg (1974).
4. On consumption see Cotton (1852, 1858).
5. Thompson (1862), p. 173.
6. Bennett (1871), p. 571.
7. Bennett (1867), p. 137.
8. Also see Thompson (1862), pp. 67–85; Bennett (1871), pp. 571–77.
9. Bennett (1867), p. 137.
10. Beale to Sanderson, 16 November 1869, University College, London, Sanderson Papers, ADD 179/1 22–23.
11. 'Discussion of the Anatomical Relations of Pulmonary Phthisis to Tubercle of the Lungs *Trans. Path. Soc. Lond.* (1873), 24:284–388.
12. Further editions were published in 1873, 1875, 1878, 1881, 1884, 1889 (Stanley Boyd, ed.), 1895, 1900 (H.M. Murray, ed.) 1905, 1911 (W.C. Bosanquet, ed.).
13. T.H. Green, *An Introduction to Pathology and Morbid Physiology* (1873 edition); cf. Shepherd (1877); Powell (1878).
14. '1. ... Of the nature of a primary morbid state; not consequent upon another disease; 2. Of the nature of a particular affection or susceptibility', 1016 N.B *Idio*—'personal, private, peculiar'.
15. *British Medical Journal* (1880), ii: 173, 318, 472–73. Acland (1878); Blyth (1878).
16. Quoted in *Veterinary Journal* (1877), 5:67. The case was lost on 'insufficient evidence', largely because the medical opinion was divided.
17. See the discussion between Prof. Axe and J. Greaves, *Veterinary Journal* (1878), 6: 282, 366; (1878), 7: 52.
18. *Lancet* (1868), ii: 88–89 and (1880), ii: 860–61.
19. Dr Wilson Fox also repeated Villemen's work independently, see: Fox (1868); *10th Report of the Medical Officer of the Privy Council*, London, 1868; 'On the Communication of Tubercle by Inoculation' *British Medical Journal* (1868), ii: 316–17; *11th Report of Medical Officer of the Privy Council for 1868*, London, 1869, 91; Sanderson (1869).
20. Bennett (1871), pp. 551–54.
21. Ewart (1882); Coupland (1882); Williams (1882), pp. 1176–80; Williams (1877).
22. Green (1878), p. 69. Unlike Burdon Sanderson, Green felt able to accommodate constitutional and diathetic ideas with inflammation.
23. Green (1878), pp. 77–79.
24. Yeo also mentioned the use of hypophosphites and local rest. The latter required the use of strapping and other mechanical restraints to reduce respiratory movements, which it was said would in turn reduce the pulmonary circulation and hence the absorption of inflammatory products. Also see: 'Antiseptic Treatment' *British Medical Journal* (1880), i: 13, 23, 72.
25. Williams (1882), pp. 1181–83.
26. Yeo (1879); 'Cohnheim on Tuberculosis as an Infectious Disease' *Medical Times and Gazette* (1880), i: 38–39.
27. On other organisms described at this time see: Saundby (1882).
28. *British Medical Journal* (1883), ii: 855–60.
29. 'Tuberculosis as a Contagious Disease' *British Medical Journal* (1880), I:704; Klein (1881), pp. 81–85.
30. C. Bastian, 'Bacteria' and 'Germ Theory', in R. Quain *Dictionary of Medicine* London (1882), pp. 98–99, 532–33.
31. Klein (1881), p. 83.
32. Hamilton (1880), p. 388.
33. Hamilton (1880), p. 280.
34. 'The Discussion at Glasgow on the Pathology of Phthisis Pulmonalis' *British Medical Journal* (1881), I:818.
35. Tyndall (1882); 'Koch on a New Tubercular Germ' *Medical Times and Gazette* (1882), I: 411–12; 'Tubercular Organisms' *Lancet* (1882), i: 655–56.
36. 'The Association for the Advancement of Medicine by Research' *Lancet* (1882), i: 542–43.

37. 'Bacilli in Tuberculosis' *Lancet* (1882), I: 797. Koch's assistant, Geltdammer came across to help Watson Cheyne, who himself made a visit to Koch's laboratories to learn bacteriological techniques and to confirm that the bacillus was the causal agent. Cheyne (1883).
38. Also see Stevens (1882, 1883).
39. Pollock, 1883, pp. 378–80.
40. 'Retrospect 1884—Etiology of Phthisis' *British Medical Journal* (1884), ii: 1294.
41. Green (1883), p. 194.
42. Worth noting are the remarks of William Ewart, a fortnight before Koch's lecture in Berlin, he said he still trusted to the naked eye and the coarser anatomy it revealed. *Lancet* (1882), i: 383.
43. Green (1883), p. 195.
44. Allbutt was replying to a paper by Williams (1882a).
45. *British Medical Journal* (1883), ii: 878–79.
46. *British Medical Journal* (1883), ii: 997.
47. Robert Saundby referred to the history of tubercle as 'a history of controversy'; Saundby (1882), p. 178.
48. The Cambridge Medical Society organised its own poll in January 1883. *British Medical Journal* (1883), i: 77.
49. 'Collective Investigation on the Contagiousness of Pulmonary Phthisis' *British Medical Journal* (1883), i: 20, 77–78.
50. 'Results of Collective Investigation Committee' *British Medical Journal* (1883), ii: 983.
51. Pollock (1883), pp. 432 and 261–63, 320–22, 378–80, 431–33, 577–79, 605–07. Cf. Andrew (1884); Powell (1884); Bennett (1884); Ransome (1884); Yeo (1885).
52. Brock (1988), pp. 169–77.
53. *Times*, 22 April, 1882.
54. 'Micro-Organisms in Disease' *British Medical Journal* (1884), I: 733–34.
55. 'Curable Phthisis' *Medical Times* (1885), i: 384–86; Yeo (1885a).
56. 'Bacillary Phthisis' *Medical Times* (1885), i: 153.
57. 'The Distribution of the Tubercle Bacillus in the Lesions of Phthisis' *Medical Times* (1885), i: 95.
58. Bryder (1988), pp. 1–44; Worboys (1992), pp. 47–54.

REFERENCES

H.W. Acland, 'Rabies and hydrophobia' *Veterinary Journal* (1878), 6: 34.
J. Andrew, 'Lumleian Lectures on the Aetiology of Phthisis' *Lancet* (1884), i: 693–98, 785–89, 833–38.
J.H. Bennett, 'The Treatment of Pulmonary Consumption' *British Medical Journal* (1867), ii.
J.H. Bennett, 'Phthisis Pulmonalis' in J. Russell Reynolds, *A System of Medicine* (London: 1871).
J.H. Bennett, 'On the Contagion of Phthisis *British Medical Journal* (1884), ii: 704–06.
A.W. Blyth, *The Prevention of Rabies in Dogs* (London: 1878).
T.D. Brock, *Robert Koch: A Life in Medicine and Bacteriology* (Madison: 1988).
L. Bryder, *Below the Magic Mountain: A Social History of Tuberculosis in Twentieth Century Britain* (Oxford: 1988), pp. 1–45.
W. Budd, 'Memorandum on the Nature and the Mode of Propagation of Phthisis' *Lancet* (1867), ii: 451–52.
W. Bulloch, 'E. E. Klein' *Journal of Pathology and Bacteriology* (1921).
C. Cameron, 'Micro-Organisms and Disease' *British Medical Journal* (1881), ii: 586.
W.W. Cheyne, 'Abstract of a Report on the Relation of Microorganisms to Tuberculosis' *British Medical Journal* (1883), pp. 507–09.
J. Coats, 'On Phthisis Pulmonalis, especially in relation to Tuberculosis' *Glasgow Medical Journal* (1881), 15: 254–69.
J. Coats, *A Manual of Pathology* (London: 1883), p. 164.
R.P. Cotton, *The Nature and Symptoms of Consumption* (London: 1852), p. 53.
R.P. Cotton, *On Consumption: Its Nature, Symptoms and Treatment* (London: 1858).
R.P. Cotton, 'Memorandum on the Nature and the Mode of Propagation of Phthisis' *Lancet* (1867), ii: 550–51.

S. Coupland, 'On Tubercle' *British Medical Journal* (1882), i: 186–87.
C. Creighton, *Bovine Tuberculosis in Man* (London: 1881).
L. Cummins, *Tuberculosis in History* (London: 1949).
A. Desmond, and J. Moore, *Darwin* (London: 1991), pp. 664–77.
J. Dreschfeld, 'Micro-Organisms in Their Relation to Disease' *British Medical Journal* (1883), ii: 1056.
R. Dubos, and J. Dubos, *The White Plague: Tuberculosis, Man and Society* (Boston: 1952), pp. 102–03.
W. Ewart, 'On Pulmonary Cavities; Their Origin, Growth and Repair' *Lancet* (1882), i: 515C.
W. Fox, *On the Artificial Production of Tubercle* (London: 1868).
J.M. Grange and P.J. Bishop, '"Über Tuberkulose": A Tribute to Robert Koch's Discovery of the Tubercle Bacillus' *Tubercle* (1982), 26: 3–17.
T.H. Green, *An Introduction to Pathology and Morbid Anatomy* (London: 1871).
T.H. Green, *The Pathology of Pulmonary Consumption* (London: 1878).
T.H. Green, 'Lectures on Phthisis' *Lancet* (1882), i: 813.
T.H. Green, 'A Lecture on the Tubercle-Bacillus and Phthisis' *British Medical Journal* (1883), i.
W. Gull, 'An Address on the Collective Investigation of Disease' *British Medical Journal* (1884), ii: 305–08.
J. Hamilton, 'Pathology of Tubercle' *British Medical Journal* (1880), ii.
D.J. Hamilton, *On the Pathology of Bronchitis, Catarrhal Pneumonia, Tubercle and Allied Lesions of the Human Lung* (London: 1883).
C. Hamilon, 'Predisposing Causes and Public Health in Early Nineteenth Century Medical Thought' *Social History of Medicine* (1992), 5: 43–70.
H. Handford, 'Tubercle and the Tubercle Bacillus' *Lancet* (1885), ii: 1038–40.
A. Hardy, *The Epidemic Streets* (London: 1993), p. 227.
G. Harley, 'Some New facts Connected with the Action of Germs in the Production of Human Disease' *Medical Times and Gazette* (1881), ii: 570–73, 596–98, 651–53, 705–06, 732–34.
G.A. Heron, 'Some of the More Recent Facts and Observations Concerning the Bacillus of Tubercle' *British Medical Journal* (1883), ii: 805–07.
G.A. Heron, *Evidences of the Communicability of Consumption* (London: 1890).
A. Hill, *The Inhalation Treatment of Diseases of the Organs of Respiration* (London: 1885).
J. Hogg, *Practical Observations on the Prevention of Consumption* (London: 1860), p. 16.
S. Jaccoud, *The Curability and Treatment of Pulmonary Phthisis* (London: 1885).
L. Kern, *Deutsche Bakteriologie im Spiegel englischer medizinischer Zeitschriften, 1875–1885* (Zurich: 1972).
E. Klein, 'Etiology of Miliary Tuberculosis' *Practitioner* (1881), 28: 81–85.
T.J. Maclagan, 'On Methods of Therapeutic Research' *British Medical Journal* (1884), ii: 260–61.
H. Neale, 'The Germ Theory of Phthisis' *British Medical Journal* (1885), i: 897.
C.T. Onions (ed.), *The Shorter Oxford English Dictionary on Historical Principles* (Oxford: 1993).
E.A. Parkes, 'Address in Medicine' *British Medical Journal* (1873), ii: 143.
W. Pirrie, *Lancet* (1882), ii: 171.
J.E. Pollock, 'Modern Theories and Treatment of Phthisis' *Medical Times and Gazette* (1883), i.
R.D. Powell, *On Consumption and on Certain Diseases of the Lungs and Pleura* (London: 1878).
R.D. Powell, 'On the Causative Relations of Phthisis' *British Medical Journal* (1884), ii: 608–704.
A. Ransome, *The Limits of the Infectivity of Tubercle* (Manchester: 1884).
R. Roberts, 'Family history' *British Medical Journal* (1882), ii: 624.
C.E. Rosenberg, 'The Bitter Fruit: Heredity, Disease and Social Thought in Nineteenth Century America' *Perspectives in American History* (1974), 7: 189–235.
J.B. Sanderson, 'On the Practical and Pathological Bearings of Recent Researches as to the Artificial Production of Tubercle' *British Medical Journal* (1869), ii: 274–75.
J.B. Sanderson, 'Phthisis Ab Haemoptysi' *Lancet* (1869a) i: 523.
J.B. Sanderson, 'The infective process of disease' *British Medical Journal* (1877), ii: 879–81, 913–15.
J.B. Sanderson, 'The infective process of disease' *British Medical Journal* (1878), i: 1–2, 45–47, 119–20, 179–83.
R. Saundby, 'Recent researches on tubercle and their bearings on the treatment of consumption' *Practitioner* (1882), 29: 178–83.

A.B. Shepherd, *Goulstonian Lectures on the Natural History of Pulmonary Consumption* (London: 1877).

E.M. Skerrett, 'Clinical Evidence Against the Contagiousness of Phthisis' *Bristol Medico-Chirurgical Journal* (1883), i: 48–70.

A.T. Smith, 'Some Questions on the Pathological Relations of Tubercle, Struma and Phthisis Pulmonalis' *Glasgow Medical Journal* (1881), 15: 110.

R.S. Smith, 'Proofs of the Existence of a Phthisical Contagion' *Bristol Medico-Chirurgical Journal* (1883), i: 1–47.

F.B. Smith, *The Retreat of Tuberculosis, 1850–1950* (London: 1988), p. 35.

W.W. Spink, *Infectious Disease: Prevention and Treatment in the Nineteenth and Twentieth Centuries* (Minneapolis: 1978), pp. 220–21.

J.L. Stevens, *British Medical Journal* (1882), ii: 735.

J.L. Stevens, 'The Tubercle Bacillus and its Relations to Phthisis Pulmonalis' *Glasgow Medical Journal* (1883), 19: 348–54.

T. Thompson, *Clinical Lectures on Pulmonary Consumption* (London: 1862).

R.E. Thompson, 'The Infection of Phthisis' *Lancet* (1880), ii: 726–28.

J. Thorowgood, 'The Communicability of Consumption' *British Medical Journal* (1885), I: 889.

J. Tyndall, 'Tubercular Disease' *Times* 22 April 1882, p. 5.

H. Weber, *Croonian Lectures on the Hygienic and Climatic Treatment of Chronic Pulmonary Phthisis* (London: 1885).

S. Wilks, *Lancet* (1867), ii: 600.

C.T. Williams, 'The Varieties of Phthisis' *British Medical Journal* (1877), ii: 405–07, 435–36.

T. Williams, 'Phthisis' in R. Quain, *Dictionary of Medicine* (London: 1882).

C.T. Williams, 'The Contagion of Phthisis' *British Medical Journal* (1882a) ii: 618–21.

C.T. Williams, 'Relations of Tubercle Bacillus to Phthisis' *Lancet* (1883), ii.

M. Worboys, 'The Sanatorium Treatment for Consumption in Britain, 1890–1914' in J.V. Pickstone (ed.), *Medical Innovations in Historical Perspective* (London: 1992), pp. 47–71.

I.B. Yeo, *The Results of Modern Research in the Treatment of Phthisis* (London: 1877).

I.B. Yeo, 'The Bacteric Origin of Tuberculosis and its Antiseptic Treatment' *British Medical Journal* (1879), i: 983.

I.B. Yeo, 'The Contagiousness of pulmonary consumption' *British Medical Journal* (1882), i: 895–98.

I.B. Yeo, *The Contagiousness of Pulmonary Consumption and its Antiseptic Treatment* (London: 1882a).

E.B. Yeo, 'Some Points of the Aetiology of Phthisis *British Medical Journal* (1885), I: 772–76.

J.B. Yeo, 'Clinical Lectures on the Treatment of Disease: Phthisis' *Medical Times* (1885a) i: 407–11.

Chapter 3

PURITY AND DANGER IN COLOR: NOTES ON GERM THEORY, AND THE SEMANTICS OF SEGREGATION, 1885–1915

JoAnne Brown

Historians of medicine in the United States have recognized how ideas about race condition the terms of public health. I want to question a key word in the history of U.S. race relations, 'segregation,' to explore how ideas about public health have conditioned the political terms of race.[2] I focus on a semantic shift, in the late 1890s, that has been duly noted but not well explained by historians of race relations.[3] Prevailing usage regarding racial policy began to change, near the turn of the century, from 'separation of the races' to simply 'segregation.' It is my argument that this salient semantic shift not only *may* be explained by referring to the medical history of the period, but that this explanation deepens our understanding of both the continuities and the discontinuities between racial slavery in the nineteenth century and racial segregation in the twentieth (Woodward, 1974). This essay explores how invidious assumptions about race persisted in medical discourse even as the hereditarian understanding tuberculosis was revised in light of germ theory. This entailed a metaphoric translation from an older vocabulary of 'race' inscribed in 'blood,' to a newer one of 'germs' figured in saliva.[4] It entailed another translation between the specialized vocabulary of medicine and the larger political language in which medicine is always situated, the two being contexts for one another.[5]

As others have pointed out, racism predates segregation, and could as well explain other forms of discrimination and violence. It follows that racism alone cannot explain any *particular* form of discrimination, including the new legal forms called racial 'segregation'.[6] The consolidation of a formal, legal system of racial segregation in the United States after 1890 demands more specific explanation.

By bringing the history of medicine and public health into this political historiography on race, however, I by no means minimize race as an explanatory category. Rather, I want to show how profoundly constructs of race were linked to matters of life and death *within* and *between* the changing, mutual contexts of medical thought and of politics. Medical authority already weighed heavily in the politics of race.[7] At a key moment in the history of medicine, the key word 'segregation' in effect translated one conjoined legacy of hereditarian and environmental thought into another, into a new ideology, ratified by germ theory, that I term 'Social Contagionism'.[8]

This semantic inquiry reorganizes some of the evidence wrought by political and economic historians as to whether formal, legal forms of segregation were, as C. Vann Woodward asserts, something new in American race relations, or whether, as revisionists would have it, these 'jim crow' laws were the effective continuance of slavery itself and its constructs of race. I take seriously the powerful rhetoric of disease not, as other scholars have viewed it, as mere pretext, but as context: as political metaphor all the more powerful because it spoke to material facts of life and death in the 1890s and beyond. Health concerns were central not only to the violent and shameful segregationist rhetoric of white supremacy, *but to its most thoughtful critics*. This paper explores the medical domain whence radical white supremacy borrowed some of its illegitimate power, capitalizing on old fears and new ideas, affecting neither a new institutions nor a continuance of old ways, but a political translation from customary 'separation' to legalized 'segregation' through the scientific authority of germ theory.

In March 1910 the Atlanta Anti-Tuberculosis Association proposed a city ordinance to 'segregate' all tuberculous patients from their families.[9] This provision would have fallen disproportionately on the classes with high TB rates: the 'Negro race,' the 'Irish race,' and the poor in general (Hoffman, 1896, 1904).[10] Yet this Atlanta segregation initiative, approved in 1913, was not the beginning of a radically new racial policy.[11] It was, rather, a late manifestation of a public health 'crusade' against tuberculosis begun in Philadelphia in 1888, given statistical credence in 1896, formalized nationally in 1904, and celebrated internationally in 1908, that translated both racial and medical precepts from an environmental-hereditarian vocabulary into a contagionist idiom (Flick, 1888; Lee, 1900; Hoffman, 1904).[12] This translation, I believe, imbued the medical term 'segregation' with its now-familiar political, racial meaning.

The proposed Atlanta ordinance was one of many legal and regulatory steps taken to control tuberculosis, once American physicians accepted its bacterial nature, following Dr. Lawrence Flick's 'crusade' to establish 'the contagiousness of phthisis' in the wake of Koch's 1882 discovery (Flick, 1888; Flick, 1891; Flick, 1903).[13] Once redefined as contagious, tuberculosis became the object of an unprecedented publicity campaign that made it recognized as the most devastating disease of the day.[14] In 1896, the year of the infamous U.S. Supreme Court decision in *Plessy v. Ferguson*, Frederick Hoffman's influential statistical study, *Race Traits and Tendencies of the American Negro*, established in a scientific idiom the post-Reconstruction association of African Americans with tuberculosis (Hoffman, 1896).[15]

What critics labelled 'phthisiophobia,' an undue fear of tuberculosis and all tubercular patients, appeared by 1900 (Baur, 1959; Adams, 1905; McClure, 1906; Spivak, 1906).[16] Earlier, in the United States, tuberculosis patients and their bodily emanations were not considered dangerous. The disease diathesis was seen as hereditary, with right living and a good environment being the safeguards of health for those predisposed.[17] The early climate therapies and sanatoria arose out of this ecological perspective (Bowditch, 1862).[18] By the turn of the twentieth century, this emphasis had been fully overlaid—not yet supplanted—by the new germ theory.

Numerous public health initiatives followed from this new understanding and wider fears. In 1900, California legislators attempted to bar consumptive health-seekers from entering the state, a bill patterned after yellow fever, bubonic plague, and cholera quarantines, and after the medical 'segregation' of leprosy sufferers, an analogy endorsed by Koch himself in 1901, and again in his 1906 Nobel lecture (Baur, 1959).[19] Quarantine isolates the sick or exposed from the healthy for a limited period of time, often forty days, whereas medical 'segregation,' as per leprosy, referred to longer-term or permanent separation. The problem of medical segregation arose in the 1890s out of discussions on registering tuberculosis patients.[20]

Dr. Edward Baldwin of Saranac Lake, New York noted in 1903, 'in the present state of unreasonable fear with which the public and many physicians are seized a consumptive postman or a butcher's or grocer's delivery clerk will be boycotted as ruthlessly as the poor seamstress, clerk or schoolteacher.' Baldwin reminded his colleagues:

> I believe it is time to pause and remind the public and that part of the profession who are lending their aid in the unfeeling campaign against the poor consumptive, that the millennium has not yet arrived; that complete protection cannot be obtained in the present state of society; and that frequently they are straining at a poor little gnat of fancied danger while swallowing, or inhaling, a large camel of infection, which they cannot avoid, try as they will (Baldwin, 1903, p. 8).

Baldwin called for 'reasonable laws, abreast of the intelligence of the people,' adding, 'let us register and disinfect, and above all, educate; but let us be humane and extremely careful how far we limit or restrict the liberty of consumptives who are decent' (Baldwin, 1903, p. 8). Despite such sane advice, exclusionary efforts mounted. 'Many of the leading winter hotels throughout the South are advertising this season—"no consumptives taken,"' noted a New York paper.[21] Others denounced the 'almost mediaeval cruelty' visited upon tuberculosis patients by a medical profession and a public who regarded them as 'lepers'.[22] A New York doctor urged political caution on his colleagues:

> I think that physicians have been laying too much stress on the bacillus, and we have raised up fanatics throughout the country in this legislation, and the consequent result has been for the public to regard the consumptive as a leper. We ought to recollect, I think, that not only have we individual cases to treat, but we have in our care the public mind in matters of legislation.[23]

Philadelphia's Dr. Lawrence Flick, who suffered from the disease himself, was particularly cautious, resisting the antisocial implications taken from his own doctrine of contagion. He too worried about 'overzealous converts to bacteriology expatiate upon the presence of the tubercle-bacillus everywhere, in the house, in church, in school, on the street, in streetcars, in railroad cars, in sleeping berths, indeed in every spot inhabited by human beings' (Flick, 1900; Rosenkrantz, 1994). The historian of medicine and public health will recognize in Flick's litany a familiar protest against the invidious social implications of quarantine, registration, and isolation (Ackerknecht, 1948).[24] But the historian of race relations will recognize the specific sites of racial segregation in the United States.

At first it appeared that hygienic exclusions in hotel accommodations, train and streetcar travel, domestic employment, and public schools would affect all tubercular invalids, without regard to race. This appeared to be the case in the North until about 1908. Nationwide, in hotels, train cars, schools, housing, and employment, something called 'segregation' was occurring, but not solely, or even primarily, upon racial bases.[25] Rather, such exclusions and discriminations were being made on the basis of individuals' behavior, 'respectability,' and visible state of health, in which race, class, and gender each played a part.[26] Meanwhile a parallel, but not identical, system of racial separation obtained in the South, but it had not the formal detail or the legal force of 'segregation,' and the courts were making mixed rulings in challenges by black people. The new public-health legislation gave discretionary police powers, in some instances, to streetcar conductors and theater managers.

The anti-tuberculosis crusade brought these two systems together, as physicians from North and South compared notes and sought professional reunion against a 'common foe,' tuberculosis. The author of a 1905 pamphlet, urging mandatory 'segregation' of 'careless consumptives' in the Virginia Sanatorium for Consumptives, predicted that tuberculosis would eventually 'exterminate' the 'colored race' as it had the 'Indian, the Hawaiian and the Bermudan.' In the meantime, while white Virginians awaited State action, the pamphlet warned, 'the lives of your friends and children are jeopardized:'

> As long as our colored people continue irregular habits, and herd together in immorality and dissipation, their homes will be hotbeds of infection, fresh from which they will enter into intimate relations with our white people, drinking from public cups, spitting around kitchens and public places, as nurses fondling and kissing children, as cooks, waiters and barbers handling food, tableware and clothing, inevitably spreading infection broadcast among all classes. Within the next few years we shall inevitably see a great increase in our [white] deathrate from this disease, which already causes more loss, individual and state, than all other diseases combined.[27]

Northern public health officials also advocated segregation for tuberculosis patients, usually without specifying race. In December 1905 the Chicago Board of Education announced 'drastic measures' for the disease now held to be 'as infectious as smallpox, scarlet fever, or diphtheria.' The Board urged that all tuberculous teachers and pupils be excluded from school.[28] By 1907 a Cleveland physician was urging a national audience of anti-tuberculosis activists that in public schools 'the open cases should be segregated' (Lowman, 1907, p. 180).[29] Pennsylvania's Commissioner of Health, Samuel G. Dixon, explained in early 1908 the necessity for state sanatoria:

> The infirmaries for advanced and hopeless cases will have the double object of segregation of the patients for the public weal and furnishing medical attendance, nursing care and the comforts of a home for those who cannot otherwise command them, in the interests of the sufferers themselves (Dixon, 1908, p. 4).[30]

None of these discussions of segregation, coming from Northerners writing before 1908, contained express racial specifications, yet they not only employed the term 'segregation,' but by 1911 involved schools, public accommodations, housing, railroads and stretcars, the very sites upon which racial separation of the social sort was already founded in the South. The problem with 'segregating' all tubercular patients was that tuberculosis was so common—particularly among medical personnel (Huber, 1904).

The problem for white Southerners, as they saw it, was that they were so dependent upon black domestic labor, and Hoffman's statistical reports made this workforce suddenly dangerous. As a doctor from Hampton, Virginia explained to the national gathering of tuberculosis activists in 1906, 'This proximity is not merely that of the Italian laborer who works on the highways or in the mines or on public buildings and dwells in a section of the city quite apart from others. The colored people . . . in a hundred ways touch the innermost parts of many homes' (Jones, 1906, p. 100).

'Hereditary consumptives' and other 'sick yankees' had for several decades 'chased the cure' in the desert climates of Florida and the Southwest (Baur, 1959; Ott, 1990; Rothman, 1994; Silber, 1995).[31] As long as the illness was not seen as contagious, the 'cons'' presence in resort towns was a source of amusement, compassion, and income, but not of danger. The anti-tuberculosis publicity of 1896–1904 changed this. Coming on the heels of the founding of the National Association for the Study and Prevention of Tuberculosis in 1904, the 1905 yellow-fever epidemic in the South renewed historical public health concerns over the hygienic dangers of rail transportation, concerns easily transferred to the tuberculosis problem (Humphreys, 1992; Ellis, 1992). In 1907 the Philadelphia *Record* carried the headline: TEXAS TO WAR UPON THE WHITE PLAGUE. Plans Strict Quarantine to Prevent Afflicted Persons Overrunning the State. TO SEGREGATE CASES. 'The new system of quarantine will involve the establishment of quarantine camps [along] the State line and the inauguration of a method of train inspection,' the *Record* explained, describing extant tent 'colonies' for tubercular patients.[32] Tuberculosis was not listed under the Federal Quarantine interstate or maritime codes, but it was a bar to immigration.[33] Dr. Harvey W. Wiley, of the U.S. Department of Agriculture, claimed in 1908 that separate train cars for tuberculosis patients were as necessary as 'detention camps,' due to the danger from cross-country invalids travelling in sleeping berths.[34]

After the success of his *Race Traits and Tendencies of the American Negro*, Hoffman carried his racial research into the study of tuberculosis in 1904 (Hoffman, 1904). Southern physicians and public health officials developed a surprisingly coherent line on the association between 'Negroes' and tuberculosis, melding old hereditarian views of race and disease with the new germ theories. Both white and black physicians agreed that TB rates among African Americans were too high, but they differed on both causes and remedies. Black physicians stressed the role of class, education, and environment in the cause and solution of the 'Negro Tuberculosis Problem,' while many white doctors increasingly favored race, though not necessarily heredity. In health as in other arenas, this focus on race eclipsed the economic and social turmoil that divided white Southerners by class and gender (Clinton and Silber, 1992).

In the 1890s many white Southerners had seen interracial sexuality as the gravest danger to white health, fearing black sickness in the transmission of 'blood' and 'seed,' especially through the so-called 'new Negro crime' of rape.[35] But the crimes of rape and lynching which seared the South after Reconstruction, cannot fully account, any more than racism can, for the precise outlines or timing of legislation and customs comprising the American system of racial segregation, given both the legal and customary forms of racial separation already in place against interracial sexuality before 1890.[36] Racism can partially explain why one group was presumed healthy and another presumed sick, and why white middle-class invalids saw themselves as 'innocent' and saw blacks seen as 'guilty,' but it was the revolution in levels of hygienic scrutiny, pursuant to the acceptance of specific germ causation in a ubiquitous and deadly disease, that raised the stakes and specified where, when, and which people would be 'set apart from the flock,' for hygienic reasons, that is, segregated.[37]

Both tuberculosis infection, and the now-dangerous habit of spitting, came to be associated by white Southern physicians with African Americans. After 1904 such deadly 'sanitary crimes' against the presumed white birthright of health were understood as much more casually committed, primarily by careless spitting and other common, unsanitary habits (Carter, 1909).[38] Racism produced the selective presumption of black guilt in these 'sanitary crimes.'

The object of white fears was unstable as the implications of germ theory for the control of the 'white plague' worked their way through the logic of public health. The 'negro' was sometimes euphemistically termed 'the criminal poor' or the 'careless consumptive,' at other times race remained a separate category in addition to economic class or sanitary disposition. In July of 1905 one of the most active publicists of the anti-tuberculosis movement, Dr. Charles Launcelot Minor of Asheville, N.C., spoke before the Alabama State Medical Association in Montgomery, a speech reprinted in the national sanatorium organ, the *Journal of the Outdoor Life*. Minor, the director of a sanatorium and later the President of the NASPT, played upon W.E.B. DuBois's familiar phrase, 'the Talented Tenth,' in asking,

> What are we to do with our Submerged Tenth, with the slumdweller, whether he be the negro of our own Southern towns or the dweller in the crowded, infected tenements of our Northern and Western cities; the ignorant, the dirty, the unsanitary masses of Italians, Poles, Bohemians, Negroes, *et id omne genus*, to whom cleanliness is unwelcome, by whom air is unappreciated, to whom hygiene is unknown [?]' (Minor, 1905, p. 134).

'What chance has the [slum] dweller ... where the many already sick cough and spit promiscuously and where his miserable rest is disturbed by the drunken carousing of those who would drown their misery in alcohol [?]' Minor wondered (Minor, 1905, p. 134).[39]

In 1905 the St. Louis Society for the Prevention of Tuberculosis issued a statement declaring the 'White Man's Plague' to be both curable and preventable, and endorsing a recent bill before the Municipal Assembly providing for the 'segregation and care' of persons afflicted,' to protect 'citizens' from this 'plague of the white race'.[40]

In a medical literature that constructed blacks not only as a race of sick people, but as 'careless consumptives' and 'promiscuous spitters', this 'news' about tuberculosis transformed the black working class into an ostensible public-health menace danger not only to each other, but to whites whose homes and bodies they touched.

The common, innocuous practice of spitting was rather suddenly rendered dangerous, then criminalized, after 1896, at the point when the contagiousness of tuberculosis, its spread through sputum, and the prevalence of the disease among African Americans, all became salient.[41] The overwhelming effect of this public-health agitation against 'promiscuous expectoration' was to specify in great detail numerous everyday practices and sites, heretofore unproblematic, which appeared to threaten the racial health of white people by exposing them to casual, indirect physical contact with black people and their sputum.

The language of 'promiscuity' reinforced an analogy between sputum, containing the 'seeds of death,' and semen, containing the seeds of race. Much as the intimate crime of rape catalyzed racial fears of hereditary 'taint' under a system of racial separation based on 'blood,' now the more ordinary, public customs of eating, drinking, kissing, and especially spitting appeared, under the new germ theories, as deadly threats to personal health.[42] Racial categories, like class, nativity, and gender categories, helped focus and contain the social anxieties generated by the broad new germ theories.[43]

Despite an association between spitting and the black population, the national public health 'crusade' against spitting was not organized around race, but around the tubercle bacillus and the behavior of 'careless consumptives.[44] Robert Koch's 1886 paper on the etiology of tuberculosis established the importance of dried sputum in spreading the disease (Koch, 1886). New York passed the first anti-spitting legislation, in 1896, followed immediately by Los Angeles and Boston. Other local and state governments followed over the next two decades (Newton, 1910).[45] Public conveyances became a prime object of these reforms; these were places where people met who otherwise lived and worked in separate spheres. As Pittsburgh physician Elmer Borland wrote in 1900 in *JAMA*, 'the question of first and second class electric and steam railway cars, to separate clean from indifferent people, will probably demand consideration, even in this democratic land' (Borland, 1900, p. 1000).[46] Men, Borland thought, were more intractable spitters than women, lower class more intractable than the bourgeois (Borland, 1900).[47]

As an adjunct of anti-vice as well as anti-tuberculosis campaigns, the regulation of spitting had a persistent racial, as well as class and gender, valence.[48] African Americans from the tobacco-growing South chewed tobacco and used snuff to a greater degree than did other groups; legislation against spitting affected them proportionately. Writing in 1903 in the *Journal of the American Medical Association*, Dr. Seale Harris insisted that spitting was a racial trait: 'Nearly all the [negro] men chew tobacco and their women use snuff, and the healthy and the tuberculous alike spit on the floors of many of their homes and churches' (Harris, 1903, p. 838). Other experts found the practice common among African Americans, but not universal. 'Infected Negroes Constant Source of Consumption,' blared the Norfolk, Virginia *Pilot*, '[T]he negro expectorates at random, never realizing that to do so is fatal. This condition, does not apply to the race as a whole, I am glad to say'.[49]

The publicity 'crusade' against spitting was nation-wide, a successful experiment in 'social advertisement' (Teller, 1988).[50] New York City Department of Health issued *'DO NOT SPIT': A Tuberculosis (Consumption) Catechism and Primer for School Children* in 1908, distributing instruction cards in Serbo-Croatian, Swedish, Italian, German, English and Yiddish.[51] In Louisville, Kentucky, St. Louis, and Paterson, New Jersey, 'Spit wardens' collected fines against unlawful spitters, sufficient to offset the cost of their own salaries.[52] In 1909, 400 arrests were made in Minneapolis, 165 in Chicago, and twenty-six in Cedar Rapids, Iowa.[53] By 1910, forty-three cities had systems of police enforcement against spitting. In Buffalo, the fine for spitting in a public place was a significant $25.00, but, as a Brooklyn newspaper noted, 'Perfectly Satisfactory Spitting May be Done in Indianapolis for 76 Cents'.[54] Activists frankly admitted that the public good derived from these campaigns stemmed more from publicity than from enforcement. Few people were arrested, but these ordinances and state laws extended the police powers of public health to new officials, and extended their mandate into new realms of everyday behavior (Newton, 1910).

In the wake of publicity over the contagiousness of tuberculosis, the germ theory was postulated for all sorts of conditions: crime, alcoholism, immorality, cancer, insanity, greed.[55] If tuberculosis—which so clearly ran in families—could be caused by a germ, then anything could. During the summer leading up to the Atlanta riot of 1906, the radical racist editor of the *Atlanta Constitution*, John Temple Graves, asserted that there was a 'germ of rape' with which black men were infected, and called for medical research toward 'inoculation' against this 'epidemic' (Baker, 1907; Williamson, 1984).[56] Graves's speculation, in which he was not alone, furthered the scapegoating of black men in this translation from 'germ-plasm' to 'germ,' from 'blood' and semen to sputum.[57]

Until about 1908, the anti-tuberculosis movement and the growing hygienic stigmatization of black Southerners remained parallel, but largely separate impulses. The 1908 International Congress on Tuberculosis in Washington, D.C., however, allowed white Southern experts on tuberculosis to nationalize their racial understanding of the disease.[58] Their position in favor of the 'segregation' of 'poor consumptives,' 'careless consumptives,' the 'Negro consumptive,' the 'criminal poor,' '*et id omne genus*' was driven not only by a legacy of white supremacy held over from the historical enslavement of black people, but also from regional resentments rubbed raw by the use of the Southland as a health resort for white consumptives throughout the nineteenth century.[59] Dr. Charles Minor, of Asheville, N.C. had become vice president of the NASPT, and Mary Lent of Baltimore led the Visiting Nurses' Association. Because so many of these same physicians and nurses also suffered from tuberculosis, it was important for them to make medical the racial discriminations that kept them from their own snare, compulsory medical segregation.[60]

At the 1908 TB Congress, graphic evidence of the merging of race and disease appeared in the award-winning exhibit of the Philadelphia Visiting Nurse's Association.[61] In three weeks' time, an estimated 200,000 people saw the exhibition.[62] The exhibit used 'Before' and 'After' photographs, a Progressive-era staple. The 'Before' picture showed a 'negress' and her tubercular son in a dark, grimy, rag-strewn tenement room. The 'After' shot depicts the same room, following the Association's intervention, cleaned, whitewashed, rearranged—and completely unoccupied.[63] Dramatically, all of the 'filth' was gone, thanks to the nurses' forceful action. Commenting on the pictures, Baltimore nurse Miss Mary E. Lent explained that dismal experience had convinced her organization that the only real constructive work that amounts to anything is in showing the conditions which we cannot cope with, and which absolutely demand segregation.'[64]

Though consumption was the most widespread danger, the public health 'crusade' it had engendered soon broadened, taking in other germ diseases. As Marion Hamilton Carter explained to the readers of *McClure's* in 1909,' 'Negro crimes of violence number dozens where his sanitary sins number tens of thousands. For one crime a mob will gather in an hour to lynch him; he may spread the hookworm and typhoid from end to end of a state without rebuke' (Carter, 1909). By 1910 many public health activists nationwide agreed with Southern publicity agent E.G. Routzahn, that '[t]heir intimate relations with the whites makes the colored consumptive [sic] the chief distributor of the seeds of death.'[65] The anti-spitting campaign criminalized, both culturally and legally, a previously innocuous hygienic custom. The absolutism of public health activists on the subject of spitting was summarized by the editor of the Wisconsin Anti-Tuberculosis Association's *Crusader*: 'to the intelligent mind open to the facts of science, spitting is murder because it spreads abroad the germs of disease and death'.[66]

By examining these still-recognizable aspects of public health regulation, we may see how the logic of public health produced what in retrospect are deemed the most absurd and arbitrary aspects of what came to be known as *racial* segregation: separate drinking fountains labelled 'colored' and 'white'.[67] This understanding does not legitimate segregated drinking fountains, however, nor any other form of racial segregation, for the very racial distinctions that obtained in these debates failed to address the problem of contagion wholly within the 'innocent' white population, much less the dangers of contagion within the black population, or the dangers of contagion from whites to blacks. It was the conviction of white superiority in the dominant culture, and the higher value placed by whites on white health, that actually confounded the understanding and control of the disease in whites as well as in blacks. This *a priori* assumption of superiority rendered the intricate hygienic logic of segregation medically invalid. It was also racism that presumed the black sufferer from tuberculosis to be uneducable, uncontrollable, impulsive, and (like the apocryphal Chinese lepers) even malicious (Jacques, 1909; La Motte, 1909).[68]

THE COMMON CUP

I.

Thou shalt not covet thy neighbor's cup,
For the cup may be unclean,
And many a germ in hiding lies
Secure in the weakness of human eyes
Where no visible taint is seen.[69]

By the mid-1890s, American physicians were largely committed to the germ theory of disease, though they differed as to how much disease the germ by itself could explain. Until this time, the common cup was the rule in town squares, schools, and railroad stations. Common cups were also used in church services. Changes in these customs grew out of attempts to write bacteriology into law.[70]

In 1894 a microscopic study of a shared chalice from a Philadelphia church yielded tubercle bacilli and staphylococci, as well as 'numerous pus cells and an abundance of pavement epithelium mixed with ovoid, purplish cells of the grape'.[71] This microscopic scrutiny of secular and sacred symbols of communion produced a train of hygienic reforms of broad social consequence.[72]

During the 1908 International Congress on Tuberculosis held in Washington's new Natural History museum, entrepreneur C.J. Ljunggren exhibited his invention, the 'Hygienic Chalice.' A brochure explained that '[t]he danger of the communication of disease germs such as the *bacilli of tuberculosis*, etc., in the administration of the wine on the Holy Communion is widely recognized today. For this reason, ... many churches have discarded the ordinary chalice or cup and adopted in its place the so called individual cups. These, however, have among other drawbacks also those of impairing the symbolism of the Sacrament...[73]

The common cup was also an historical icon of political danger. As Dr. Samuel G. Dixon imagined in 1909, the backward courtesy of the common cup, endangered the unsuspecting modern citizen no less than medieval assassination plots:

> We read with horror the story of the poisoners of the Middle Ages and the Crimes of the Borgias. We shudder when we think of the days when the preparation of deadly poisons with murderous intent was a profession, and the goblet proffered under the guise of courtesy was too often the instrument of death (Dixon, 1909, p. 5).

A Virginia Public Health Report for 1910 illustrated the contemporary danger with a drawing of the town pump, surrounded by a well-cared-for white little girl, an emaciated consumptive white man, and, between the white girl and the healthy 'citizen', a thick-necked black man; behind him, a cigar-smoking carpetbagger.[74]

By 1910, seven states and twenty-five cities had altogether banned the 'poisoned cup' from railroad stations, trains, schools, and places of public amusement.[75] As officials sought sanitary alternatives to the and common cup, manufacturers of 'Dixie' cups, new sanitary 'bubblers,' and collapsible metal travellers' cups, readily obliged.[76] The new public drinking fountains allowed 'that the water, gently bubbling upward, may be drawn in by the thirsty lips without contact with ought else' (Dixon, 1909).

Legislation banning the common cup followed the social map of Jim Crow segregation, focusing in particular on railroad cars and stations, schools, theaters, and restaurants.[77] In its place, the individual white paper 'dixie' cup symbolized white separateness:

> [T]he war against public drinking cups typifies the whole movement on behalf of more hygienic conditions. The child, or for that matter, the careless or ignorant adult who sees a clean white individual drinking cup and is made to understand that he should use it for the protection of his health has received a valuable lesson in general hygiene.[78]

As anti-tuberculosis publicity began to appear on billboards, in streetcars, hotels, schools, and other public sites after 1904, these places were also officially marked off as sites of potential danger from contagion. While marking an advance for bacteriology, and likely for disease prevention, this wave also marked a retreat from symbolic forms of general sociability, and created small, daily reminders of contagionist fears. Moreover, the map drawn by public health regulations had the sanction of the state and the motives of commerce backing it (Brown, 1999).

The very innocuousness of artifacts like the Dixie cup, and the powerful martial rhetoric of the public health crusade, masked the broad pattern formed by these small constraints and scrutinies. A few health activists recognized what was happening, however. Responding to the trend exhibited at the International Congress to merge 'The Negro Problem' with 'The Tuberculosis Problem,' the Dean of Arts and Sciences at the historically black Howard University, Professor Kelly Miller, A.M., argued in classic progressive language for the primacy of social bonds over fears of contagion, in the face of a growing public health crisis (Shaler, 1884; DuBois, 1903).[79] He noted that, 'Although the negro race, as such, had little part in this great International Congress, it enjoyed no small part of attention throughout the proceedings,' he noted dryly, 'and must be the chief beneficiary of its deliberations' (Miller, 1909, p. 129). 'The health of a community depends upon the sickest individual in it,' Miller wrote in the *Journal of the Outdoor Life* (Miller, 1910).[80] 'Tuberculosis germs,' he affirmed, 'can not be confined to one race. The Jim Crow law can not be made to apply to disease' (Miller, 1910, p. 259).[81] Miller, who chaired the legislative committee of the Negro Anti-Tuberculosis Society of Washington, was resisting the wholesale translation of old social 'separations' into the new tenets of public health 'segregation'.

Meanwhile, national attention within the anti-tuberculosis movement shift Southward. Baltimore initiated residential segregation by race in 1911, in part around a designated area known as the 'lung block' (Magnum, 1940, p. 140).[82] Atlanta, the site of a racial massacre in 1906 and a center for organized white supremacy as well as black progressivism, continued in the *avant garde* of public health segregation and regulation. In 1913, city officials passed what was heralded as 'one of the most radical ordinances on tuberculosis passed in any city in the United States.'[83] The editor of the *Journal of the Outdoor Life* observed that 'while the ordinance is so worded that it applies to any person afflicted with tuberculosis, it is primarily designed to provide for compulsory removal, segregation, and detention of colored servants and others, who by their habits may endanger the lives of others'.[84]

Kelly Miller continued to resist such moves, as he reaffirmed in 1916:

> The germs of disease have no race prejudice. They do not even draw the line at social equality, but gnaw with equal avidity at the vitals of white and black alike, and pass with the greatest freedom of intercourse form the one to the other. One touch of disease makes the whole world kindred, and kind' (Miller, 1916).

Despite the dangerous confounding of race and disease, to which Professor Miller and others responded, black sufferers were not the only people being 'segregated' as part of the quintessentially progressive, social-scientific logic of Social Contagionism.[85] This breadth may have made these early turns of the segregation screw difficult to interpret, particularly for black physicians concerned above all with health. As community leaders, professional men like Dr. Miller were forced to articulate a public-health philosophy that ran against the social and medical tide. Recognizing the real health problems of African Americans, they nonethless had to resist what had become the skewed racial onus of prevention. Earlier municipal sanitary improvements of the nineteenth century had bypassed black neighborhoods and rural counties. 'Give the negroes plenty of water and better sewerage in this section where they live. Make it easy to keep clean. This is far more necessary than segregation laws,' explained R.R. Wright of the *Christian Recorder* at a conference devoted to 'the New Chivalry—Health.' Dr. Charles V. Roman of Nashville, editor of the black physicians' *National Medical Association Journal*, concurred, directly challenging Frederick Hoffman's 1896 data, to which he ascribed much of the problem:

> ...it is the type of overzealous agitation which has laid undue emphasis upon the negro's susceptibility to disease and his high death rate, with the result of alarming the public instead of stirring the health authorities to activity. Segregation and not sanitation has been the response—destroying the negro's opportunity instead of improving his health (Wright, 1915; Roman, 1915; Gilbert, 1915).

THE HISTORICAL SEMANTICS OF 'SEGREGATION'

Prior to 1890, the term 'segregation' was a technical one, used almost exclusively in medicine and public health contexts to describe the removal or separation of any sick person or animal from any healthy population, or the removal of healthy people from the site of contagious disease, particularly for long periods of time and in cases of chronic illness.[86] Race as an explicit medical concern moved in and out of this medical discourse.[87] Segregation continued as a medical term of art, but after 1899, the term acquired a political connotation with a racial charge that has been conventional for most of the twentieth century.[88]

Historian John Cell argues that for the practices of separation to become a system of segregation, 'some new set of circumstances, some hitherto unacknowledged challenges must have come into existence' (Cell, 1982, p. 103). It is just at this point, after 1897, that the language of medical segregation takes on a racial charge around the problem of tuberculosis, by analogy to other contagious diseases.[89] But what significance might this new usage have, coinciding as it did with what historian Joel Williamson calls the 'new organism, Radicalism?' (Williamson, 1984, p. 182).

'That something new was happening in the separation of the races was indicated by the fact that a new word was required for such occasions', Williamson notes (Williamson, 1984, pp. 253–54).[90] 'The word "segregation"' Williamson observes, 'was not much used before 1899, and when it was used it had no special racial connotations. In and after that time, it was used to refer to the separation of the races, and it seemed to carry with it the idea that the separation referred to was effected by law.' Williamson adds that he cannot explain this new term.[91] Rabinowitz notes that *de jure* segregation was not widespread until after 1890, and the term did not come into national usage in a racial sense until 1913 or 1914.[92]

'Segregation,' however, was not the 'new' term that political historians understand it to be. In medical dictionaries it is found as early as 1864, meaning 'to separate from the flock,' reflecting an even older religious meaning that was in part hygienic: 'Clean *segregate* from all kinds of uncleanness.'[93]

Medical usage of the term 'segregation' was characteristic of colonial encounters from at least the 1840s forward.[94] Embedded in these health provisions, as Curtin and others have pointed out, was the arrogance of white supremacy.[95] European and American medicine gave priority to white health even at the expense of Africans, Indians, and black Americans, rationalized through anthropological premises that figured non-whites or non-Europeans as always, already diseased or, conversely, as curiously immune.[96] American usage in matters of racial and public health policy continued to parallel colonial examples, where colonial authorities presumed their own health primary. Some contemporary observers noted this colonial parallel too. In his muckraking articles on ‘the color line’ in America, Ray Stannard Baker remarked upon the international similarity of these practices: ‘In much the ways that in the United States “Negroes are being segregated [,] [s]o are the Chinese segregated, and the blacks in South Africa, and certain classes in India” (Baker, 1907, p. 299).[97]

Historians of race relations have noted the sudden appearance of the term ‘segregation’ at the moment when *de facto* separation of the races becomes ‘crystallized’ into *de jure* ‘segregation,’ in the 1890s.[98] They have recognized patterns of emulation between the American South and contemporary colonial regimes such as South Africa.[99] John Cell, following Maynard Swanson, even noted the medical usage: ‘the reader must know that the “sanitation syndrome” was frequently used as a **pretext** for creating exclusive living and recreational facilities for Europeans throughout the colonial empires ... (Cell 1982, pp. 1–2).[100] While Cell does not pursue the term “segregation” from a medical context into a broader political one, seeing that earlier medical context as pretext, he nonetheless concludes that “the primary origins of segregation are not in slavery or on the frontier but in the modern conditions of the 1890s and after” (Cell, 1982, p. 20). Among the most important “modern conditions” of the 1890s, I argue, was advent of germ theory, and the framework of Social Contagionism that was its logical, invidious extension.

Political historian Joel Williamson sees 1889 as a pivotal year in the transformation of customary forms of racial separation and even idiosyncratic accommodation, to a decade of extreme violence accompanied by ‘virulent’ rhetoric. He offers political events—President Harrison’s election and the appointment of 16,000 black people to federal jobs, and the economic depression of 1892–1893—as triggers for white backlash. But he admits that ‘[p]recisely why Radicalism appeared with such dramatic suddenness in 1889 and swept so powerfully through the South in the years that followed is beyond easy explanation ...’ (Williamson, 1984, p. 112). Williamson is correct; the explanation is highly complex.[101] It is my contention that the sea-change in popular understandings of disease in general, and of specific diseases that were already racially inscribed, raised the mortal stakes of casual social contact between the races and produced the exquisitely detailed, overwhelmingly powerful, racist system known as ‘segregation’.[102]

'Segregation' translated hereditarian views of race and disease, which focused on intimate contact, into a new contagionist vernacular, centered around casual contact. Its hygienic purpose was never in its racial form valid, but nonetheless in the terms of the new germ theory it was quite persuasive. These forms outlasted in practice the rationale that gave them meaning. The 'bacteriomania' and 'phthisiophobia' of the tuberculosis crusade having been largely forgotten by the time historians encountered segregation as a historical problem in the antibiotic era of the 1950s, the practices of segregation appeared so perversely irrational and purposefully degrading as to have been the arbitrary invention of white psychosis. Once the threat of germ diseases was contained by antibiotics, and unknown by a new generation of political historians, separate drinking fountains appear as petty humiliations without rhyme or reason. These were, as Cell rightly observes, arbitrary expressions of power, but they were not only this.[103] Understood in the medical context of 'bacteriomania,' such provisions have a certain logic whose basic hygienic premise—that deadly contagious disease must be prevented—was accepted by segregationists and critics alike, who shared a new 'progressive' belief in germ theory. The logical and ethical flaw was not their belief in contagion, but the selective application of that belief. Segregation proceeded as a translation of a much older legacy of racism, to put in place laws on the premise that, solely because of race, contagious disease among non-Aryan people was somehow more threatening to the (white) body politic than the same disease among whites. It thus reified a long-standing analogy between whiteness and purity, jeopardizing black health. This was the point that C.V. Roman, Kelly Miller, and R.R. Wright were trying to make.

More recently, historians of race relations puzzling over the sudden rage of Radicalism have turned attention to the wave of lynching that lashed the South in the 1890s. The large exception to nineteenth-century patterns of social proximity between the races was, of course, the strong prohibition against miscegenation.[104] While actual practices in plantation societies violated these rules, subjecting black women to rape, the prohibitions themselves have been legitimated by that powerful precept of nineteenth-century medical and racial thought among whites, namely, that racial superiority was a matter of inheritance, of 'blood.' The 'purity' of white 'blood' was to be absolutely guarded. A white Southerner in 1870 explained, '[If] I sit side by side in the Senate, House, or on the judicial bench, with a colored man, how can I refuse to sit with him at the table? What will follow? ... If we have social equality we shall have intermarriage ... Sir, it is a matter of life and death with the Southern people to keep their blood pure' (Macrae, 1870)[105] The converse of white purity was the long-standing construction of blackness itself as a diseased state.[106]

The prohibition against racial 'amalgamation' also guarded, theoretically, against the transmission of those diseases that were historically understood as 'hereditary': consumption (tuberculosis), leprosy, and syphilis.[107] There was not, in general, a very clear distinction between sexually transmitted infectious disease and hereditary conditions. The 'seed and soil' metaphor obtained in 'germ theory' and 'germ-plasm.'[108] To make matters worse, syphilis, leprosy and consumption were strongly associated with African Americans.

These three diseases tended to stand in as heuristic models for one another in medical thought, and as each was found to be specifically contagious, this racial association of each disease further tightened the equation between race and all germ diseases. Once a disease was known to be contagious, its sufferers became candidates for medical segregation. In 1901 Robert Koch addressed the issue of 'segregation' for the tuberculous in the *Journal of the American Medical Association*.[109] Koch argued, by analogy to leprosy, that his bacteriological research mapped a new direction in health work against tuberculosis, to replace vague efforts at environmental sanitation and social betterment with strict measures of 'segregation.'[110] As germ theory spread, from 1881 through 1900, it carried with it this new understanding of danger from persons, across geographic and social boundaries.[111] Detached from locale, class, and even in many instances the invented biological category of race, germs transcended the boundaries previously deemed sufficient to protect privileged groups from diseases associated with inheritances of poverty or blood.

HISTORIANS AND DISEASE METAPHORS

The unwillingness of American political historians to take seriously the medical contexts of segregation may be a figment of the artificial separation that has obtained generally between medical history and general political history. This separation has had peculiar implications for the rhetoric of historiography among scholars of race relations. General historians, by underestimating health and disease as political factors embedded in the language, have also freed up disease metaphors for their own evaluative uses.

Political and economic historians of segregation, in their well-founded ethical position as opponents to the American system of racial injustice and violence, have turned segregationist rhetoric against itself. In classical debating form, they have co-opted the medical, diagnostic idiom of nineteenth- and early-twentieth-century white supremacy and used it against itself, much as the black-nationalist minister Alexander Crummell did when he termed 'caste the canker of diseased souls' (Crummell, 1883). Documenting 'virile strains' of racial hostility, the 'feverish' spread of segregation, the 'wasting away' of more Conservative racial ideologies in the face of Radicalism, Cell, Williamson and others borrow disease metaphor from their distant interlocutors, to align themselves rhetorically with the voices of contemporary resistance.[112] They share this moral-medical vocabulary with progressive-era public-health reformers, such as one who called poor housing 'the cancer that sends its poison to the finger tips of the social body' (Bacon, 1912). Participating in an ancient political tradition of body-politics, this language is both banal and powerful—its metaphors are almost inexhaustible, their applications infinitely malleable.[113] These metaphors capitalize politically and historiographically on our most basic concerns and judgments about life and death, borrowing medical authority to raise the cultural capital of historians, by analogy.

Yet in this particular historiographic literature on the politics of racial segregation, these disease metaphors peculiarly eclipse the material history of medicine, public health, and contagious disease that belongs among the contexts in which politics, including the politics of race, may be understood. If racism is a 'disease', then what is disease? Too easily the answer has come, as the reciprocal logic of metaphor dictates, 'disease is racism'.[114] The vital point is not that this equation is baseless (it is not), nor that historians' language can be purged of metaphor, as Sontag would have it (Sontag, 1978). By thinking instead about the language that comprises our historical evidence as *translation* (the definition of metaphor), we can recognize the contexts of refutation in which a congeries of *de facto* separations became a coherent, overwhelmingly powerful propaganda of social contagion, with profound consequences for the health of its subjects (Pocock, 1960; Brown, 1986; Edelman, 1977; Lakoff and Johnson, 1980). The suasive weight of Social Contagionism, at the beginning of the century, was such that it at first took in even those who stood to lose the most.

Finally, I fear I 'preach to the converted' in proclaiming the opportunities for medical historians to use their knowledge about matters of life and death into historiographic debates among political historians. Yet as this essay suggests, the results of 'bringing health back in' can be startling.

I want to suggest that the history of race relations, or more exactly, of ideologies of white supremacy in this country, in the period between Reconstruction and the Great War, comprises at least three separate strands of thought tied together not just by concepts of race but also by concepts of health and disease, of the individual and the 'body politic'. The most heinous form of white supremacy, lynching, as a violent sexual crime, revolved around the issues of rape, voluntary interracial sexual intercourse, miscegenation, and 'blood'. Political disfranchisement hinged on the rhetoric of temperance, moral and medical, as applied to the race, gender, and class of African American men. The regulations that comprised segregation depended upon germ theory, overtly with regard to tuberculosis, more subtly with regard to venereal disease. Moreover, through the nexus of syphilis with phthisis, understood as a contagious *and* sexually-transmitted disease, and through eugenical conflations of sexual and casual 'transmissions' that confused 'germs' and 'germ-plasm', sexual transgression implied disease transmission, linking the old fears of miscegenation with the new fears of contagion. This gave urgency to the arguments in which African Americans resisted segregation—not, as whites often presumed, because they preferred the 'better' company of whites, but because segregation was literally, in the context of knowledge about contagious disease, a matter of life and death. Progressive black leaders, like white progressives, accepted the public health propaganda of germ theory and, like physicians with tuberculosis, did not want to be 'bottled up' with sick people by compulsory segregation. As anyone subject to these provisions knew, there is no more abject 'subject' than the invalid.[115]

Racism, despite its permeation of American life, itself does not explain fully the development of formal, legalized segregation. Something new must have entered into the old ideology of white supremacy to create segregation's characteristic, gerrymandered forms. By imagining a pre-antibiotic social world in which contagious disease meant death, it is possible to perceive how the principles of germ theory, newly applied to a disease heretofore understood as inherited or sexually transmitted, could have produced a panic. The alarm around tuberculosis, and the clamor for 'segregation', was far broader than race, but the high salience of the disease among blacks, their specific social and economic position in the labor force, and the racial arrogance of white reformers reorganized both the legal basis and social geography of racial separation in what came to be known as 'segregation'.

NOTES

1. This speculative and very preliminary exploration of the connections between medical and racial segregation was originally prepared for the conference on Transmissions at INSERM, Paris, in May 1996. I have subsequently developed it in my forthcoming book, *Matters of Life and Death*, and presented a substantially new account, based on new evidence on housing segregation in Baltimore, at the American Association for the History of Medicine Meetings in Bethesda, MD, USA, in May 2000, the text of which is currently under review. Thanks to Ilana Lowy and her colleagues at INSERM, and to John Warner, Naomi Rogers, Frederick Holmes, Ed Mormon, Harry Marks, and Philip Curtin for their readings of earlier drafts; thanks also to Nancy Tomes, Katherine Ott, and Tera Hunter for generously sharing unpublished work, and to my teacher, Murray Edelman.
2. On language, politics and history see Pocock (1960), Banta (1983), Toews (1987), Scott (1991), and Brown (1990, 1992).
3. Cell (1982), p. 4; Williamson (1984), p. 115.
4. On blood see Jordan (1968), pp. 165–66. See also Brown, *Matters of Life and Death*, chapter 1, 'The Parable of the Sower' (forthcoming).
5. Pocock (1960), p. 15.
6. Cell (1982), p. 4; Williamson (1984), p. 115.
7. Jordan (1968), pp. 53, 88, 165, 258–260, 263–264, 377, 423, 501, 518–521, 528–529.
8. Cell (1982), p. 4, citing Williams (1958, 1976); Daniel Rogers, Contested Truths ([**cte**]). See also Rogers (1982). I emphasize metaphor in Brown (1986, 1992, 1999).
9. 'Movement to Segregate All Tuberculosis Patients,' clipping, n.d., dated in text ca. 3/1910, MSS 322 Box 8 file 2, Atlanta Anti-Tuberculosis League, Atlanta Lung Assn. Coll., Atlanta Historical Society Archives, Atlanta, GA.
10. See also Harris (1903); Jones (1906), pp. 97–113 esp. p. 97; Hubbard (1896); Powell (1896); Mays (1897); Faison (1899); Coleman (1903); Rawlins (1904); 'Tuberculosis the Negro's Problem,' Louisville, KY *Courier-Journal* (July 26, 1909), clipping, vol. (1/2/09–10/3/09), Henry Barton Jacobs Scrapbooks, Chesney Archives, Johns Hopkins University Medical Institutions (JHMI), Baltimore, MD; Galishoff (1985); Gamble (1989); Edson (1895).
11. 'Radical Tuberculosis Measure in Atlanta' *Journal of the Outdoor Life* (February 1913), X: 2 p. 57. See also 'Save Babies from Kisses, Her Plea' *Philadelphia Press* (June 19, 1907); 'Determined Warfare on Tuberculosis,' *Atlanta News* (February 18, 1907), all, Jacobs Scrapbook, vol. (1/27//07–1/28/07), JHMI; and Hunter (1990).
12. *NASPT Transactions* (1904), 1:1, entire. Histories of tuberculosis include Dubos and Dubos (1952), Bryder (1988), Teller (1988), Ott (1990), Bates (1992), Rothman (1994), Barnes (1995) and Tomes, (1998).
13. See also Webb (1878) as cited in Bates (1992); Paquin (1896); Benedict (1898).
14. On the TB 'crusade' historicized as a 'holy war,' see Brown (1993, 1994, 1996).
15. Plessy v. Ferguson, 163 U.S. 537, 16 Sup. Ct. 1138 (1896). On the Plessy case see Meier (1963); Murray (1952); Magnum (1940).
16. 'Tuberculosis Panic' *Outlook* (February 8, 1908), 88:328–9; 'Consumptives Like Outcasts ... Shunned Like Lepers' *Boston Herald* (May 14, 1913), Jacobs Scrapbook vol. (11/4/12–10/22/14), JHMI.
17. Warner, *The Therapeutic Perspective*; Rosenberg, *The Care of Strangers*, p. 71. On the persistence of hereditarian views of TB see Flick (1900) and de Schweinitz (1912).
18. Ott (1990), Chapter 3; Rothman (1994), pp. 31–56.
19. Ott (1990), Chapter 3; and Rothman (1994), p. 190.

20. 'The Segregation and Supervision of the Tuberculous' *Boston Medical and Surgical J.* (1897), 136:75; Koch (1901), p. 260; and Knopf (1910); Bryder (1988).
21. 'No Consumptives' *NY Press* (January 6, 1905), Jacobs Scrapbook vol. (12/1/04–10/3/05). On 'phthisiophobia' see also 'Urge School War on Tuberculosis' *NY Herald* (December 20, 1904); 'Taxing Arizona's Hospitality' *Brooklyn Eagle* (December 19, 1904), both, Jacobs Scrapbook vol. (12/1/04–10/3/05); 'Consumption on Railways' *NY Tribune* (October 3, 1905), Jacobs Scrapbook vol. (10/3/05–1/26/07).
22. 'The Segregation and Supervision of the Tuberculous' *Boston Medical and Surgical J.* (1897), 136: 75; and Koch (1901). See also the later comparison between 'careless spitters' and 'vicious lepers in China' who allegedly 'jostled' healthy people to spread the disease, 'Careless Expectoration' *SPUNK* (November 1913), V: 8 p. 10.
23. Dr. Ira van Geison, discussion of Knopf (1910).
24. On germ theory see Kramer (1948); Richmond (1980); McClary (1980); Rosenkrantz (1979); Brown (1984); Tomes (1990); Rogers (1992); Kraut (1994); Brown (1999).
25. Smoking was one criterion for separating passengers; the 'ladies' car' was the 'no smoking' car, and black women and prostitutes were generally excluded from it: Roback (1986). The medical term 'segregation' also described the municipal districting of prostitution and other vice: see Rosen (1982). On 'relatively fluid' race relations prior to enactments of segregation laws see Rabinowitz (1978), pp.192–94. On segregated housing for the tuberculous see e.g. 'Liberty [N.Y.] Excludes License in Tuberculosis' *Boston Medical and Surgical J.* (1901), 145: 689, as cited in Ott (1990), p. 182. Residential 'segregation' by race began in California in the 1890s against the Chinese, and in Baltimore against blacks in 1911, the latter struck down by the Supreme Court in 1917: Meier (1963), pp. 162–63.
26. Roback (1986), pp. 896–7; Matthews (1974); Campbell (1986); Folmsbee (1949).
27. 'Virginia Sanatorium for Consumptives' Ironville, Virginia, 1905, 2–3, TB Pamphlets, vol. 46, no. 18a, Flick Papers; Jones (1906), p. 175.
28. 'Fighting Tuberculosis' Washington [D.C.] *Star* (December 23, 1905), Jacobs Scrapbook, vol. (10/3/05–1/26–07), JHMI.
29. See also Jacobs (1909).
30. See also untitled newspaper article, Washington *Herald* (October 28, 1907), Jacobs Scrapbook, vol. (1/27/07–11/28/08), JHMI.
31. **'Sick Yankees'**
32. 'Texas to War upon the White Plague' Philadelphia *Record* (August 11, 1907); 'Texas Bars Consumptives' Austin *Transcript* (June 24, 1907); 'To Consider Quarantine. Health Officials May Meet in Houston during August' Houston *Chronicle* (July 4 (?) 1907); see also the editorial opposing the quarantine; Bernard (1907).
33. 'Texas Will Quarantine against White Plague' Atlanta *News* (June 24, 1907), Jacobs Scrapbook vol. (1/27/07–11/28/07).
34. 'Danger from Consumptives. The Necessity of Isolating them on Railroad Trains' clipping, dateline Washington, D.C., December 5, 1908), Jacobs Scrapbook, vol. (9/5/08–1/1/09).
35. On rape and lynching see Wells-Barnett (1895); McGuire and Ldyston (1893); Baker (1907); Williamson (1984), pp. 117, 214–5; Hall (1983); Carby (1986); Hodes (1992, 1993); Weigman (1993).
36. Cell (1982), p. 4; Pascoe (1996).
37. 'Segregation', in Gould (1899).
38. On disease as crime, see Brown (1999); Rogers (1989); Leavitt, 'Typhoid Mary Strikes Back', 'Careless Expectoration' 10; 'Object-Lessons in White Plague War. Striking Exhibition. Consumption Declared a Crime' Philadelphia *Record* (November 21, 1907) Jacobs Scrapbook vol. 1/27/07–11/28/07.
39. Minor's emphasis on darkness, overcrowding, immorality and alcohol are consistent with the major tenets of racist discourse in this period. See Williamson (1984), Chapter 5.
40. 'Consumption, If Not Retarded, May Kill One in Every Seven in City' *St. Louis Republic* (February 23, 1905), Jacobs Scrapbook vol. (12/1/04–10/3/05).
41. Naomi Rogers recounts a very similar transformation of the house fly in Rogers (1989).

42. On 'blood' and 'seed' in the discourse of race, see Jordan (1968), pp. 165–66. On venereal disease see Brandt (1985); Atkinson (1877); Carter (1883); Jones (1904); Murrell (1910). On 'Syphilitic Phthisis' see Roy (1881); Edgar (1906).
43. See Brown (1999).
44. Cartoon, 'Dr. Hurty is Here,' in the South Bend, Indiana *Tribune*, (July 25, 1906) 4; 'Plain-Clothes Men to Look After "Spitters",' Louisville, KY *Courier Journal* (December 23 1909), Jacobs Scrapbook vol. (11/3/09–1/7/10) p. 66; n.a., 'Want a Spit Warden' *Journal of the Outdoor Life* (May 1909), VI:5.
45. See also Feldberg, *Disease and Class*, pp. 86–89.
46. Borland reported from his survey of regulations that in some regions change was not conceivable: 'Women can, but men cannot, change their filthy spitting habits,' wrote one respondent; and from St. Augustine, 'Can not stop a Florida cracker from spitting'.
47. See also Barnes (1995), pp. 82–91 and *passim*, on France's antispitting 'crusade'.
48. Allen (1903) compared the new 'Moral standards [against] expectoration' to the 'immorality of drunkenness'. This was a powerful rhetorical link: on liquor as the Southern populists' rhetorical link between the twin threats of the black man and big business, see Campbell (1986).
49. 'Infected Negroes Constant Source of Consumption—One Infects One Hundred Before Death—Difficulties Confronting Health Departments in Crusade against Plague' Norfolk, Va. *Pilot* (October 27, 1907), Jacobs Scrapbook 1/27/07—11/28/07.
50. Orlando Lewis, Social Advertising, Trans NCCC 37 (1910), pp. 537–547.
51. City of New York, Department of Health, '*DO NOT SPIT*': A Tuberculosis (Consumption) Catechism and Primer for School Children (New York: Department of Health, 1908), Flick Papers, vol. 65, no. 103; 43–48, 'TB-Int. Congress 1908'.
52. 'Plain-Clothes Men to Look After 'Spitters' Louisville, KY *Courier-Journal* (December 23, 1909), Jacobs Scrapbook, vol. 11/3/09–1/7/10, 66; 'Want a Spit Warden' *Journal of the Outdoor Life* (May 1909), VI: 5 p. 146.
53. Kerr and Moll, *op. cit.*
54. 'Don't Spit in Buffalo. Costs $25 There, but Perfecty Satisfactory Spitting May be Done in Indianapolis for 76 Cents,' Brooklyn *Times* (May 5, 1910), Jacobs Scrapbook 1/3/10–7/30/10, n.p.
55. See Brown (1999).
56. 'Contagion of Unnatural Crime' *Independent* (August 8, 1907), 63: 346–8. Graves also wrote an earlier book on Southern health-resorts for consumptives (1883).
57. See chapter 1, 'The Parable of the Sower' in Brown, 'Matters of Life and Death', (forthcoming).
58. 1908 was also the year of the Springfield, Illinois race riot, touched off on August 14 when Mrs. Earl Hallam, a white woman, accused a well-known black man, George Richardson, of attempted rape.
59. Minor (1905), p. 12.
60. **[fll cte nec?]** Dr. W.B. Stanford of the College of Physicians and Surgeons, Memphis, TN, 'Curse of the Negro Race. Yet Few Blacks Died of Consumption Before the Civil War. Southern Physicians Speak' Baltimore *Sun* (September 26, 1908), Jacobs Scrapbook vol. (9/5/08–1/1/09), 12, JHMI.
61. 'Tuberculosis War to Begin Monday' Washington *Herald* (September 17, 1908); 'Monster Exhibit on Tuberculosis,' Augusta, GA *Herald* (November 10, 1908), both, Jacobs Scrapbook, vol. 9/5/08–1/1/09; 'Ready for the Study of Tuberculosis' Philadelphia *Inquirer* (January 21, 1909), Jacobs Scrapbook, vol. 1/2/09–10/30/09, 9. An earlier TB exhibit had been part of the 1904 Louisiana Purchase Exposition.
62. 'Mrs. Sage and Mr. Carnegie as "Crusaders"' Chicago *Inter-Ocean* (November 29, 1909), Jacobs Scrapbook, vol. 11/3/09–11/7/10, 40.
63. 'A Most Interesting Section,' Jacobs Scrapbook, vol. 9/5/08–1/1/09; see also Washington *Herald* (October 12, 1908); 'Great Health Fair Ready to Open' *NY Times* (November 27, 1908); '10,000 Attend the Tuberculosis Show. Visitors Throng the Natural History Museum on the Exhibit's Opening Day,' *NY Times* (December 1, 1908); '16,000 People See Tuberculosis Show' *NY Times* (December 3, 1908); 'Phonographs Cry Out Dangers of Tuberculosis' NY *American* (December 1, 1908), all, Jacobs Scrapbook, vol. (9/5/08–1/1/09), 77.
64. 'Object Lessons in the Plague War'; 'A Most Interesting Section'.

65. 'Tuberculosis the Negro's Problem,' *op. cit*; Northen (1909); Minor (1905), pp. 132–34; Williamson (1984), pp. 288–91.
66. Untitled pamphlet of the Wisconsin Anti-Tuberculosis Association, No. 15 (June 1911), p. 4., Flick Pamphlets, College of Physicians, Philadelphia, vol. 114, No. 51.
67. On martial metaphor borrowed from the Civil War in the Great War-era 'holy war' against tuberculosis, see Brown (1994). See also Emily Martin, *Flexible Bodies*, on post-1945 immunology.
68. La Motte (1909) includes among the unteachables 'the shop girl, the day laborer, the drunken negro'; among the elect, 'the millionaire, the professional man, the bank clerk'.
69. 'The Crusader's Decalogue,' (First Commandment) *The Wisconsin Crusader* (June 1912), reprinted in *Journal of the Outdoor Life* (December 1913), X: 12, p. 360. Commandment VII forbade spitting.
70. For a more thorough discussion of the common cup and communion, including religious aspects and medieval precedents see Brown, *Matters of Life and Death* (forthcoming).
71. Kerr and Moll, 'Common Drinking Cups,' 6, citing Dr. Howard S. Anders and Dr. C.L. Furbush, *New York Medical J.* (October 27, 1894) p. 538; Dr. Charles Forbes, [Public Drinking Cups], *J. of the American Medical Assn.* (October 16, 1897) 790 & ff.; 'Individual Drinking Cups: The Deadly Public Cup Gives Way to the New Sanitary, Paper Cup' *Journal of the Outdoor Life* (August 1909), VI:8, p. 237; Kerr and Moll, 'Common Drinking Cups,' 5–6.
72. On Microscopy see John Harley Warner, 'Exploring the Inner Labyrinths of Creation,' *op. cit.*
73. Brochure, 'Hygienic Chalice,' C.J. Lunggren, inventor, 'Pamphlets: TB-International Congress 1908,' vol. 65, no. 63, Flick Papers.
74. *Virginia Health Bulletin* (March 1910), II:3, pp. 52, 63.
75. 'The Arrival of the Individual Cup' *Journal of the Outdoor Life* (December 1909), VI: 12, p. 371; 'The Public Drinking Cup' *Journal of the Outdoor Life* (September 1910), VII: 9, p. 281.
76. Illustration, '[The Individual Drinking Cup]' *Journal of the Outdoor Life* (August 1909), VI: 8, n.p.
77. 'The Public Drinking Cup' *Journal of the Outdoor Life* (September 1910), VII: 9, p. 281. See also 'The Common Drinking Cup' *The Survey* (1909), 22: 695; 'Dangers of Public Cups' *SPUNK* (November 1911), 3: 8, p. 11.
78. 'The Arrival of the Individual Cup' *Journal of the Outdoor Life* (December 1909), VI: 12, p. 371–72.
79. See also Brown, *Matters of Life and Death* (forthcoming), for other plays on language: the 'White Plague' with the 'White Man's Plague,' and the 'White Man's Plague' with earlier 'White Man's Grave,' and the 'White plague' with the 'black (i.e. bubonic) plague.' See e.g., 'Consumption, if not retarded, May Kill One in Every Seven in City' *St. Louis Republic* (February 23, 1905); 'Consumption His Theme' *Pittsburg Leader* (January 13, 1905), both, Jacobs Scrapbook (vol. 12/1/04–10/3/05), JHMI; Curtin (1961); Curtin (1985) citing Rankin (1836); Painter, *Standing at Armageddon*, pp. 153–155, citing Kipling, 'The White Man's Burden' *McClure's Magazine* (February 1899); Meier (1963), p. 22.
80. Miller founded the Negro Anti-Tuberculosis League, and wrote on the history of black physicians. See also Rogers (1982); Brown (1992), p. 195.
81. See also 'White Man's Burden', same issue.
82. The Baltimore ordinance was declared unconstitutional in 1917. It is usually cited as the first residential segregation act, but Magnum (1940), p. 140 also cites a city of San Francisco ordinance directed at relocating Chinese people in 1890, *in re* Lee Sing, 43 Fed. 359 (C.C.N.D. Cal. 1890). A quarantine against bubonic plague was placed around the city's entire Chinese neighborhood in 1900, also found unconstitutional. See McClain (1988).
83. 'Radical Tuberculosis Measure in Atlanta' p. 57.
84. 'Radical Tuberculosis Measure in Atlanta' citing 'An Ordinance to Protect the People of Atlanta from the Dangers of Infection with Tuberculosis when the Disease Exists in Washer-women, Cooks, Servants, or other Employees'; on a similar public-health measure in New Jersey, signed by then-Governor Woodrow Wilson, see 'Compulsory Segregation Law Passed' *Journal of the Outdoor Life* (May 1912), IX: 5, p. 115. The *Journal of the Outdoor Life* commented: 'Without

doubt the most advanced State legislation on tuberculosis that has ever been enacted in this country, if not in the world'.

85. John Cell associates early segregation, initially, with moderate, modernizing, 'progressive' influences: Cell (1982), xii, pp. 17–18.
86. See Koch (1901); Spitzer (1968), esp. p. 55.
87. That is, I do not argue that this earlier medical usage was 'free' of racial assumptions.
88. Lent, 'The True Function of the Tuberculosis Nurse' 265–269. Historians have also read the word 'segregation' back into nineteenth-century discussions of race which typically employ the term *separation*, obscruring these medical connections: Kelly (1993). Only in 1904 does a racial definition appear in the *Oxford English Dictionary* (Cell, 1982, p. 1). When Baker published *Following the Color Line* in 1907, he still found it necessary to explain what kind of segregation he meant: 'segregation by race', p. 70.
89. 'The Segregation and Supervision of the Tuberculous' *Boston Medical and Surgical J.* (1897), 136:75; Koch (1901), p. 260.
90. The author also notes that he cannot explain the new term, 'segregation'.
91. Williamson (1984), p. 254.
92. Rabinowitz (1978), p. 182; see also Rabinowitz, 'The Not-So-Strange Career of Jim Crow'.
93. Latin., '*se*, by oneself' and '*grex*, a flock', Thomas (1864); see Gould (1899); *Oxford English Dictionary* (1970), 9:398.
94. *Oxford English Dictionary* (1970), 9: 398; Swanson (1977); Spitzer (1968); Curtin (1985). A British physician in 1849, for example, noted the 'utter impossibility of complete segregation [of fever patients] even in the most roomy vessels' Bryson (1849) cited in 'segregation' *Oxford English Dictionary* (1970), 9: 398.
95. See on this large subject Curtin (1985, 1989); Packard (1989, 1989a); Swanson (1977); Anderson (1992, 1992a, 1993); Fanon (1978); Sullivan (1988); R.C. Ileto, 'Cholera and the Origins of the American Sanitary Regime in the Philippines', in D. Arnold (ed.), *Imperial Medicine and Indigenous Societies* (Manchester): Alfred W. Crosby, *Ecological Imperialism*; Davin (1978); Arnold (1993).
96. Jordan (1968), pp. 518–521, 536, 583–584.
97. The first edition of the *Oxford English Dictionary* offered this 1904 usage: 'segregation[is] the first law of hygiene for the Europeans in the tropics,' citing *British Medical J.* (September 17, 1904) p. 631, *Oxford English Dictionary* (1970), Vol. 9:398; see also Cell (1982), pp. 1–2. Humphreys (1992), p. 165, notes new racial meaning in the 'healthy carrier' *in re.* yellow fever in the same period, 1895–1905.
98. Cell (1982), citing Williams (1958, 1976).
99. Frederickson (1981); Cell (1982), entire.
100. Emphasis mine; Swanson (1977). On the historians' error of reading medical metaphor as 'mere' metaphor, and the erasure of contagion as a political factor, see the final chapter of Brown, *Matters of Life and Death* (forthcoming), and Brown (1996a).
101. For reasons of space, I have omitted a longer discussion of colonial segregation from 1840 through 1900; of the disease analogues of leprosy and syphilis that linked tuberculosis to African Americans; and the epidemics of yellow fever that established train inspections. On the second, see Rosenberg (1974); Brandt (1985), p. 14. For a more detailed analysis of these issues, see Brown, *Matters of Life and Death* (forthcoming).
102. Cell (1982), pp. 18–20.
103. Cell (1982), p. 18 and *passim*.
104. See Jordan (1968), p. 164; note the comparison between smallpox and miscegenation.
105. As cited in Rabinowitz (1978), p. 187.
106. See e.g. Rush (1799), as cited in Jordan (1968), pp. 15, 17, 518–21.
107. See Rosenberg (1974); Brandt (1985), p. 14.
108. Brown (1999); see also e.g. Babcock (1901).
109. Koch (1901), p. 260.
110. 'The Segregation and Supervision of the Tuberculous' *Boston Medical and Surgical J.* (1897), 136: 75.

111. Galishoff (1985); Gamble (1989), introduction.
112. Campbell (1986); Cell (1982), pp. 83, 85, 118 and *passim*; Williamson (1984), p. 182 and *passim*.
113. Burbick, *Healing the Republic*; Banta (1983).
114. See e.g. Swanson (1977).
115. Banta (1983), pp. 85–86.

REFERENCES

E. Ackerknecht, 'Anticontagionism between 1821 and 1867' *Bull. of the History of Medicine* (1948), 22: 562–593.

S.H. Adams, 'Tuberculosis, the Real Race Suicide' *McClure's* (January 1905) 24: 234–49.

W.H. Allen, 'Sanitation and Social Progress' *American Journal of Sociology* (1903), 8: 635.

W. Anderson, 'Climates of Opinion: Acclimatizationi n Nineteenth-Century France and England' *Victorian Studies* (Winter 1992) 35: 135–157.

W. Anderson, '"Where Every Prospect Pleases and Only Man is Vile"; Laboratory Medicine as Colonial Discourse' *Critical Inquiry* (Spring 1992a) pp. 506–529.

W. Anderson, 'Immunities of Empire: Race, Disease and the New Tropical Medicine, 1900–1920' presented to the Twentieth-Century History of Medicine Seminar, Johns Hopkins University, November 9, 1993.

D. Arnold, *Colonizing the Body* (Berkeley: 1993).

I.E. Atkinson, 'Early Syphilis in the Negro' *Maryland Medical Journal* (1877), 1: 135–46.

R.H. Babcock, 'The Limitation of Drug Therapy' presidential address, *Trans. of the American Climatological Assn.* (1901), 17: 10.

A.F. Bacon, 'What Bad Housing Means to the Community' *SPUNK* (July 1912) IV: 4, pp. 3–13.

R.S. Baker, 'Following the Color Line' *American Magazine* (April (1907), 63: 564–66.

E.R. Baldwin, 'The Tuberculous Patient: When and to What Extent Shall His Liberty be Limited' reprint from *Medical Review of Reviews* (January 1903) TB pamphlets, vol. 30, no. 34, p. 8, Flick Papers.

M. Banta, 'Medical Therapies and the Body Politic' *Prospects* (1983), 8: 59–128.

D.S. Barnes, *The Making of a Social Disease* (Berkeley, CA: 1995).

B. Bates, *Bargaining for Life* (Philadelphia: 1992).

J.E. Baur, *The Healthseekers of Southern California* (San Marion, CA: 1959) pp. 150–173.

A.L. Benedict, 'Consumption Considered as a Contagious Disease' *Popular Science* (November 1898) 48: 33–39.

J.T. Bernard, 'The Consumption Quarantine' *Houston Chronicle* (July 28, 1907) all, Jacobs Scrapbook, vol. (1/27/07–11/28/07).

E.B. Borland, 'Municipal Regulation of the Spitting Habit' *J. of the American Medical Assn.* (October 20, 1900) pp. 999–1001.

H.I. Bowditch, *Consumption in New England* (Boston: 1862).

A.M. Brandt, *No Magic Bullet* (NY: Oxford, 1985).

J. Brown, 'Policing the Body: Criminological Metaphor and the Popularization of Germ Theories of Disease' paper presented to the American Assn. for the History of Medicine, May 1984.

J. Brown, 'Professional Language: Words That Succeed' *Radical History Rev.* (January 1986) 34: 33–51.

J. Brown, 'The Rhetorical Force of Numbers' in J. Brown and D. Van Keuren (eds.), *The Estate of Social Knowledge* (Baltimore: 1990).

J. Brown, *The Definition of a Profession* (Princeton, NJ: 1992).

J. Brown, 'Tuberculosis: A Romance' presented at the Berkshire Conference of Women Historians, Vassar College, Poughkeepsie, NY, 1993.

J. Brown, 'Playing the Game: Tuberculosis, Medievalist Nostalgia, and the Great War' presented at the Organization of American historians, Atlanta, 1994.

J. Brown, 'Losing Battles: Feminism, Militarism, and the Everyday Language of War,' presented to the Berkshire Conference, Chapel Hill, North Carolina, June 1996.

J. Brown, 'Death Takes a Holiday: Historical Writing in the Antibiotic Interregnum, 1945–1990', parts of which were presented to the Committee for the Study of Political and Social Thought conference, 'The Politics of History' Tulane University, New Orleans, LA, May 1996a.

J. Brown, 'Crime, Commerce, and Contagionism' in R.G. Walters (ed.), *The Authority of Science and Reform in the United States* (Baltimore: 1997).

L. Bryder, *Below the Magic Mountain* (Oxford: 1988).

A. Bryson, *Med. Statist. in Man. Sci. Enq.* (1849), p. 45.

W.E. Campbell, 'Profit, Prejudice, and Protest: Utility Competitions and the Generation of Jim Crow Streetcars in Savannah, 1905–1907' *Georgia Historical Q.* (Summer 1986) 70: 2, p. 213.

H. Carby, '"On the Threshold of Woman's Era": Lynching, Empire, and Sexuality in Black Feminist Theory' in H.L. Gates (ed.), *'Race,' Writing, and Difference* (Chicago: 1986) pp. 301–16.

H.R. Carter, 'Manifestations of Syphilis Among Negroes; a Statistical Inquiry' *Report of the Surgeon General of the Marine Hospital* (Washington, D.C.: 1883) pp. 131–15.

M.H. Carter, 'The Vampire of the South' *McClure's Magazine* (October 1909) 33: 631.

J.W. Cell, *The Highest Stage of White Supremacy* (NY: 1982).

C. Clinton and N. Silber, (eds.), *Divided Houses: Gender and the Civil War* (New York: Oxford, 1992) introduction.

T.D. Coleman, 'The Susceptibility of the Negro to Tuberculosis' *American Climatological Assn. Transactions* (1903), 19: 122–32.

A. Crummell, 'The Need of New Ideas and New Aims for a New Era' in *Africa and America* (orig. 1883; Springfield, MA: 1891).

P.D. Curtin, '"The White Man's Grave": Image and Reality, 1780–1850' *Journal of British Studies* (1961), I: 94–110.

P.D. Curtin, 'Medical Knowledge and Urban Planning in Tropical Africa' *American Historical Review* (1985), 90: 594–613.

P.D. Curtin, *Death by Migration* (Cambridge: 1989).

A. Davin, 'Imperialism and Motherhood' *History Workshop* (Spring 1978) 5: 9–65.

S.G. Dixon, 'The Lesson of a State' *Journal of the Outdoor Life* (February 1908) V: I, p. 4.

S.G. Dixon, 'The Poisoned Cup' *Spunk* (October 1909) 1: 7, p. 5.

W.E.B. DuBois, *The Souls of Black Folk* (1903).

R. Dubos and J. Dubos, *The White Plague*, intro. by B.G. Rosenkrantz (orig. 1952; New Brunswick, NJ: 1987).

M. Edelman, *Political Language* (New York: 1977).

J.L. Edgar, 'Syphilis as a Cause of Phthisis' *Georgia Physician* (1906), 4: 219.

C. Edson, 'The Microbe as a Social Leveller' *North American Review* (1895), 161: 421–426.

J.H. Ellis, *Yellow Fever and Public Health in the New South* (Lexington, KY: 1992).

J.A. Faison, 'Tuberculosis in the Colored Race' *Medical Record* (1899), 55: 375.

F. Fanon, 'Medicine and Colonialism' in J. Ehrenreich (ed.), *The Cultural Crisis of Modern Medicine* (1978), pp. 229–251.

L. Flick, 'The Contagiousness of Phthisis (Tubercular Pulmonitis)' *Trans. of the Medical Society of the State of Pennsylvania* (June 1888) 20: 164–182.

L. Flick, 'The Duty of the Government in the Prevention of Tuberculosis' *J. of the American Medical Assn.* (August 22, 1891) 17: 287–290.

L. Flick, 'The Registration of Tuberculosis' (June 2, 1900) TB Pamphlets, vol. 30. no. 5, Flick Papers, Philadelphia College of Physicians.

L. Flick, *The Crusade against Tuberculosis* (Philadelphia: 1903).

S.J. Folmsbee, 'Notes and Documents: The Origins of the First "Jim Crow" Law' *J. of Southern History* (May 1949) 15: 235–247.

G. Frederickson, *White Supremacy* (NY: 1981).

S. Galishoff, 'Germs Know No Color Line: Black Health and Public Policy in Atlanta, 1900–1918' *J. of the History of Medicine and Allied Sciences* (January 1985) 40: 23.

V.N. Gamble (ed.), *Germs Have No Color Line* (NY: 1989).

J.W. Gilbert, 'City Housing of Negroes in Relation to Health' in, *The New Chivalry—Health* (Nashville: Southern Sociological Congress, 1915) p. 405–12.

G.M. Gould, *An Illustrated Dictionary of Medicine, Biology and Allied Sciences* (Philadelphia: P. Balkiston's Sons & Co., 1899).
J.T. Graves, *The Winter Resorts of Florida, South Georgia, Louisiana, Texas, California, Mexico, and Cuba* (NY: C.G. Crawford, 1883).
J.D. Hall, '"The Mind That Burns in Each Body": Women, Rape, and Racial Violence' in Ann Snitow *et. al.* (eds.), *Powers of Desire* (New York: 1983).
S. Harris, 'Tuberculosis in the Negro' *J. of the American Medical Assn.* (1903), 41: 834–38.
M. Hodes, 'Wartime Dialogues on Illicit Sex: White Women and Black Men' in Clinton and Silber (1992).
M. Hodes, 'The Sexualization of Reconstruction Politics: White Women and Black Men in the South after the Civil War' *J. of the History of Sexuality* (1993), 3:3, p. 402–17.
F.L. Hoffman, *Race Traits and Tendencies of the American Negro* (NY: 1896) pp. 95, 312.
F.L. Hoffman, 'The Statistical Laws of Tuberculosis' in *Report of the Tuberculosis Commission of Maryland 1902–1904* (Baltimore: 1904).
G.W. Hubbard, 'Consumption among the Colored Population of the Southern States' *Medical and Surgical Reports* (1896), 75: 423–5.
J.B. Huber, 'The Great White Plague' *Popular Science Monthly* (November 1904) 65: 298–305.
M. Humphreys, *Yellow Fever and The South* (New Brunswick, NJ: 1992) Chapter 5.
T.W. Hunter, *Household Workers in the Making* (Ph.D. thesis, Yale University, 1990).
P.P. Jacobs, 'National Association Meeting' *Journal of the Outdoor Life* (June 1909) VI: 6, p. 175.
M. Jacques, 'The Visiting Nurse in Tuberculosis: Her Importance as an Educational Agent' *Journal of the Outdoor Life* (May 1909) VI: 5, pp. 136–37.
F. Jones, 'Syphilis in the Negro' *J. of the American Medical Assn.* (1904), 42: 32–4.
T.J. Jones, 'Tuberculosis Among Negroes' *National Assn. for the Study and Prevention of Tuberculosis Transactions 2* (1906).
W.D. Jordan, *White Over Black* (orig. 1968; NY: 1977).
R.D.G. Kelly, 'We Are Not What We Seem: Rethinking Black Working-Class Opposition to the Jim Crow South' *J. of American History* (June 1993) 80: 1, p. 103.
S.A. Knopf, 'State Phthisiophilia and State Phthisiophobia, with a Plea for Justice to the Consumptive' *National Assn. for the Study and Prevention of Tuberculosis Transactions* (1910), 6: 155.
R. Koch, 'The Etiology of Tuberculosis' in W.W. Cheyne (ed.), *Recent Essays by Various Authors on Bacteria in Relation to Disease* (London: The New Sydenham Society, 1886).
R. Koch, 'The Fight Against Tuberculosis in the Light of the Experience Gained in Successful Combat of Other Infectious Diseases' *J. of the American Medical Assn.* (1901), 37: 259–65.
H.D. Kramer, 'The Germ Theory and the Early Public Health Program in the United States' *BHM* (1948), 22: 233–47.
A. Kraut, *Silent Travellers* (NY: 1994).
G. Lakoff, and M. Johnson, *Metaphors We Live By* (Chicago: 1980).
B. Lee, 'State Provision for the Treatment of the Consumptive Poor' *J. of the American Medical Assn.* (1900), 35: 989–991.
J.H. Lowman, 'TB in the Schools: Plans Suggested for Detecting the Disease and Caring for Its Victims' *Journal of the Outdoor Life* (June 1907) IV, 5.
D. Macrae, *The Americans at Home, II* (Edinburgh, 1870).
C.S. Magnum, *The Legal Status of the Negro* (Chapel Hill, NC: 1940).
L.M. Matthews, 'Keeping Down Jim Crow: The Railroads and the Separate Coach Bills in South Carolina' *South Atlantic Q.* (1974), 73: 117–129.
T.J. Mays, 'Increase of Insanity and Consumption among the Negro Population of the South since the War' *Boston Medical and Surgical J.* (1897), 136: 537–40.
C. McClain, 'Of Medicine, Race, and American Law: The Bubonic Plague Outbreak of 1900' *Law and Social Inquiry* (Summer 1988) 13: 447–513.
A. McClary, 'Germs are Everywhere: the Germ Threat as Seen in Magazine Articles, 1890–1920' *J. of American Culture* (1980), 3: 33–46.
W.K. McClure, 'Boycott of Consumptives' *Living Age* (December 8, 1906) 251: 624–9.
H. McGuire (President of the AMA) and G.F. Lydston (M.D., of Chicago), pamphlet, 'Sexual Crimes Among the Southern Negroes' (Louisville, KY: Renz and Henry, 1893).

A. Meier, *Negro Thought in America, 1880–1915* (Ann Arbor, MI: 1963).

K. Miller, 'The Negro Anti-Tuberculosis Society of Washington' *Journal of the Outdoor Life* (May 1909) VI, 5.

K. Miller, 'The Negro and Tuberculosis' *Journal of the Outdoor Life* (September 1910) VII: 9, p. 259.

K. Miller, 'The Historic Background of the Negro Physician' *J. of Negro History* (April 1916) 1: 2, pp. 99–109, esp. pp. 108–9.

C.L. Minor, 'Cooperation of the Public: Indispensable in the War on Tuberculosis' *Journal of the Outdoor Life* (July 1905) II, 6.

E.N. La Motte, 'The Unteachable Consumptive' *Journal of the Outdoor Life* (April 1909) VI: 4, pp. 105–7.

P. Murray, *States' Laws on Race and Color* (Cincinnati, OH: 1952).

T.W. Murrell, 'Syphilis in the Negro, a Medico-Sociologic Study' *J. of the American Medical Assn.* (1910), 54: 846–49.

R.J. Newton, 'The Enforcement of Anti-Spitting Laws' *National Assn. for the Study and Prevention of Tuberculosis Transactions* (1910), 6: 110; 'Progress of the Crusade'.

W.J. Northen, 'Tuberculosis Among Negroes' *Journal of the Southern Medical Association* (October 1909) 6: 415.

K. Ott, *The Intellectual Origins and Cultural Form of Tuberculosis* (Ph.D. thesis, Temple University, 1990).

R. Packard, 'The "Healthy Reserve" and the "Dressed Native": Discourses on Black Health and the Language of Legitimation in South Africa' *American Ethnologist* (November 1989) 16: 686–703.

R. Packard, *White Plague, Black Labor* (Berkeley: 1989a).

P. Paquin, 'Inherited Wretchedness: Should Consumptives Marry?' *Arena* (September 1896) 16: 605–11.

P. Pascoe, 'Misegenation Law, Court Cases, and Ideologies of "Race" in Twentieth-Century America' *J. of American History* (1996), 83: 44–69.

J.G.A. Pocock, 'Languages and Their Implications,' in Pocock (ed.), *Politics, Language and Time* (orig. 1960; Chicago: 1989).

T.O. Powell, 'The Increase of Insanity and Tuberculosis in the Southern Negro since 1860, and its Alliance and Some of the Supposed Causes' *J. of the American Medical Assn.* (1896), 27: 1185–1188.

H.N. Rabinowitz, *Race Relations in the Urban South* (NY: 1978).

F.H. Rankin, *The White Man's Grave* 2 vols. (London, 1836).

J.S. Rawlins, 'What Can We Do to Prevent the Spread of Consumption by the Negro Race in the Southern States?' *Memphis Medical Monthly* (1904), 24: 630–633.

P.A. Richmond, 'The Germ Theory of Disease' in A.M. Lilienthal (ed.), *Time, Places, and Persons* (Baltimore: 1980) pp. 84–93.

J. Roback, 'The Political Economy of Segregation: The Case of Segregated Streetcars' *J. of Economic History* (December 1986) XLVI: 4, pp. 893–917.

D. Rogers, 'In Search of Progressivism' *Rev. in American History* (1982), 10: 113–132.

N. Rogers, 'Germs with Legs: Flies, Disease, and the New Public Health' *BHM* (Winter 1989) 63: 4, pp. 599–617.

N. Rogers, *Dirt and Disease* (New Brunswick, NJ: 1992).

C.V. Roman, 'The Negro Woman and the Health Problem,' *The New Chivalry—Health* (Nashville: Southern Sociological Congress, 1915) pp. 392–405, 399.

R. Rosen, *The Lost Sisterhood* (Baltimore: 1982) pp. 78–85.

C.E. Rosenberg, 'The Bitter Fruit: Heredity, Disease, and Social Thought' *Perspectives in American History* (1974), pp. 189–235.

B.G. Rosenkrantz, 'Cart Before Horse: Theory, Practice, and Professional Image in American Public Health, 1870–1920' *JHMAS* (1979), 29: 55–73.

B.G. Rosenkrantz (ed.), *From Consumption to Tuberculosis* (NY: 1994).

S.M. Rothman, *Living in the Shadow of Death* (NY: 1994).

G.G. Roy, 'Syphilitic Phthisis in the Negro' *Southern Medical Record* (1881), 11: 175–77.

B. Rush, 'Observations Intended to Favor a Supposition That the Black Color (As It Is Called) of Negroes Is Derived from the Leprosy' *American Philosophical Society Transactions* (1799), 4: 289–297.

K. de Schweinitz, 'Consumption not Inherited' *SPUNK* (August 1912) IV: 8.
J.W. Scott, 'The Evidence of Experience' *Critical Inquiry* (Summer 1991) 17: 773–797.
N.S. Shaler, 'The Negro Problem' *Atlantic Monthly* (1884), p. 703.
N. Silber, *The Romance of Reunion* (Chapel Hill: 1995).
S. Sontag, *Illness as Metaphor* (NY: 1978).
L. Spitzer, 'The Mosquito and Segregation in Sierra Leone' *Canadian J. of African Studies* (1968), 2: 49–61.
C.D. Spivak, 'Isolation of Advanced Consumptives' *Charities* (May 26, 1906) pp. 279–80.
R. Sullivan, 'Cholera and Colonialism in the Philippines, 1899–1903' in R. MacLeod and M. Lewis (eds.), *Disease, Medicine, and Empire* (1988), pp. 284–300, and entire volume.
M. Swanson, 'The Sanitation Syndrome: Bubonic Plague and Urban Native Policy in the Cape Colony, 1900–1909' *J. of African History* (1977), 18: 387–410.
M. Teller, *The Tuberculosis Movement* (NY: 1988).
J. Thomas, *A Comprehensive Medical Dictionary* (Philadelphia: 1864) p. 498.
J. Toews, 'Intellectual History After the Linguistic Turn: The Autonomy of Meaning and the Irreducibility of Experience' *American Historical Review* (1987), 92: 879–907.
N. Tomes, The Gospel of Germs : Men, Women and the Microbe in American Life, (Cambridge, Mass, 1998).
W.H. Webb, 'Is Phthisis Pulmonalia Contagious, and Does It Belong to the Zymotic Group?' (Philadelphia, 1878).
R. Weigman, 'The Anatomy of Lynching' *J. of the History of Sexuality* (1993), 3: 3, pp. 445–467.
R. Williams, *Culture and Society, 1780–1950* (London: 1958).
R. Williams, *Keywords* (NY: 1976).
J. Williamson, *The Crucible of Race* (NY: Oxford, 1984).
I.B. Wells-Barnett, *A Red Record: Tabulated Statistics and Alleged Causes of Lynchings in the United States 1892–1893–1894 (Chicago: 1895).*
C. Vann Woodward, *The Strange Career of Jim Crow* 3rd. rev. ed. (NY: 1974).
R.R. Wright, 'Health the Basis of Racial Prosperity' *The New Chivalry—Health* (Nashville: Southern Sociological Congress, 1915) pp. 437–446.

Part 2
ETIOLOGY AND EXPERIMENTAL PRACTICES

Chapter 4

STANDARDIZING EPIDEMICS: INFECTION, INHERITANCE, AND ENVIRONMENT IN PREWAR EXPERIMENTAL EPIDEMIOLOGY

Olga Amsterdamska

Concluding the final chapter of his heroic history of *The Conquest of Epidemic Disease*, C.-E.A. Winslow wrote:

> There is today a wholesome reaction against exclusive emphasis on the germ and recognition of the importance—even in many germ diseases—of factors of constitutional resistance (diathesis) and of the influences of climate and season and nutrition upon vital resistance (Winslow, 1943, pp. 379–80).

Winslow's book is a celebration of the contribution of bacteriology to public health. But while he hails 'the practical application of the principles developed by a series of clear thinkers and brilliant investigators—from Fracastorus to Chapin—[which] has forever banished from the earth major plagues and pestilences of the past,' Winslow also admits that

> the practical triumphs of bacteriology did indeed tend to oversimplify the problem and to cause medical men for nearly half a century to ignore the true many-sidedness of disease. It is well that today we recognize once more—as in the pre-Pasteurian era—the importance of such factors as diathesis and nutrition and climate and season in the prevalence of even germ diseases. No clinical case of any disease is caused solely by the entrance of a germ (Winslow, 1943, p. 335).

Indirectly, Winslow's comments suggest that the common assessments of the influence of the bacteriological revolution on the understanding of infectious diseases as a public health problem deserve more detailed historical examination. Most historians agree that by the first decades of the twentieth century, bacteriology had radically transformed the concerns and organization of public health, and with it, of epidemiological research. The wide-ranging advocacy of social and hygienic reform had given way to more specific efforts to control of specific sources of infection and the transmission of pathogenic bacteria. A reform movement, which appealed to social and political values, had given way to a profession whose legitimacy was to be guaranteed by science.

> The powerful new methods of identifying the causes of diseases through the microscope drew attention away from the larger and more diffuse problems of water supplies, street cleaning, housing reform and the living conditions of the poor. The approach of locating, identifying and isolating bacteria and their human hosts seemed a much more efficient way of dealing with disease than environmental improvements. By focusing on the diagnosis of infectious diseases, the public health laboratory demonstrated the scientific and diagnostic power of the new public health, and began to cast doubt on the objectivity and credibility of those public health professionals who continued to insist on the necessity of social reform.
>
> Like the new bacteriology, epidemiology now became firmly oriented to the control of specific diseases (Fee, 1994, p. 237).

This change in the focus of public health concerns meant also that the bacteriological laboratory with its microscopes, stains, and biochemical and immunological diagnostic tests became a necessary and often a central setting for the investigation and control of public health problems. When the laboratory emerged as 'the premier location for the production and ordering of medical knowledge ... epidemiological inquiry and clinical observation came to exert a decreasing influence on the shape of etiological reasoning and immunological thought. To understand disease one needed a laboratory—or at least a microscope' (Anderson, 1996, pp. 105–06).

What these accounts neglect is that the dominance of bacteriological reasoning was not all-encompassing. Already in the 1890s, bacteriologists were acutely aware of the fact that the germ theory failed to explain certain phenomena and that epidemics of infectious diseases continued to be a difficult public health problem which the bacteriological theory could not solve. Epidemiology had never moved entirely into the laboratory: in addition to the microscopes, petri dishes, and laboratory animals, epidemiologists continued using morbidity and mortality statistics, surveys, and historical reconstructions of disease patterns to study correlations among factors involved in the spread and mode of transmission of infectious diseases, the waxing or waning of epidemics. Laboratory research did not replace all field investigation of infectious diseases, nor eliminate all attention to other biological 'predispositions' or to the cultural or social or environmental conditions in the causation and transmission of disease. In fact, during the period immediately before and especially after World War I, the relationships between bacteriology and epidemiology—between the laboratory and the field, microbes and epidemics, the seed, the soil and the environment—became subjects of intense debate.

In his essay 'Explaining Epidemics' Charles Rosenberg draws a distinction between the configuration and contagion perspectives which have alternately dominated attempts to understand the spread of epidemic disease. Rosenberg identifies bacteriology with a contamination perspective focusing on contagion and the specific mode of transmission of morbid matter from one individual to another, contrasting it with a configuration view which is more interactive and contextual, and emphasizes the multiple relationships between disease and the natural and social environment. Whereas the contamination perspective tends to be reductionist and monocausal and to focus on individuals, the configurational perspective is holistic, emphasizes system, interconnection, and balance, and is more likely to focus on groups or populations (Rosenberg, 1992, p. 295). The bacteriological revolution, he claims, 'helped swing medical opinion toward contamination' (Rosenberg, 1992, p. 299) and the reductionist assumptions of this perspective fit well with the reductionist methods and approaches of laboratory research. Rosenberg does acknowledge that the dominance of the contamination perspective brought about by the germ theory of disease 'did not banish the configurational impulse. A continuing concern for what is often called social medicine and an interest in environmental determinants of health and disease remained alive and in continued dialogue with the new bacteriology' (Rosenberg, 1992, p. 299). What Rosenberg does not acknowledge is the extent to which this dialogue was taking place within bacteriology itself, particularly in those locations where laboratory research 'intersected' with either clinical management of diseases[1] or with investigations of epidemics conducted in the field. For a variety of reasons, the contamination perspective seemed always to fare better in the laboratory than in the world outside it. Conversely, it seems that—at least in the cases examined here—bacteriologists and epidemiologists who attempted to bring the configurational perspective into the lab ended up providing reductionistic explanations of epidemics that were typical of the contamination view.

This article attempts to illustrate this point by following some of the developments in 'experimental epidemiology,' an area of research initiated in the immediate post-World War I era with the explicit purpose of disentangling the various factors governing the waning and waxing of epidemics by studying the spread of infectious diseases in animal populations, especially mice, in the laboratory. Experimental epidemiologists tried to model (or imitate) epidemics in the laboratory in order to study them from what would at first sight appear to be a configurational perspective. Epidemics were to be studied as multicausal or multifactorial group phenomena, but they were to be brought into the laboratory where appropriate tools and procedures would allow scientists to discipline these multiple factors. Laboratory control and precision were to go hand in hand with an examination not just of the microbial pathogen and its interaction with a single host organism, but also of the multiple web of causes governing the transmission of an epidemic within a group, its morbidity and mortality, its rise and fall. This at least was the hope of W.C.C. Topley in London and of the researchers at the Rockefeller Institute when in the immediate post-World War I period they began studying epidemics among laboratory mice.[2] Topley, who since 1910 was director of clinical pathology at the Charing Cross Hospital in London, began his experiments on mice epidemics shortly upon his return from Serbia, where during the war he served as a bacteriologist to the British Sanitary Commission. He continued working on the subject when he moved to the chair of bacteriology in Manchester in 1922, and when he returned to London to assume a position at the London School of Hygiene and Tropical Medicine in 1927.[3] Convinced of the importance of combining bacteriological with biometric methods, in the early 1920s he secured the collaboration of the epidemiologist Major Greenwood, a student of Karl Pearson and an advocate of the use of statistical methods in medicine (though not an advocate of eugenics, about which he was critical already in 1913).[4] The Rockefeller Institute research, developed on the initiative of the Institute's director, Simon Flexner. Initially conducted by Clara Lynch and Harold Amoss, it was continued throughout the interwar period under the direction of Leslie T. Webster—a graduate of Johns Hopkins Medical School who joined the Rockefeller Institute group after having spent a year conducting research at the Department of Pathology at Hopkins.[5] From the mid-1920s, studies in experimental epidemiology were also conducted at the Koch Institute in Berlin by F. Neufeld and his associates. Although my analysis compares British and American research on experimental epidemiology, this article concentrates primarily on the program pursued by the Rockefeller group of researchers in an attempt to trace the transformation of their initial, multifactorial formulation of the problem of the rise and fall of epidemics into a bi-causal model of an epidemic as an encounter between a pathogenic microorganism and a genetically susceptible host.[6]

EXPERIMENTAL EPIDEMIOLOGY AS BACTERIOLOGISTS' REVENGE?

The bacteriological translations required to make infectious diseases subject to laboratory investigation reduced the messy clinical reality of an infectious disease, with its variety of symptoms and individual manifestations, its regional, seasonal and environmental variations, and its often unpredictable outcomes, to relatively predictable model diseases of laboratory animals infected with specified amounts of standardized and pure bacterial cultures grown on ever more precisely composed substrates under carefully-controlled conditions.[7] By manipulating bacteria in test tubes and on petri dishes and then injecting them into animals, bacteriologists were able to study various properties of pathogenic microorganisms, the host's (induced) immunity and immunological reactions, the pathological changes accompanying individual infection, etc. In all these studies, infectious disease was an artificially transmitted individual event, a preferably uniform result of an interaction between an increasingly well characterized microorganism and an increasingly well controlled animal host. Perfect uniformity of outcome (such that, for example, all animals injected with a specific amount of X would invariably die—or survive without signs of disease—during a precisely specified time period) was a definite experimental ideal even if it was rarely achieved, as the meticulous records of the fate of each individual laboratory-infected mouse or rabbit in late nineteenth- and early twentieth-century bacteriological articles testify. And, to achieve this ideal, bacteriologists isolated, purified, measured, weighed, and attempted to control or eliminate all possible intervening factors, while the very fact of studying human infection in the laboratory by focusing on bacteria and individual infected animals excluded other environmental, social, and biological factors from consideration. But did these 'other' factors really need to be excluded or ignored? Could they not be subjected to the same laboratory discipline and manipulation?

Even at the height of the excitement about the germ theory of disease, bacteriologists often worked in the field, where many of the factors programmatically excluded from laboratory consideration were impossible to avoid and where full control of either the host or the organism was impracticable. In these field-based epidemiological studies of endemic and epidemic disease, it was also impossible to ignore some of the well-known facts which were either put aside or made to disappear in the laboratory. It was known, for example, that even during a severe epidemic not everyone got sick, and that even among those who had intimate contact with cases of the disease some remained healthy; it was also known that epidemics came and went, and that they followed a characteristic curve, that the severity of infectious diseases was changeable, that the incidence of some infectious diseases was subject to seasonal variation, and that some diseases affected particular social groups more than others, or were more prevalent in particular locations. In the early twentieth century, the germ theory of disease—even when armed with the concept of healthy carriers of infection, knowledge of animal vectors, and ideas about acquired immunity—could not explain these facts. As Andrew Mendelsohn has shown, they were deeply problematic for bacteriologists and epidemiologists in the last decade of the 19th and in the early decades of the 20th century. The seemingly simple etiological model according to which the presence of pathogenic bacteria necessarily brought about infection and disease came increasingly to be regarded as untenable, and bacteriologists began examining the 'terrain'—the constitutional and environmental factors involved in the spread of infectious diseases—alongside the microbe.[8]

The problems with the simple models of bacterial etiology of infectious disease were forcefully brought to the attention of the medical profession by the 1918/1919 influenza epidemic which claimed more than half a million lives in the U.S. and more than 150,000 in England and Wales alone. Bacteriologists argued about the identity of the causative organism. Epidemiologists disputed the novelty of the disease and the factors involved in the rise and decline of its prevalence and virulence.[9] In Britain, where statistical and multifactorial approaches to epidemiology were well entrenched and never replaced entirely by the purely bacteriological models, the divisions were particularly sharp. As the successive discussions in the Section of Epidemiology and State Medicine of the Royal Academy of Medicine in London testify, epidemiologists invoked a whole variety of factors and influences to explain both the influenza pandemic and epidemics of encephalitis, meningitis, and polio, all of which seemed to defy the available public health control measures. Thus, while some epidemiologists, e.g., John Brownlee, appealed to the Pasteurian idea of variability of the virulence of bacteria in different stages of an epidemic, others (e.g., Ronald Ross) tried to explain the rise and fall of epidemics in terms of mathematical models of changes in the composition of the population, and still others called for long-term historical investigations and the return to the seemingly forgotten Hippocratic notion of epidemic constitutions and its seventeenth-century elaboration by the English physician Thomas Sydenham. Historical epidemiologists such as Crookshank, Hamer, and Creighton argued that 'during natural periods of time, the epidemiological happenings in any stated area tend to exhibit peculiarities and particularities that are more or less distinctive' and that 'observation, if pursued by the historical method over long periods of time, will show that there is a tendency for the periodic recurrence of like epidemic constitutions.'[10] Although Crookshank and Hamer claimed that their position did not imply 'either disregard of what bacteriology has to say, or belief in "atmospheric influences" as causes of epidemics,' they chided bacteriologists for adopting simplistic models of causation:

> Bacteriologists [...] finding certain bacilli to be generally the immediate agents in particular *cases* during an epidemic declare these bacilli to be the cause of the *epidemic*; a position as illogical as to declare that bullets and poison gases were the cause, not merely of deaths during the late war, but of the war itself. The bacteriologists evidently forget the complexity of causation, and forget, too, that, however important from the point of view of the pathologist may be the exact nature of this or that bacillus in this or that form of disease, the real problem before the epidemiologist is to make out the nature of the factors that favor the propagation and dissemination of these special microorganisms at periodic intervals (Crookshank, 1921, pp. 548–49).[11]

The first proponents of experimental epidemiology regarded the rift between bacteriology and epidemiology as profoundly unfortunate and detrimental to the understanding of epidemics. They acknowledged and emphasized the insufficiency of monocausal bacteriological explanations, but regarded appeals to 'epidemic constitutions' as scientifically futile and defeatist since they relegated the rise and fall of epidemics to some 'occult' or 'telluric' factors. Acknowledging that 'the mere occurrence of potent microorganisms does not suffice to produce an outbreak of epidemic disease' and that the available epidemiological and bacteriological knowledge was incapable of accounting for many features of epidemics, Simon Flexner 'questioned whether progress is to be gained by a return to the indefinite concepts of epidemic constitutions or of the interdependence of diseases as various as epidemic meningitis, poliomyelitis, influenza, etc. united by an inevitable nexus of events the nature of which cannot even be surmised' (Flexner, 1922, p. 11). In his initial statements on experimental epidemiology Topley did not refer to conflicts within epidemiology, though he did argue that 'In any attempt to form a bacteriological conception of the processes involved in the epidemic spread of infection, the ascertained facts of epidemiology must be kept constantly in mind' (Topley, 1919, p. 1). By the mid-1920s, however, Topley, now working together with Major Greenwood (who had been one of the proponents of the return to the Hippocratic ideas of Sydenham), was just as adamant as Flexner in his rejection of the concept of 'epidemic constitutions' as 'occult,' and ridiculed Sydenham's followers for appealing to 'inexplicable change in the very bowels of the earth' and 'super-meteorological phenomena, whether variations of terrestrial magnetism or ... other high cosmic phenomena' (Greenwood and Topley, 1926, p. 35).[12] Topley and Greenwood saw Crookshank's plan for epidemiological research as 'apocalyptic rather than feasible' and declared it 'not likely that the doctrine of Epidemic Constitutions will prove a generally valuable organon in the investigation of human epidemic disease' (Greenwood and Topley, 1925, p. 50).

When the arguments of the experimental epidemiologists are read against the background of epidemiological disputes, they appear as attempts of the laboratory-based bacteriologists to reaffirm the primacy of bacteriological methods and modes of understanding epidemics by correcting their earlier shortcomings: excessive concentration on the microorganism, exclusion of other factors, the belief that an understanding of epidemics will follow naturally from the understanding of individual diseases, etc. By using experimental laboratory methods, they claimed, one could not only study individual infections but also address some of the acknowledged epidemiological problems which neither the germ theory of disease nor epidemiology alone had been able to solve. Monocausal, individualistic models of epidemics were seen as misleading and insufficient, but there was no need to reject proven scientific methods and to minimize the advantages of the laboratory; the challenge was to use these methods so as to disentangle the multiple factors involved in rise and fall of epidemics without reducing them to simple germ/host interactions.

In the Goulstonian lectures delivered at the Royal College of Physicians in London in 1919, Topley announced his new research program and the first results of his studies on a deliberately induced epidemic of typhoid among mice. The aim was to address precisely the issues which epidemiologists claimed the germ theory of disease could not explain. Experimental study of epidemics had to 'explain the constant presence of a specific cause of disease through long periods of time, the periodic reappearance of the disease in epidemic form and the characteristic form of each such wave of disease in its rise, crest, and subsidence, leading to another disease free period' (Topley, 1919, p. 2). Although Topley believed that these problems could be solved by focusing primarily on the interactions between bacteria and the hosts, he did not deny the role of environmental factors. The influence of the environment, however, was indirect: 'alterations in the environment may be the determining cause initiating an outbreak of bacterial disease; but they will almost certainly act through the variations which they bring in the other two factors,' namely, the bacteria and the host (Topley, 1919, p. 3). Thus, although from the very beginning Topley limited the 'configurational' aspect of his research by (temporarily) excluding direct observations of environmental influences, he emphasized another aspect of the configurational view: the necessity to look at epidemics as events affecting collectives rather than individuals. In contrast to traditional bacteriological approaches, Topley stressed the fact that an epidemic was a biological phenomenon *sui generis* and, like the epidemiologists, he regarded it as a herd phenomenon that could not be reduced to individual cases of disease.

Just like Topley, the Rockefeller Institute researchers saw experimental epidemiology as the bacteriologists' answer to epidemiological difficulties. According to Flexner, traditional epidemiological methods of (*post facto*) statistical analysis did provide some 'valuable and highly suggestive if only tentative conclusions.' Laboratory studies with their precision and control could certainly do better than this, and Flexner was certain that 'whatever facts can be determined with accuracy concerning the factors governing the incidence of epidemic diseases among animals will apply in some measure to epidemic diseases of man.'[13] Flexner listed a number of considerations that led him to propose an experimental study of epidemics:

> the growing importance of the knowledge of laws of epidemicity, the difficulties in the way of obtaining complete data in the field and then of interpreting these data; the imperfection of our knowledge of the extent and then of the significance of the carrier state, and of the ordinary modes of communication of the inciting microbe from animal to animal; and also, the influence of the contributing factors—food, temperature, humidity, preexisting pathological states, etc., etc. … One finds that the variation in

> the mortality statistics of epidemic influenza in certain American cities is related seriously to such indefinite sources as so-called 'constitutional diseases,' that is, that the primary factor in causing the observed discrepancy in different communities in respect to the influenza epidemic was 'the biologic constitution of organic fitness of the people making up the population of these communities.'[14]

Starting from the proposition that 'the factors involved in a given epidemic wave are doubtless intricate,' Flexner seemed to acknowledge the importance of a configurational perspective. To understand the fluctuations of epidemics, their intensity, severity, spread, etc., he argued that, it was necessary to consider not only 'two sets of living organisms, themselves complex, namely the microbic parasite and the larger animal host,' but also their interactions and the less definite agencies which—once properly investigated—will no longer fall into the undefined and speculative class of 'telluric' influences or 'epidemic constitutions.'[15] For Flexner then, the epidemiologists' critique of bacteriology implied that laboratory methods needed to be used to study not the microbe alone, and not even just the microbe and its host alone, but rather their complex relations and the multiple factors that shape them. Flexner's plan 'to reduce the study of certain phases of epidemiology to an experimental basis'[16] certainly involved important elements of the configurational view of epidemics and explicitly rejected monocausal explanations; but, as opposed to Topley, Flexner and his associates did not emphasize the fact that epidemics should be regarded as collective rather than individual phenomena.

It remains to be seen how these two related but different attempts to introduce elements of a configuration perspective into the laboratory fared in practice.

MODELING EPIDEMICS

The distinctive emphases of the two groups were directly reflected in their experimental arrangements, in how they understood the models they constructed, in how they posed their questions, and in how they interpreted their results.

From the very beginning of their experiments, both the British and the American researchers had to make trade-offs between the laboratory control believed to be necessary to isolate the roles of the various factors which could be involved, and the desire to make experimental epidemics simulate as closely as possible 'the natural' course of epidemic infectious disease in human populations. Thus, on the one hand, Topley claimed that in his experiments 'the conditions were arranged to reproduce, as closely as possible, the sequence of events which must actually occur in the epidemic spread of disease' (Topley, 1919, p. 91), and Amoss justified his procedures by claiming that they were 'chosen to simulate the conditions of epidemic outbreaks of disease not only among carefully segregated small domestic animals but also those which occur among human beings' (Amoss, 1922, p. 26). On the other hand, the specific advantage of the experimental method was the fact that it allowed control over numerous factors that could not be easily controlled in field investigations. Thus, Topley and Greenwood claimed that in comparison with an epidemiologist they had a clear advantage because in their experiments 'The social and economic milieu has not changed, the rate of increase has been strictly controlled and, excepting the victims of cannibalism, all deaths have been certified with pathological accuracy' (Greenwood and Topley, 1926, p. 35). Similarly, Amoss and Webster emphasized that the experimental method allowed them to 'control and keep constant as many factors as possible' (Webster, 1922, p. 71), that in the laboratory 'an epidemic [can] be started under fixed conditions,' and 'by providing a more homogeneous material yield results of greater consistency, while the two major factors of host and parasite would be placed under highly favorable conditions of control' (Amoss, 1922, p. 25). In both cases, the initial control involved such environmental factors as nutrition and temperature, and above all, both groups of researchers evolved complex routines (quarantine, sterilization of cages, implements, and food; rules about handling the animals, etc.) to prevent extraneous infections of the experimental population by other microorganisms, introduction of already infected mice into the colony, or introduction of insect vectors. Thus, if the experimental approach was to deliver relevant knowledge that was not to be had from the field investigations of actual human epidemics, the model epidemics had to be carefully controlled, eliminating all possible 'disturbing' factors:

> Great care was taken throughout the experiment to exclude roaches, ants, flies, and other insects and vermin from the room. The temperature of the room was kept constant at 68 F. [...] The cages were thoroughly cleaned once a week, always in the same order, and the mice were fed in the same order daily as follows: grain in the morning and bread moistened with milk in the afternoon. Precautions were taken to exclude extraneous disease. The mice used came from a carefully controlled healthy stock

> bred at the Institute and free of communicable disease. (Amoss, 1922, p. 26).

We shall see that as the experiments went on, the number of factors which Webster standardized or controlled grew: the microbes, the mice, and the environment in which the experiments were conducted were made ever more uniform. And yet, Webster continued to insist that 'the method of experimental infection most apt to yield critical evidence on epidemiological problems is one which simulates natural conditions by bringing about the clinical, pathological, and mass phenomena of natural disease' (Webster, 1932, p. 322).

But while both groups agreed on the necessity to model natural conditions and on the fact that the experimental method offered advantages by making it possible to control the various factors involved in an epidemic, the British and American researchers disagreed on what constituted an adequate experimental model of an epidemic. What should and what should not be controlled or manipulated? How should the model and the results obtained be evaluated and what comparisons could legitimately be made?

The most obvious difference between the two experimental set-ups was that while Topley and his collaborators kept the experimental mice in a single large cage and allowed for the infection to spread directly from mouse to mouse, at the Rockefeller the mice were kept in separate jars or cages in groups of five and the infection was carried from cage to cage on the hands of the assistants responsible for cleaning the cages and feeding the mice. In some experiments, Webster was actually inoculating all the individual mice involved in the experiment with the bacillus, and still claiming to be studying epidemics. This seemingly minor technical difference reflected an important difference in theoretical emphasis: each group wanted its model epidemic to include what they saw as the major (irreducible) feature of the phenomenon. Flexner, Amoss, Webster and their Rockefeller associates saw an epidemic as a sum of individual mice succumbing to the same disease; a herd was a collection of individuals. (Accordingly, controlling the dosage, for example, meant controlling the number of bacteria each individual mouse received). Keeping the mice in small isolated groups (or even infecting them individually and thus controlling the dosage of the pathogen) did not essentially affect the phenomenon to be investigated and served as an experimental expedient: it facilitated control, simplified the monitoring of individual mice, and allowed for careful record keeping. It was also intended to prevent or at least minimize cannibalism (when kept in large groups, mice devoured their dead companions, making it difficult to ascertain the cause of death or monitor the spread of the disease). Thus, despite the fact that mouse typhoid is transmitted by ingestion of infected materials, the experiment was set up so that 'infection could be transferred only by the hands and implements of the person cleaning the cages and feeding the mice' (Amoss, 1922, p. 26). Interestingly, this limited—and highly artificial—mode of transmission was justified as one that was more natural than the method used by Topley, who, according to the Rockefeller scientists, attempted to 'achieve the greatest likelihood of infection occurring by providing the greatest possible concentration of contaminated materials [...]' According to Amoss, the Rockefeller experiments did not aim to provide 'optimal conditions for infection to take place in mice, but the imitation, if only roughly, of those occurring naturally in man and in laboratory animals, in connection with which epidemics of disease occur' (Amoss, 1922a, p. 67).

In contrast, Topley and Greenwood emphasized the fact that an epidemic is a herd phenomenon so that the factors affecting spread of disease in a group might not be reducible to those responsible for each individual mouse dying of the infection. Explaining the waxing and waning of epidemics was not equivalent to explaining why each individual mouse succumbed or survived. Instead, the collective mortality rate was to be explained by some change in the conditions in the herd. One of their principal questions was how the composition of the herd and its changes—for example, not only the presence of individually susceptible mice, but also the relative proportion of such mice in the herd—affect the spread of infection. Accordingly, housing the mice in a single cage was a means of 'ensuring as uniform an admixture as possible of the individuals within the experimental herd' (Topley, 1926, p. 482) while eliminating possibilities for 'random variation.' This does not mean, of course, that Topley's method was any less 'controlled' or 'artificial' than that of Amoss and Webster: the cage in which experimental mice were housed was sterilized daily, all contact with external contaminants was avoided, new mice were added to the cage following a careful quarantine protocol, etc. But the decisions about what should be controlled and what should be carefully 'imitated' before the model could be considered an appropriate one were shaped by both theoretical and practical considerations, and when theoretical assumptions differed, they could be always subject to dispute.

In some respects, the attempt to imitate 'natural' conditions was basically a rhetorical device. For example, the Rockefeller mice were housed in a 'village' in which 'healthy mice in jars of five each, arranged in the manner of houses along streets, were placed.'[17] But while the rhetoric suggested human settlements, the arrangement precluded direct communication between individual 'houses' in which only a limited number of mice were kept, and there was no obvious analogue for the hands of the assistant that carried the infection from one house to another. The epidemic in a mouse 'village' was taken as equivalent to the incidence of infections in a number of virtually quarantined 'houses.' It is tempting to see Amoss's and Webster's 'mouse village,' in which there is no direct contact among the inhabitants and where only the 'service personnel' spreads the infection, as a peculiar model of a human community organized either as a prison or as an affluent American suburb. The arrangements, however, appear to have been structured more directly by theoretical ideas about epidemics and the exigencies of the laboratory than by conscious imitation of specific social conditions.

The British researchers also—though somewhat more reluctantly—invoked social analogues to their laboratory epidemics. At one point they compared the conditions in their herd to those obtaining in 'a strict monastery, receiving each year the same number of novices, losing no members save by death.' But they immediately rejected this analogy as inexact and admitted that their population was 'essentially artificial,' not even an analogue of a natural mouse community (since the experimental population did not increase by birth—whatever young were born were either devoured by the other mice or removed). At best, they claimed, the experimental herd was to be likened to 'a collection of savages, men and women, practicing infanticide and cannibalism and receiving in each unit of time a constant number of other savages of both sexes and between the ages of puberty and early middle life' (Greenwood and Topley, 1925, pp. 56–57). And yet—even though these comparisons to monasteries and 'collections of savages' were somewhat ironic—the fresh, unexposed mice which were regularly added to the experimental population were repeatedly and seriously referred to as 'immigrants' and, even though they were carefully preselected and screened, and their ages and weights were more or less standardized, their movement was taken as analogous to movements of uninfected human populations and to the growth of such population through births. These changes in the population—equilibria between bacteria and populations composed of varying mixtures of susceptible and resistant hosts—were the main focus of attention in Topley's and Greenwood's studies and the grounds on which they attempted to establish the practical relevance of their models. The claim that their mouse 'microcosm' was at least in this respect directly analogous to the human 'macrocosm' in which actual epidemics took place was the basis on which their models rested.

Ultimately, however, the validity of experimental models was established neither by the claim that the laboratory epidemics were taking place in a 'mouse village' nor by the assertion that the new mice which were continuously added to the cages to keep the epidemic going were in some respects comparable to 'immigrants' or the 'growth of population by reproduction.' For the Rockefeller researchers, this validity was decisively established by a comparison of two epidemic curves; for Topley and Greenwood it rested on statistical comparisons between the laboratory epidemics and biometric epidemiological data collected about human epidemics.

At the Rockefeller Institute, the immediate stimulus for undertaking a study of an animal epidemic in the laboratory was a spontaneous outbreak of mouse typhoid in the animal room of the Institute where mice were being bred for the study of cancer. From the end of 1918 until January 1921, some 7000 mice were affected and some 3000 mice died of typhoid in two separate epidemic waves. The course of the epidemic, its spread among the mouse colony, the characteristics of the bacteria involved, and the reactions of mice to attempted vaccination with killed cultures were monitored and charted by Clara Lynch (Lynch, 1922). Lynch's study was primarily observational—the epidemic was followed and charted, but there were no attempts to induce new outbreaks of the disease, nor to prolong or control the infection of the mice or the spread of the disease. The data gathered in this 'observational' study or in the early experiments conducted by Amoss served later as a base line against which the data generated in the more strictly experimental investigations of mouse typhoid epidemics were to be evaluated. The success in the establishment of the model did not depend on a demonstration that all features of the laboratory epidemic resembled those found in 'natural' epizootics or epidemics; given the advantages of control and manipulability, such a full imitation of nature would have been useless as a basis of experimentation anyway. To establish the relevant similarity, Webster needed hardly more than a guarantee that the epidemic curves constructed by Lynch and Amoss on the basis of the mortality rates in their study of the initial 'spontaneous' epidemic of the laboratory mice were similar to his own curves constructed to portray mortality in 'experimental' epidemics. Since the mortality rates were similar in both cases, the model constructed in the laboratory was deemed a faithful one.[18] By demonstrating this parallel, Webster claimed to have established that the source of variation lay within his model.

In the first pages of his 1932 Harvey lecture, Webster summarized the various means by which epidemic curves could be faithfully reproduced in experimental epidemics. By this time, he had induced and studied experimental epidemics caused not only by the mouse typhoid organism, but also by the Friedländer bacilli responsible for pneumonia in mice, and by bacterium lepisepticum, the agent of rabbit 'snuffles' pneumonia, both of which were respiratory diseases.[19] In all these cases, 'a technique has been evolved which is quantitative and yet reproduces the natural phenomena of infection, which permits critical observations on the questions of whether microbic and host factors account for all the mass phenomena of infectious disease ...' (Webster, 1932, p. 324). Following the review of the methods used to induce epidemics and adducing as evidence the comparisons of mortalities in spontaneous and experimental epidemics, Webster concluded that

> The results thus far described demonstrate that various endemic and epidemic phenomena of natural enteric and respiratory infection can be elicited experimentally solely in bringing together susceptible hosts and pathogenic microorganisms (Webster, 1932, p. 328).

In other words, the relationship between the 'model' epidemic in the laboratory and the 'naturally occurring' laboratory epidemic was interpreted as a warrant for disregarding the role of all factors other than 'host susceptibility' and the bacterium. If, under controlled and constant environmental conditions, epidemics still followed the characteristic curve and some mice survived while others succumbed, then the explanation for the observed variation had to be found in those factors internal to the experiment whose variation was—as yet—uncontrolled and unexamined. Accordingly, Webster turned his attention first to the possibility of variations in the virulence and in the quantitative distribution of the microbe and then to variability in the 'susceptibility of the host.' With this formulation the 'configurational' study was, for all practical purposes, reduced to three biological factors.

As Webster retreated to the study of bacteria and experimental animals as the two sources of variance, rhetorical references to 'villages,' 'houses,' and 'streets' were no longer necessary and could be safely deleted.

Topley and Greenwood strenuously objected to such appeals to the similarity between the two epidemic curves as grounds for validating an experimental model or the results of experiments. Not only were they fond of repeating that 'comparaison n'est pas raison' (Greenwood and Topley, 1926, p. 35) and of warning their readers about 'too adventurous an application of the dangerous argument from analogy' (Topley, 1926b, p. 648), but they also objected to Webster's comparisons between the two epidemic curves as indicative of some underlying similarity in causation. A similar curve could be produced by any number of underlying causal processes, they argued, and a mere similarity in the rates of mortality over time is no proof that the mechanisms involved were similar in the two cases. Topley objected particularly strongly to one of Webster's experiments in which a large number of mice were individually injected with a dose of bacteria and their mortality, charted over time, provided data such that the 'typical' epidemic curve could be reproduced. Such an experiment, he claimed, could not be compared to or serve as a model of any actual epidemic, since it is impossible that in an epidemic each individual member of the population would be infected with an identical dose of the infectious agent simultaneously with all the other individuals.[20]

Throughout their investigations, Topley and Greenwood insisted that the only valid comparisons were those between the statistically significant data gathered on the basis of experiments and actual epidemiological data subjected to similar statistical treatment:

> If, from the results of such experiments as we are carrying out, it should be possible to formulate any general description of the spread of epidemic disease, it will clearly be necessary to test the validity of such generalisations by statistical analysis of the data of epidemic diseases as they occur in man and animals under natural conditions, and so to develop the inquiry on its purely biometrical side (Topley, 1926, p. 478).[21]

Topley and Greenwood regarded their experimental epidemics as simplifications of the conditions obtaining outside the laboratory. Since in the field the 'multiplicity of factors, certainly or probably modifying the evolution of epidemic disease is so great that nobody yet succeeded in unraveling the tangled skein,' model epidemics should allow only for the isolation of the most important causal determinants of the waxing and waning of epidemics. If the periodic rise and fall of epidemics similar to those found in human populations could be reproduced when certain specific factors were controlled, these factors were not likely to play a major role in the process. But the possibility that they contributed to it in some way was by no means excluded, especially if they were found to play some role in non-experimental epidemiological studies. Thus, for example, Topley and Greenwood argued that despite the fact that everyone knows that seasonal variation is important in influencing the epidemicity of some diseases and that it stands to reason that dietary variation may have some influence on susceptibility to infection, these factors could not be decisive, both because in laboratory situations epidemics arise also when these factors are kept constant, and because actual epidemiological data on human disease show them to be at most of secondary importance (or as Greenwood put it, the correlations between season or diet and an epidemic are stochastic rather than functional). Ignoring these factors in the laboratory was thus a justifiable simplification.

But, as we have seen, not all simplifications could be so easily justified. Those simplifications which reduced an epidemic as a herd phenomenon to the (sum of the) individual cases of disease were inadmissible, as were those which in their opinion made the structure of the herd or the conditions in it incomparable to any 'normal' population. Given this emphasis on the herd and on an epidemic as an irreducible biological phenomenon, Topley and Greenwood re-defined a number of variables which had to be measured in their models: dosage of bacteria, for example, was no longer a quantity of bacteria which an individual received but 'a probability, the chance that any one individual among the population at risk will, within a given time, receive a given number of bacteria' (Topley, 1926a, p. 531). This was complicated further by the fact that in a herd not only total probability of ingesting or inhaling a dose of bacteria had to be considered, but also the probability of receiving repeated doses over varying amounts over time, and by the possibility of differences arising out of the manner in which bacteria entered the host organism. As a result, 'dosage' in a herd could not be accurately measured and comparisons became difficult. Similar considerations governed the concept of resistance, since in the study of epidemics it referred to 'herd' rather than individual resistance. A herd could, for example, increase its immunity by 'selection' (the death of susceptible individuals), or by better sanitary conditions preventing the spread of bacteria, or by the elimination of animal vectors.[22] Throughout their experiments, Topley and his co-workers focused on the manipulation of factors that affected the structure of the herd (for example, by admitting varying numbers of immigrants, dispersal, or immunization), but they objected to any conclusions about epidemics being drawn from the investigation of individual infections or even from comparisons with phenomena observed in individual infections. They also insisted that before any other contributing factors are considered one must account for the periodic recurrence of epidemics in herds in which the only source of variation was the addition of new unexposed 'immigrants.'

STANDARDIZING VIRULENCE AND STANDARDIZING BACTERIA

The theory that the rise and fall of epidemics could be attributed to increases and decreases in the virulence of the bacteria was initially proposed by Pasteur. It was well known that bacterial virulence changes under different methods of cultivation in the laboratory, and animal passage of a bacterial strain was the traditional method of increasing the virulence of cultures used in laboratory experiments. The apparent ability of the bacteriologist working in the laboratory to vary bacterial virulence was then assumed to be analogous to changes taking place under natural conditions.[23] Variations in virulence of the microbe were also invoked by epidemiologists (for example, Brownlee) to account for the waxing and waning of epidemics. At the time of his Goulstonian lectures, Topley also regarded bacterial variability as the most likely explanation for why new epidemics arise and subside. Similarly, when Amoss began to study epidemics he expected to find variations in the virulence of the mouse typhoid bacteria:

> The evidence at hand is to the effect that the degree of infectivity of 'mouse typhoid' bacilli is highly fluctuating, and it appears that all the bacilli which are included under that name, classed variously as *Bacillus enteritidis*, Gärtner bacillus, *Bacillus paratyphosus B*, *Bacillus suipestifer*, and *Bacillus pestis caviae*, infect mice in a similar, possibly indistinguishable manner, including self-limited outbreaks of disease reaching at times epidemic proportions (Amoss 1922a, p. 69).

Webster set out to test this idea of variations in bacterial virulence quite early in his research and continued testing it throughout the 1920s. To do so, he relied on inoculations with pure, single cell cultures of one of the strains of the bacillus isolated by Amoss; Webster tested and re-tested the virulence of this particular strain in a long series of experiments conducted between 1922 and 1930. Bacteria were isolated from the animals at various stages of the artificially induced disease, from healthy carriers as well as diceased victims, and in various seasons; in periods immediately preceding an artificial epidemic, at the height of the epidemic, and in the post-epidemic period. Once isolated, these bacteria were cultured and given in precisely specified doses to random lots of mice. The mortality rates of these mice were then compared. The differences in any one of these comparisons were so small that Webster judged them insignificant. Microbes isolated from carriers killed approximately as many new mice as those isolated from very sick individuals; injections of strains taken during the height of the epidemic resulted in as many deaths in the new susceptible lots of mice as those taken from animals which had survived the infection.[24] The mortality rates in mouse typhoid infections in the lots of 20 mice hovered around 75–80% (in one case, it was as low as 60 and in another as high as 100 percent). Similar, though less numerous, experiments were made with the pasteurella infection and the Friedländer bacillus infections.

Webster also examined the relations between virulence and bacteriophage infection and between virulence and dissociation.[25] Although he was able to manipulate bacterial virulence *in vitro* by artificially inducing bacterial dissociation, he did not detect any such dissociation occurring 'spontaneously' in the animals used in his study. Topley's studies on bacteriophage also produced negative results (Topley, Wilson and Lewis, 1925; Topley and Wilson, 1925).

Webster concluded that through the entire period of his experiments, and independently of the source of the culture and the stage of the epidemic at which bacteria were tested, the virulence of any given strain was constant and dissociation played no significant role in the process.

In 1924, when the experiments in question were still in the relatively early stages, Webster argued:

> It was found that several paratyphoid-enteritidis strains, isolated from man and animals and related antigenically, differed markedly in pathogenicity, but that the inherent virulence of each strain remained constant, uninfluenced by repeated animal passage and varied natural environmental circumstances. Stock cultures kept in the ice-box and carrier cultures did not differ in virulence from cultures isolated from acute cases. Hence, individual strain virulence among the ubiquitous paratyphoid-enteritidis bacilli may be viewed as relatively a fixed quality (Webster, 1924, p. 137).

The coexistence of the different strains and variations in their pathogenicity or virulence were, however, quickly excluded from consideration as Webster attempted to standardize and control the conditions in his experiments. To test whether a strain changed its virulence, Webster had to assure himself that the mice were initially free from all other contact with mouse typhoid microorganisms, and then use a single, well defined, pure strain to induce the epidemic:

> To control these two variables, dosage and host susceptibility, and thereby measure virulence changes more accurately, experiments were performed with known doses of mouse typhoid bacilli under conditions in which the individual differences in host susceptibility were eliminated as far as possible through the use of large numbers of mice of similar breed, living conditions, and of corresponding age and weight (Webster, 1924, pp. 136–37).

Given this degree of control of the microbe and the host, and the maintenance of constant environmental conditions, Webster did not detect any variations in bacterial virulence.

When Webster came to review his results in 1930, the bacillus of mouse typhoid (now referred to as B. enteritidis) was no longer said to be composed of strains of different though unchanging virulence, but simply of organisms of constant virulence: in the process of examining whether a single strain of specified virulence undergoes modifications in the course of experimental epidemics, Webster had so standardized mouse typhoid that it was now seen as the result of an interaction between different numbers of bacilli of constant virulence and a variously susceptible host:

> Studies of mouse typhoid, a disease resembling paratyphoid fever in man, show that in this infection, microbes from animals apparently well, from others moderately ill, and from still others dying after a brief and serious illness are all equally virulent and retain their precise power of producing disease under ordinary conditions prevailing in nature. Microbes from cases taken in times of severe epidemics, and from animals during relatively 'healthy' times prove equally pathogenic (disease-producing), whereas any given cultures of the microbe administered to a group of animals lead to all gradations of reaction from apparent immunity on the part of the host to acute fatal illness. It seems, therefore, as if the resisting power of the mouse and not variation in the bacteria is mainly responsible for different manifestations of the disease, and that factors other than supposed changes in the virulence of the microbes bring about an epidemic.[26]

In Webster's rejection of the hypothesis of variations of virulence as one of the factors responsible for the epidemic waves, we encounter yet again the conflict between laboratory standardization and control of all variables, which was deemed necessary to isolate the relevant causal factors, and the wish to consider the interaction of variables. Bacterial variation could be measured in the laboratory only under conditions in which all the other variables—whether environmental or linked to the microorganism or to the host—that were thought to be implicated in the process were kept constant; but insofar as all the variables were said to be interrelated, keeping all these conditions constant often seemed to standardize also the very factor whose changes were to be examined. This experimental dilemma brings to the fore the apparent difficulty of adopting standard bacteriological techniques and their methodological constraints to a study that was initially conceived as an attempt to examine the configuration of factors involved in epidemics. For in the process of rendering such configuration experimentally manageable, Webster attempted to disentangle them systematically and consider them 'one by one,' assuming that their variations were independent of the factors that were being controlled. This difficulty was compounded by the fact that many of the variables (such as virulence) could only be measured by using the same types of animal models which, when seen as models of epidemics, were also supposed to bring about their variation. At one and the same time, Webster had to guarantee that 1) when he measured virulence, the different animal tests were directly comparable and measured a property of the well-defined microbe rather than some property of the hosts or the environment or some result of their interactions, and that 2) when he used these animal models as settings in which to provoke changes in virulence, they provided optimal conditions for variation of this one variable, while keeping constant all other known variables—including those known to affect virulence when bacteria were cultured in vitro. For example, when a significant decrease in the mortality rate was found in one of the experiments, Webster promptly attributed it to the excellent state of resistance of the hosts at that particular period—but the measurement of host resistance was at that time procedurally the same as the measurement of microbial virulence.

In this manner, Webster systematically modified his experimental model, bringing it ever closer to the monocausal 'contamination' perspective in which epidemics were once again considered as direct results of the interaction between an individual susceptible host and a stable bacterial pathogen.

Topley and his co-workers, particularly G.S. Wilson, were reluctant to accept Webster's conclusions regarding the stability of virulence. Like

Webster (and using comparable methods), they were generally unable to demonstrate that changes in virulence actually took place during the epidemics under the experimental conditions in their herds. Nevertheless, Topley and Wilson considered the question open as late as 1936.[27] On many occasions they referred back to the few instances in which they detected such changes in their experiments, appealed to the fact that bacterial virulence was often manipulated under laboratory conditions, and argued that even single-cell cultures of any particular strain exhibit heterogeneity, and that a variety of strains of different degrees of virulence could be isolated during any given epidemic. Even when he admitted that variations in virulence are unlikely to account for the rise and fall of any single epidemic, Topley held on to the opinion that such changes might play an evolutionary role over long periods of time (Topley, 1942). Moreover, in line with his general opinion that all variables change their meaning when applied to a herd rather than individuals, Topley attempted to define separately two aspects of virulence—pathogenicity and infectivity—which in his views needed to be distinguished in experimental epidemiology: 'we have no right to assume,' he claimed, 'that an organism which is capable, once it has gained access to the tissues, of causing a rapidly fatal septicaemia or toxaemia is also endowed with those properties which will enable it to spread readily from host to host' (Topley, 1926a, p. 531).[28] Although in some of his experiments with various strains of mouse typhoid bacteria he was able to show that certain strains were highly 'infective' while not being highly 'pathogenic,' Topley also despaired of anyone's ability to measure virulence and its various biological components with precision. Given the possibilities of variations in the susceptibility of mice and the possible interference of many 'chance' factors, he regarded all measurements of virulence as only statistically valid. And, as opposed to Webster, he was acutely aware of the fact that measurements of virulence were not independent of measurements of susceptibility, no matter how carefully one standardized all conceivable variables. Given these reservations, Topley was far more reluctant than Webster to accept the claim that variations in bacterial virulence played no role in epidemics.

CONTROLLING DOSAGE

In a series of ingenious experiments, Webster showed that the amount of the pathogen available within the population is directly proportional to the likelihood and severity of an epidemic. The number of bacteria present in the population increased directly prior to the epidemic and declined when the epidemic was waning. Higher doses of virulent bacteria resulted in more explosive and severe epidemics.[29] Despite his more probabilistic and dynamic definition of dosage, Topley reached similar conclusions and showed a similar pattern of changes by measuring the excretion of bacteria in the infected herd (Topley, Ayrton and Lewis, 1924). The earlier findings on carrier rates preceding epidemics of encephalitis in human populations gave additional support to Topley's and Webster's conclusions. The dosage factor, however, was not regarded by either Webster or Topley as an independent variable. Webster regarded it as a function of the number of susceptible animals present in the population, since the number of bacteria present in the population, which itself was kept under standard conditions, no longer depended on any of the environmental conditions but exclusively on the multiplication of bacteria in individual animal hosts. Thus, although dosage was found to be one of the factors influencing the course of the epidemic, the question 'what leads to the increase in dosage?' was immediately reduced to the question of how to account for the varying susceptibility of individual mice.[30] In the process, Webster also redefined the environmental condition of crowding—known to promote the spread of infections—into a biological condition of the presence of increased numbers of bacteria.

Topley also regarded the increase in dosage as a function of the number and proportion of susceptible animals in the herd. But he disagreed profoundly with Webster about what constituted susceptibility and resistance.

STANDARDIZING HOSTS: HEREDITY OR ACQUIRED IMMUNITY?

Having constructed 'dosage' as a result of host susceptibility, and having eliminated variations of virulence as a factor responsible for the production of epidemic waves in his experimental model, Webster turned his attention to what he believed to be the last uncontrolled variable, 'host susceptibility':

> Hitherto, the weight of action has been placed on the 'microbe'; from now on more weight will have to be placed on the 'host'. When large outbreaks of disease occur, these two factors: microbe vs. host, will be considered jointly as determining the result ... Since the virulence of germs is fixed, it remains to be discovered how widely host susceptibility varies with natural conditions.[31]

In the early 1920s, Webster considered host susceptibility to be a complex variable in need of further analysis. At that time, he distinguished racial immunity, acquired immunity, and a non-specific general immunity that included the influences of environmental factors such as diet or season as well as such characteristics of the organism as age and weight. Until the mid-1920s, susceptibility, and especially 'genetic' or 'racial' susceptibility, was just one factor among many.[32]

To investigate these multiple factors, Webster again systematically searched for a system in which all possible sources of variation would be under experimental control. Accordingly, from the very beginning of his research Webster attempted to standardize the mice used in his experimental epidemics by selecting only mice of a specific age and weight, feeding them a standard diet, keeping the mice free of any health problems, and trying to insure that the mice introduced into the experimental population had no prior contact with the typhoid organism (and thus no prior acquired immunity). In fact, the first experiments conducted by Webster after he began working together with Amoss and Flexner were attempts to induce specific acquired immunity by inoculating mice with dead cultures of the bacillus and cultures of low virulence (Webster, 1922). Although Webster succeeded in inducing such immunity in some mice, the results were variable and the acquired immunity only relative: not all mice were successfully immunized when given a standard dose, and some immunized mice exposed to massive doses of the typhoid bacillus still developed an infection. The variability of results, moreover, was taken as an immediate indication that the immunized mice were not homogenous: the effect of specific immunity could not be estimated unless one knew the general, nonspecific resistance of the immunized individual.

The attempts to control this underlying variability, and to select mice with identical inherent, genetic susceptibility, were thus, at one at the same time, supposed both to estimate the role of the 'genetic' factor, and to provide Webster and his associates with a more homogenous set of animals which he believed were needed for more systematic tests of the influence on susceptibility of the other, more easily controllable factors (such as diet, light, temperature, or acquired specific immunity).

In the initial experiments designed to test 'racial' susceptibility, Webster and his associate Ida Pritchett employed inbred stocks of mice from various suppliers. These inbred stocks were found to succumb to the infection to a differing extent suggesting some inherent genetic differences between them. As the mice of the relatively more susceptible strains were added to the more resistant populations, their mortality rates were also consistently higher (Webster, 1923a; Webster, 1924a).

These initial experiments on genetic resistance and susceptibility were conducted simultaneously with attempts to investigate the effects of different diets. Here, variations in resistance could also be experimentally induced: one of the diets clearly conferred a great deal of resistance to the population, since the mortality rates among mice fed on these diets were much lower than among those kept on the regular Rockefeller Institute diet (Webster and Pritchett, 1924).

As we have seen, the importance of innate susceptibility as a factor increased as Webster solidified his conclusion that changes in microbial virulence were not responsible for the rise and fall of experimental epidemics, and as the dosage factor came to be seen as a result of differing host susceptibilities. Already in 1926, when the first comparisons of the different strains of mice were made by Pritchett, Flexner described plans to hire a geneticist to assist in 'experimental investigations into the nature of epidemics' (Flexner, 1926, pp. 24–25).[33] In 1929, definite plans were being made for 'a genetic study of the inheritance of resistance to infectious disease' and Webster presented an estimated budget to Flexner.[34] In his report for 1930, Webster concluded that 'host resistance' was the one factor playing 'the most significant role in the spread of infectious disease'[35] and by that time he clearly considered the control of genetic factors to be a first step in the attempts to analyze the composite category of 'host resistance.'

To analyze the role of these genetic factors, Webster devised an ingenious system of selective breeding in which the offspring of the most susceptible and the most resistant mice stemming from a particular strain were bred brother to sister for a number of generations. The standardization was successful: by 1933, Webster had at his disposal both highly susceptible and highly resistant lines of mice (Webster, 1933), and had shown that the hereditary susceptibility/resistance were not specific to the mouse typhoid organism but extended to several other (though not all) pathogenic bacterial infections that he tested, as well as to certain forms of poisoning (Webster, 1933a). Webster was also ready to subject these lines to further genetic investigation by back-crossing, observing epidemics in populations composed of mice of known resistance, and, above all, testing the role of acquired immunity, of diet and of other environmental factors, now that the last source of uncontrolled variations had been practically eliminated 'through strict selective inbreeding' which allowed Webster to develop 'a line of mice with 95% of the individuals highly susceptible to *B. enteritidis* typhoid and another line from similar parentage with 95 per cent of the individuals highly resistant.'[36]

Addressing the question 'what determines survivorship of an epidemic?' Webster distinguished his own experimental technique from that of Topley and Greenwood, who had, according to him, erroneously assumed that 'constituents of an uninfected population [were] alike with regard to their initial resistance to an infection.' In contrast, Webster 'considered the possibility that [individuals] might have been innately resistant at the outset,' confounding all attempts to test specific immunity or the effects of other factors. Given this possibility, it was thus necessary to find out whether individuals 'might differ widely in their initial resistance.' The possibility of different genetic endowment, however, had important methodological consequences, for 'the question of survivorship could not be tested directly until not only the average resistance of a population at the outset but the resistance of each individual was known.'[37]

Equipped with strains of mice with very high and very low inherited resistance, Webster turned his attention to acquired immunity only to discover 'that it is extremely difficult to render a naturally susceptible mouse resistant by a "natural" exposure to small doses, and readily accomplished with a naturally resistant mouse.'[38] Thus, on the basis of experiments using the 'naturally resistant' and 'naturally susceptible' mice whose 'natural' susceptibility was a result of years of experimental manipulation and inbreeding, Webster expressed doubts about the commonly accepted belief that 'survivorship during an epidemic results from acquisition of specific immunity following non-fatal doses' and proposed his own 'experimentally developed' epidemiological theory which explained epidemics 'on the basis of only one remaining assumption—differences in individual resistance' which were now regarded as largely genetic, though possibly modifiable by 'environmental factors.'

Webster firmly believed that he was now in possession both of a new epidemiological theory and of an adequate experimental model (see Fig. 4.1). The rise of epidemics and their mortality rates depended on the presence of a microbe whose virulence was stable and on a sufficient number of inherently susceptible individuals in a population. The proportion of inherently susceptible and inherently resistant hosts was now *the only factor* necessary to account for the shape of the epidemic

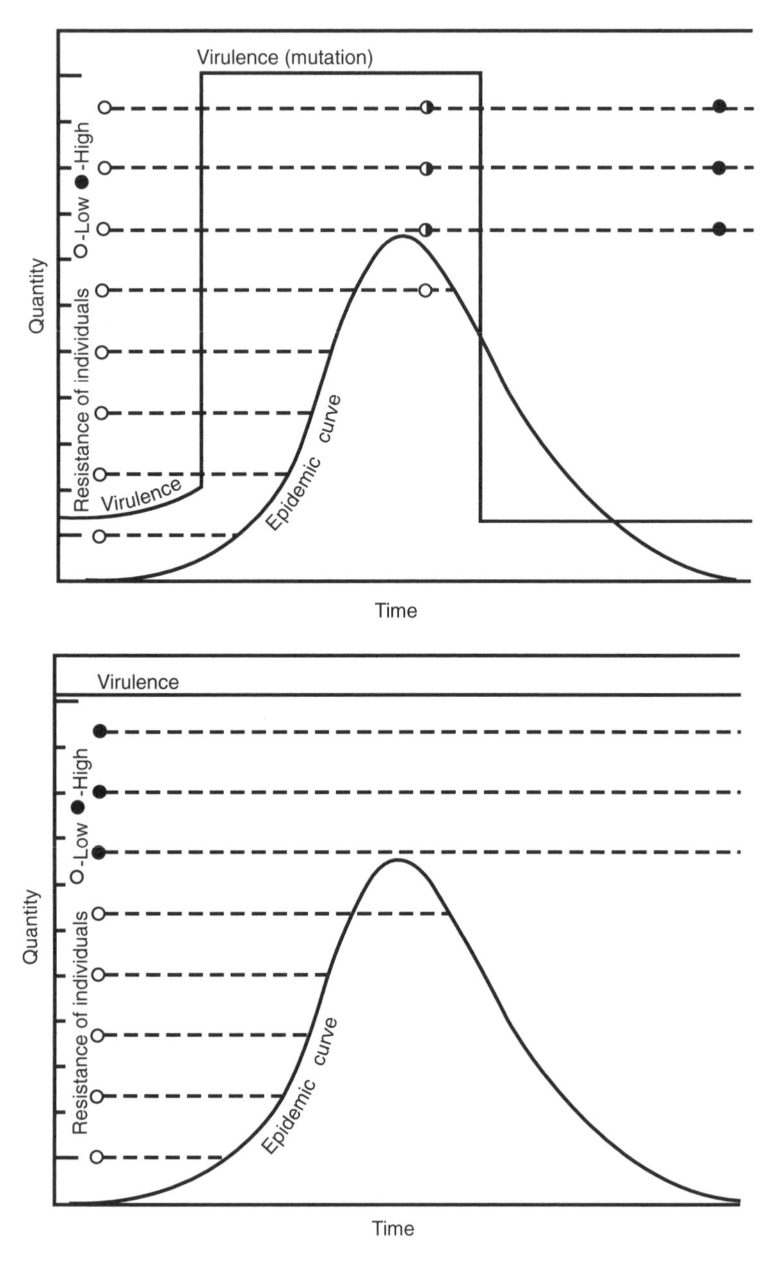

FIGURE 4.1 WEBSTER'S VISUALIZATION OF THE DIFFERENCE BETWEEN HIS MODEL OF EPIDEMICS, IN WHICH NEITHER THE MICROBE NOR THE RESISTANCE OF THE POPULATION CHANGES, AND THE TRADITIONAL MODEL IN WHICH CHANGE CAN BE OBSERVED BOTH IN THE VIRULENCE OF THE MICROORGANISM AND IN THE RESISTANCE OF THE HOST THROUGH IMMUNIZATION. SOURCE: 'REPORT OF THE DIRECTOR OF THE LABORATORIES TO THE CORPORATION,' OCTOBER 28, 1938, *SCIENTIFIC REPORTS*, BOX 6, VOL. 26, P. 304, ROCKEFELLER UNIVERSITY ARCHIVES, ROCKEFELLER ARCHIVE CENTER, TARRYTOWN, N.Y.

curve, for the survival of some individuals and the deaths of others, and for the waning or waxing of epidemics.[39]

Thus formulated, Webster's theory fits all the characteristics of Rosenberg's contamination model: it is monocausal, reductive, and focused on individuals. It was made so, however, not through a commitment to a monocausal explanation, or an interest in genetic explanations, or even faith in a contamination view of epidemics, but by continuous attempts to control experimental conditions in which, in order to measure the role of any one factor, all the others had to be carefully standardized, kept constant, or eliminated.

Topley, Greenwood, and their collaborators criticized Webster's conclusions about heredity from the very beginning. In 1925, after Webster's first comparisons of susceptibility in different inbred stocks of mice which were at that time not yet bred specifically for resistance or susceptibility to mouse typhus, Topley, citing the statistical reanalysis of Webster's and Pritchett's data by his collaborator, Lockhart, objected that the results obtained on the group-specific mortality rates were not statistically significant. At the same time, he argued that one could never be sure whether the differences observed were innate, rather than due to some 'chance factors or to some transient physiological condition, so that a mouse which showed more than average resistance when tested at one time might show less than average resistance if tested at another.' According to Topley, an experimental test which would exclude the possibility that the observed variations were transient rather than innate was impossible, since 'successive trials would not be independent': after all, a mouse that resisted infection at one time would, as a result of the experience, possess an altered resistance to subsequent inoculation' (Topley, 1926a, p. 535). Similar objections were raised by Topley's long-time collaborator, G.S. Wilson (Wilson, 1929).

Topley and his collaborators did not deny that there might be an innate component in the immunity of individual mice, but they argued that given the constraints of the experimental situation in laboratory epidemics, the contribution of this component could not be estimated independently of the contribution of such factors as acquired immunity and chance. To illustrate this claim, Topley resorted to an analogy, comparing microbes infecting variously resistant mice to stones randomly hitting stakes of different thickness:

> Suppose we take a number of stakes of different thickness, plant them in the ground and expose them to bombardment with stones of varying size flung

> by catapults of varying strengths. After certain time we shall find that a number of the stakes have been broken. This will not have happened to many of the thicker stakes, but other survivors will consist of thinner stakes, around which ineffective missiles have formed protective armour. Survivors of the latter class are in a precarious state; subsequent bombardment may displace the protective heap, and perhaps add its impetus to that of the new missile. Survivors of the former class may eventually be destroyed by missiles of sufficient momentum. At any given moment many of the stakes still standing will owe their survival to chance escape from fatal impact (Topley, 1926a, p. 532).

The problem was of course that one could not measure the thickness of individual stakes other than by throwing stones and seeing how many stakes fall, and the momentum of the stones could only be estimated by observing how many stakes have fallen. As we have seen, Webster's solution to this dilemma was to try to make both the stones and the stakes as standardized as possible. Topley objected—at first because Webster's method of selecting the 'thick' and 'thin' stakes did not discriminate between the various ways of escaping the missiles, and later because the very thick and very thin sticks he created were too artificial to tell us what happens when thickness is more 'naturally' distributed.

The objection of 'artificiality' and 'inapplicability to human epidemics' was used once Webster began to breed mice specifically for resistance and susceptibility. Topley and Greenwood did not question the existence of innate differences, nor did they believe—as Webster occasionally accused them of believing—that 'all mice were alike,' but they saw hereditary differences as randomly distributed in the population and innate resistance as only one component of immunity. Webster's pure strains of mice, they believed, had no natural analogues in human populations, and there was no reason to believe that conclusions about disease in populations of mice which were specifically bred for resistance and susceptibility would be relevant for the understanding of factors shaping epidemics among men:

> None of us working on experimental epidemiology in England at all doubt the existence of innate, probably inheritable, variations of resisting power.
>
> We can conceive—with some difficulty—that it might be practicable, by judicious genetic selection, to reach a population of mice none of whom would die of the infections with which we have worked. We cannot conceive that the variations of incidence and mortality of zymotic diseases [...] are explicable in terms of a genetic hypothesis. But since we study mice

> not for their own sakes but for the sake of the light their doings and sufferings may throw upon the fate of social man, we are not greatly interested in the elucidation of factors which have not, we think, played a very important part in human epidemiological experience (Greenwood, 1932, p. 55).

Throughout his experiments Topley was interested in herd rather than individual susceptibility, and understood immunity as a frequency distribution of individuals with different levels of susceptibility. Just like dosage it could only be expressed as a probability.[40] In fact, in some formulations, the question about the role of innate resistance was posed in terms of whether successive waves of epidemic increase herd immunity by active immunization or by the selection (i.e., elimination) of susceptible members of the herd. In effect, while Webster was asking the question of why some individuals survive an epidemic while others die, Topley and Greenwood wanted to know to what extent the herd's resistance to infection is a result of selection processes and to what extent it is due to immunization.

To disentangle the two factors contributing to herd immunity Topley and Greenwood tried to resort to statistical techniques. In their 1925 paper, for example, they constructed a mathematical model to show that the assumption that innate resistance is responsible for the observed skewness in the rates of mortality over time does not satisfactorily account for the data obtained from mouse epidemics, while the assumption that immunity increases with exposure might explain the curves in a more satisfactory manner.[41] Similarly, in his Herter Lectures at Johns Hopkins, Greenwood rejected both a pure selectionist and a pure environmentalist hypothesis and hoped to isolate the two factors by statistical analyses of data from epidemics in which the mortalities of previously exposed survivors and previously unexposed mice were compared when they were both put in an epidemic environment. The experiment was, as he himself acknowledged, far from decisive. All that Greenwood was able to conclude was that 'since the vast majority of the old and seasoned inhabitants ultimately die of the specific infection, mortuary selection [was] not all powerful' and that 'weighing the probabilities, the balance seems to incline in favour of immunisation rather than weeding out' (Greenwood, 1932, p. 69). But then, Topley and Greenwood based their conclusion not only on their statistical treatment of laboratory data, but also on statistics from real epidemics. Topley and Greenwood had always treated such data as a touchstone for their own conclusions. It was known, after all, that in some infections, a single attack protects one against future attacks of the same disease, and that at least in some cases, vaccination seemed to have significantly lowered the incidence of disease. If Webster's conclusions about hereditary differences were to be seriously applied to human epidemics, immunization would not occur naturally and all efforts to induce immunity by artificial means would be doomed to failure. For Topley and Greenwood this was clearly both an irrational and a politically unacceptable conclusion. As opposed to Webster, in all their work, Topley and Greenwood spelled out the direct practical implications of their findings: they were eager to demonstrate the inefficacy of such measures as isolation of cases, and to show that if an epidemic of a disease appears once in a population it does not need to be reimported from outside to appear again. It is of course impossible to say to what extent their opposition to Webster's conclusions about the minor importance of acquired immunity was based upon different kinds of political, scientific, 'methodological,' and 'common-sense' considerations.

Having set themselves the task of disentangling the various factors involved in the waxing and waning of epidemics by combining experimental data with laboratory analysis and comparing the results to those available from human epidemiological data, Greenwood and Topley seemed unable to move beyond their early finding that if an infectious disease has been introduced into a population, it will continue indefinitely in some proportion to the number of new susceptible animals introduced into the herd. They continued to vary the numbers of new entrants, the timing of such additions, the overall structure of the herd, and so on. The crucial factors relating to the microbe or the host (virulence, herd immunity, innate susceptibility, acquired immunity) seemingly could not be measured experimentally without the standardization of all the other variables, but when standardization optimized some factors, it destroyed the analogy with human epidemics. Human beings, after all, were not selectively bred for resistance, and microbes did not enter human bodies all at the same time in single and precisely measured doses. On the one hand, Greenwood and Topley worried that the standardization and optimalization of variables artificially distorted the equilibrium among the various factors in the model, making it impossible to measure their relative and interdependent contributions to the overall balance. On the other hand, neither the laboratory methods, nor statistical analysis of the data from the appropriately simplified models (i.e., those models which controlled some of the secondary factors but left the primary ones approximating 'natural conditions') provided tools by which the individual factors contributing to such equilibria could be measured independently. Redefining all variables in terms of the herd and in terms of probabilities and frequencies made the problem even more intractable.

In a 1926 paper read before the Section of Epidemiology and State Medicine, Greenwood and Topley presented what they saw as the likely—and not all together misplaced, though perhaps short-sighted—objections against experimental epidemiology:

> Perhaps it will be said: Your positive contribution to knowledge is very meagre. After years of labour, holocausts of mice, much arithmetic, much expenditure of money, you have only proved that in communities of mice a particular microbial disease will continue indefinitely if you introduce fresh susceptible individuals, and that the disease will wax and wane, being at times replaced by another microbial infection, in a fashion similar to what occurs in human experience. You have made it probable that the form of the secular curve of epidemicity is dependent upon the rate of addition of susceptibles and that an acquired immunity as well as selective immunization have a part in the advantage survivors of one epidemic enjoy when exposed to another. All these things any reasonable man might have inferred from human experience [...] Was it really worth while to spend so much time and money in demonstrating these notorious facts?' (Greenwood and Topley, 1926, p. 40).

Already at that time their defense against such potential criticism was rather modest, as they asserted that experimental epidemiology allowed clearer formulation of problems.[42] But in his obituary of Topley, Greenwood also suggests disappointment: 'Perhaps Topley might not have lost courage—he would never have done that; in his scientific work he bore cruel disappointments with perfect good temper—but have become bored, the word sounds trivial, but expresses my meaning, possibly "frustrated" is the appropriate term. "Nature" is harsh, answering the questions put and only those [...]' (Greenwood, 1944, p. 709).

In so far as Topley and Greenwood were indeed willing to follow the laboratory logic of control and standardization, they reduced the problem just as Webster was doing. In this light, their conclusion that the waxing and waning of epidemics was the result of a disturbance in equilibrium between a microbe and the number of susceptible hosts in a herd was indeed a reduction of the multifactorial, configurational perspective. The controlled, excluded factors (diet, temperature, season, etc.,) were relegated to the background, even if Topley and Greenwood did not claim these factors were totally irrelevant. But having made this move towards a 'contamination' perspective, Topley and Greenwood seemed unwilling to go as far as Webster had done, and to cease thinking in terms of collectives and equilibria. They were not prepared to follow the logic of laboratory control when it seemed to transform all the collective factors into individual ones, or when standardization upset what they regarded as an irreducible 'equilibrium' of various factors. This rather limited configurational perspective—their emphasis on the balance and interdependence of factors and on the importance of collective variables—was undoubtedly strengthened by their continuous reference to the more traditional epidemiological research. When criticizing Webster, they called attention to the danger and the illusion involved in the logic of 'testing one factor at a time while controlling all the others' to which the Rockefeller group tenaciously subscribed, but—as they explicitly acknowledged in some of their later articles—maintaining such a configurational perspective in the laboratory did not allow them to move beyond confirming what had already been found in the field:

> I have communicated to you no discoveries the immediate application of which might lessen the sum of human misery, no methodological principles which might swiftly revolutionise human thought. I have only, very unskillfully, described a biostatistical technique, which, so far, has yielded little but a confirmation of the harsher inductions from human epidemiological experience [...] (Greenwood, 1932, p. 79).

And while this might have been less that they had hoped for when they begun their studies in experimental epidemiology, it did seem to indicate that the marriage of a configurational perspective and the laboratory was no simple matter.

> In biology, above all in *human* biology, the laboratory man has been too impatient, has aspired to the short cut of an *experimentum crucis*. Yet [...] 'of all methodological procedures in biology the *experimentum crucis* is the most dangerous. [...] [N]othing emerges more clearly from the history of biological thought than that, almost without exception, the crucial experiments which have been most loudly hailed at the time they were made, as for ever settling the problem under discussion, have been subsequently found to have led to quite erroneous conclusions' (Greenwood, 1932, p. 79).

Though this remains implicit in the text, Greenwood's warning does seem to have been directed at Webster and his associates.

THE LIMITS OF STANDARDIZATION

When Leslie Webster died in 1943 he was collaborating with a biochemist, H.A. Schneider, in research whose goal was to isolate and specify the nutritional components of the diets which had been found in the 1930s to influence the 'nonspecific' resistance of mice to typhoid. Diet was by then the only factor other than inheritance whose variation had been associated with variations in the epidemic curves.

Not surprisingly, the researchers planned to capitalize on the existing experimental model:

> we have adhered to the techniques of experimental epidemiology formerly employed and have added certain procedures calculated to control other ecological factors. [...] In using this technique [stomachal instillation of fixed doses of bacilli], which reproduces natural conditions closely, we are aware of the importance of controlling all possible variables which may influence the reaction of the mice to the dose of bacilli. We are confident that bacterial factors are adequately controlled and are fortunate in having available mice of known high and low innate resistance to the test infection. Moreover, it has now become possible to control the effects of temperature and humidity [...]. Light is partially controlled by excluding all daylight [...] Each animal is housed separately so that it has no access to excrement or to bedding, and so that it is exposed at all times to the controlled temperature and humidity and light.[43]

(One cannot help but wonder whether any irony was intended when this set up was characterized in the same report as 'natural yet controlled technique.')

Although the initial results reported in 1942 and 1943 seemed to be encouraging, the first report written by Schneider after Webster's death acknowledged that 'nutritional effects on natural resistance *cannot* be demonstrated in inbred, pure-line mice selected for either resistance or susceptibility to mouse typhoid, or in their hybrids.'[44] Schneider was still able, however, to demonstrate such effects on 'unselected' mice. Within a year, he was conducting his experiments on wild house mice that were apparently more sensitive to dietary changes than any of the inbred laboratory stock.

A year later, a laboratory accident led Schneider to deconstruct not only the usefulness of Webster's 'genetically homogenous' mice, but also that of his model pathogen. Following an attempt to re-purify a standard culture of B. enteritidis that had accidentally become contaminated during a routine transfer, Schneider was no longer able to demonstrate any effects of diets on the resistance of any mice.[45] And when Schneider proceeded to grow B. enteritidis cultures on artificial media, it also appeared that 'in Webster's strains of resistant and susceptible mice, identical mortality rates occur in both kinds of mice when they are infected with one of the synthetically grown virulent cultures. When these synthetically grown cultures are the source of infection, in other words, the relative designations of these mice as "resistant" or "susceptible" are without meaning.'[46]

By 1946, then, Schneider insisted that in order to test the effects of nutrition on resistance, it was necessary to use not only a 'randomized, outbred stock,' but also 'an invading pathogen population which in itself presents an array of variation.' By that time, Schneider was also ready to question the very methodology, which in his view departed too far from 'the world of nature' to allow for meaningful conclusions:

> Ostensibly the way to study the effect of nutrition on resistance to infection is to feed different diets to identical samples of hosts in a controlled environment and then infect these hosts with a pathogen under controlled conditions of dosage, route of administration, etc., and observe the results. Yet the whole history of our investigations in this field over the past six years drives us irresistibly to the conclusion that [. . .] due to the influence of certain present-day laboratory traditions in methodology, the probability is great that such studies will reveal no dietary effect at all.[47]

Armed with such arguments, Schneider proceeded to stress the importance of a more configurational model in which phenomena of infection result from the multiple interactions and interdependencies of heterogeneous factors which cannot be reduced to single, independently controlled variables affecting only individuals. Or, according to Schneider, 'the relationship of nutrition to resistance to infection has its seat in phenomena of an integrative nature wherein populations, both of hosts and of pathogens, have properties not possessed by individual hosts and individual pathogens.'[48]

Schneider's findings, and his insistence that in order to imitate nature properly one not only had to cope with variations but explicitly to include them in the experimental design,[49] bring to the fore Webster's methodological dilemmas as he attempted to construct his model epidemics. For in order to measure the variables Webster was interested in, he continuously had to standardize and control all conceivable sources of variation. Moreover, the same standardization processes were being used to develop both the model in which the variation to be measured was supposed to occur and to develop techniques by which these measurements could then be performed. In the process, Webster developed a perfectly self-contained—or as Ian Hacking would put it—self-vindicating (Hacking, 1992) laboratory system in which variation was standardized to such an extent that only those factors which were purposefully varied experimentally (such as genetic resistance) were found to produce variation in the model itself. And, as the control of the variables was increased and only one factor was allowed to vary in any particular experiment, the results precluded the observation of any configurational effects (for example, he could observe and measure the effects of the diet, or the effects of genetic resistance, or of the differences in the virulence of bacteria, but once he maximized and standardized the effects of any one of these factors, the effects of the others became invisible).

CONCLUSIONS

Webster described his ideal of a medical-school department of epidemiology in a memorandum sent to Alfred Cohn, a cardiologist working at the Rockefeller Institute Hospital. Webster asserted that epidemiology engages in 'investigations of disease as social phenomena,' and claimed that 'in order to understand disease we need to [...] pay more attention to the correlation of the many factors which appear germane to a synthetic appreciation of the concept disease.' He continued:

> A diseased individual is not a 'hospital' case but an individual whose misfortune is related to a setting, his natural environment in a family and a community, living in certain social and economic conditions and, of course, at a certain period of time.
>
> It is desirable, furthermore, going beyond the individual, to consider the diseased human population as a whole, in the light of its varied hereditary background and its geographic and climatic environment. [...]
>
> The meaning of this reach beyond the contemplation of the immediate phenomena of a single diseased individual, is the belief that there attaches to a proper understanding of his situation the need for great attention to the correlation of the various factors which are involved in his disease [...][50]

In his 'Memorandum' Webster stressed the importance of 'accumulating relevant clinical psychological, bacteriological and sociological data in individual families' and the need to 'gain further light on the etiological role of specific microbic agents, the predisposing role of heredity, social and environmental factors, and finally on the effect of well-regulated hygienic regimes in the health of individual and communities.' This plea certainly seems at odds with the claim that the rise and fall of epidemics can best be explained as the result of an interaction between a stable bacillus and an innately, genetically susceptible host. The views expressed in the memorandum could almost serve as a definition of the 'configurational' view of epidemics, yet at the very time he wrote the memorandum Webster was deeply engaged in demonstrating experimentally that hereditary endowment is the only factor necessary to explain why some individuals survive an epidemic while others do not. In the laboratory, a configurational ideology seemed repeatedly to give way to a methodology embodying a contamination model.

The research in experimental epidemiology can be understood as an outcome of a continuous tension between laboratory control believed to be necessary to isolate the role of the various factors which could be involved, and the desire to make experimental epidemics simulate as closely as possible the 'natural' course of an infectious disease in human populations. In this context, the choice between the contamination and the configuration view of epidemics can be seen as an unintended—although hardly a politically or socially neutral—consequence of the practical exigencies of laboratory research, in which standardization, control, and manipulability serve as prerequisites of 'do-ability.'

As the comparison of the two research programs shows, it was difficult, if not impossible, to maximize the advantages of the laboratory while upholding the configurational view. The choice of whether to abandon 'configuration' or to forgo some of the advantages of 'laboratory control' was shaped in some important respects by the institutional and political context in which the two groups worked. On the one side, there was the commitment to experimentation as the basis of medical knowledge which the founders and directors of the Rockefeller Institute never ceased to emphasize, Webster's background in pathology, and his collaboration with geneticists but not with epidemiologists. As against this configuration of factors, Greenwood and Topley had close links to epidemiology and to public health discussions in Britain, relied heavily on the statistical expertise and style of thought which Greenwood had inherited from Pearson, and the biometric tradition which had remained strong in the U.K. throughout this period.

But we should not overestimate the power of the individual ideological preferences of researchers: Webster's transformation of the mouse typhoid epidemic into a genetic problem of host resistance and susceptibility was much more a result of the possibilities and constraints of the available biomedical laboratory techniques, and of his commitment to the laboratory ideals of measurement, manipulation, and control, than a result of his commitment to a particular social or medical ideology in which problems of epidemics were to be treated as problems of individual biological endowment rather than as broader socioeconomic and biological phenomena. And even Greenwood's and Topley's stronger commitment to configurational perspectives was not enough to overcome the reductionist and mechanistic consequences of standard bacteriological laboratory approaches and methodologies. Insofar as they followed this laboratory logic, they did indeed move away from a configurational perspective, but when they resisted it, their studies brought little beyond the confirmation of the older, field-based epidemiological knowledge.

I have argued earlier that experimental epidemiology could be understood as the bacteriologists' answer to an epidemiological critique of the simplifications of laboratory science. As against the epidemiologists' invocations of peoples, times, seasons, airs, waters, and places, experimental epidemiologists proposed establishing laboratory models and controlled experiments which would enable them to study the interactions of a variety of factors. But the constraints of the laboratory as understood by the bacteriologists seemed stacked against such an attempt; to make their models tractable they either indeed reduced the complexity down to a very few biological variables and created a model and a theory that stood or fell together, or else they found themselves without a means to disentangle the very configuration they were intent on understanding.

ACKNOWLEDGEMENTS

The archivists at the Rockefeller Archive Center in Tarrytown, especially Lee Hiltzik and Thomas Rosenbaum, have provided invaluable help in the gathering of materials for this article. I would also like to thank the participants of the discussions at the seminar on Rules, Norms and Standardization at the Centre de Recherche en Histoire des Sciences et Techniques in Paris and at the progress conference in the Department of Science and Technology Dynamics in Amsterdam. Ilana Löwy, Steve Sturdy, Ton van Helvoort, Jean-Paul Gaudillière, Jean Gayon, and Theodore Porter deserve special thanks for their insightful comments on the various versions of this article.

NOTES

1. See, for example, Michael Worboys's studies of the clinical management of pneumonia and tuberculosis in Britain (1993), and his article in this volume.
2. According to Flexner, the two studies were begun independently of one another, though at approximately the same time. In October 1922, Flexner asserted that 'While the experiments were in progress, a series of papers by Topley appeared in England' ('Report of the Director of the Laboratories to the Corporation, for 1921–22,' October 20, 1922, *Scientific Reports*, Box 3, vol. 10, p. 234, Rockefeller University Archives, Rockefeller Archive Center, Tarrytown, N.Y.) and indeed Topley's name does not come up in Flexner's first report on experimental epidemiology (October 15, 1920, Box 3, vol. 8, p. 293–96). Corner, however, claims that Flexner visited Topley in 1919 at the London School of Tropical Medicine. I do not have any confirmation of this, but Corner is certainly wrong about the place of the visit, since in 1919/1920 Topley worked at the Charing Cross Hospital (cf. Corner, 1964).
3. For details of Topley's life, see Dean and Wilson (1944) and Greenwood (1944).
4. On Greenwood, see Matthews (1995) and 'Major Greenwood, 1880–1949,' *Obituary Notices of the Fellows of the Royal Society* (1950), pp. 139–141.
5. For a short biography of Webster, see Olitsky (1943) and 'Leslie Tillotson Webster, 1894–1943. Memorial Minute,' Rockefeller University Archives. Faculty, L.T. Webster, Biographical Material (450 W395 (1)), Rockefeller Archive Center, Tarrytown, N.Y.
6. Although the two groups repeatedly criticized each other's work, they seemed—at least at the beginning—to have maintained good personal relations: 'It is of some interest to record that quite independently of the study in experimental epidemiology begun by Drs. Flexner and Amoss, a similar investigation was begun at about the same time by Dr. Topley in London. The two studies have often converged, and served to support and extend each other. Through a correspondence which has always been of the most friendly character, the work has been forwarded, and unnecessary repetition of experiment has been avoided. On the other hand, the stimulus of conflicting observation has not been lost, but rather has led to the more rapid elucidation of active factors in this highly complex subject than might otherwise have been secured' ('Report of the Director of the Laboratories to the Corporation, 1922–23,' October 19, 1923, *Scientific Reports*, Box 4, vol. 11, p. 216, Rockefeller University Archives, Rockefeller Archive Center, Tarrytown, N.Y.).
7. Of course, not all infectious diseases affecting humans could be modeled on animals, and even in those cases in which such infections were possible, the symptoms from which animals suffered and the course of the disease were often different from those encountered in humans; see Löwy (1992, 1995). On the difficulties and problems of animal modeling of a viral infection such as polio, see Grimshaw (1996) and Rogers (1992).
8. Andrew Mendelsohn in this volume.
9. See, for example, *Report on the Pandemic of Influenza, 1918–19, Reports on Public Health and Medical Subjects*, 4 Ministry of Health (London: His Majesty's Stationery Office, 1920).
10. Crookshank quoted by Hamer (1928).
11. The disputes among British epidemiologists and bacteriologists during this period were quite complex and cannot be discussed here in detail. Some points of dispute and the various positions taken in them are discussed in Mendelsohn (1996).
12. The paper includes a satire of both the traditional public health proponents of isolation as a means of combating epidemics and of the neo-Sydenhamian historical epidemiologists, portrayed as erudite mice reflecting on the course of (experimental) epidemic in their colony: ' "Have you," they would ask, "paid attention to the events of forty years ago, the strange sickness—it is very unphilosophical to talk about diseases—which were prevalent in the time of our ancestors, all far wiser mice than we; there is nothing *new* about these prevalences. Cease to chatter idly about novelty. When you fully comprehend the nature of the vast genii who at roughly periodic intervals transport is, in a manner still obscure, from one habitation to another, when you fully and exactly grasp the whole cosmos, embracing us mice, the genii themselves, and the larger genii who no doubt control *them*, you will have a right to call yourselves epidemiologists, and be sure that

whatever you *do* discover will be no more than a tedious amplification of what our incomparable ancestors believed, as we—when you *have* made the discovery—shall not fail to mention.' (p. 35).

13. 'Report of the Director of the Laboratories to the Members of the Corporation,' October 15, 1920, *Scientific Reports*, Box 3, vol. 8, p. 197. Rockefeller University Archives. Rockefeller Archive Center. Tarrytown, N.Y.
14. 'Report of the Director of the Laboratories to the Board of Scientific Directors,' January-October 1920, *Scientific Reports*, Box 3, vol. 8, p. 12. Rockefeller University Archives. Rockefeller Archive Center. Tarrytown, N.Y.
15. 'Report of the Director of the Laboratories to the Corporation for 1920–21,' October 21, 1921, *Scientific Reports*, Box 3, vol. 9, p. 205. Rockefeller University Archives. Rockefeller Archive Center. Tarrytown, N.Y.
16. 'Report of the Director of the Laboratories to Board of Scientific Directors,' January-October 1921, April 16, 1921, *Scientific Reports*, Box 3, vol. 9, p. 6. Rockefeller University Archives. Rockefeller Archive Center. Tarrytown, N.Y.
17. 'Report of the Director of the Laboratories to the Corporation,' 1920–21, October 21, 1921, *Scientific Reports*, Box 3, vol. 9, p. 207, Rockefeller University Archives. Rockefeller Archive Center. Tarrytown, NY.
18. In a 1923 paper, for instance, Webster compared the 'standard control curve with the epidemic curve as given by Amoss for mouse typhoid epizootics' (p. 269) and claimed that although 'it is unwise to relate causally independent series of phenomena merely because they may be described by the same types of curve,' such procedure was justified in this case because 'conditions are [...] closely allied and [...] we are comparing controlled phenomena with like uncontrolled phenomena,' so that the conclusion that there exists 'a similar causal relation on the basis of similarity of curves seems more justifiable' (Webster, 1923).
19. Flexner had been particularly eager to study infectious diseases spread by respiratory means which he considered a more urgent public health problem than the enteric infections which in his opinion were both better understood and easier to control through sanitation and personal hygiene. He might have also been particularly interested in respiratory infections because he considered polio, on which much of his research concentrated, to be spread in this manner.
20. Topley (1926b), pp. 647–48.
21. Elsewhere, Topley and Greenwood resort to a metaphor: 'We sail some of our model yachts in the children's pond and invite the real sailors to tell us whether their motions have any instructive resemblance to those of real ships.' (Greenwood and Topley, 1925, p. 32; see also p. 46).
22. (Topley, 1926a, pp. 531–32). Topley went as far as to assert that 'A herd may be immune to a particular disease—in the logical sense that it will resist the introduction of infection from without—although each of its members is fully susceptible' (Topley and Wilson, 1936).
23. 'The assumption that the virulence of a microorganism, its capacity to harm its native host, is a highly labile variable is derived from the early observations of Davaine and Pasteur. Supported by later bacteriologists, it has come to form the basis of modern epidemic theory. The experimental epidemiologist, however, recognizing that it is derived from tests which are now regarded as inadequate, has interested himself in virulence of bacteria under conditions as nearly natural as possible' (Webster, 1932).
24. See for example Webster (1923a), Webster (1930). For a summary of Webster's experiments on virulence, see Webster (1932) esp. pp. 328–33, and Webster (1946), pp. 91–94.
25. For example, Webster and Burn (1927).
26. 'Report of the Director of the Laboratories to the Corporation for 1926/27,' October 28, 1927, *Scientific Reports*, Box 4, vol. 15, p. 449, Rockefeller University Archives, Rockefeller Archive Center, Tarrytown, N.Y.
27. In the second edition of *The Principles of Bacteriology and Immunity* (London: Arnold & Co. 1936), W. W. C. Topley and G. S. Wilson concluded their discussion of this subject by claiming on the one hand that 'variations in virulence in a bacterial parasite during an epidemic prevalence have often been sought for with negative results,' and on the other hand reasserting that 'Taking the experimental evidence as a whole, it would seem to accord quite well with the view, expressed by many epidemiologists, that variations in the characters of the parasite are of major importance

in the spread of human infections, and that an outbreak of disease may be initiated by the evolution, or importation, of an 'epidemic strain' of the causative organism' (p. 985).

28. Webster could not really follow Topley's reservations about equating virulence with 'disease-producing power' for individual mice: 'Topley states that theoretically virulence may be a very complicated series of factors and that disease-producing power, as measure by the natural inoculation of individual mice, may not measure disease-producing power as determined by the ability of the organism to cause diseases in herds. We know of no experimental basis for assuming the operation of additional mysterious properties, no can we conceive of any factor inherent in the agent which would operate in the herd and not in the individual' (Webster, 1946, 77ff, p. 92).
29. Dosage, as carrier rate was initially explored by Amoss (1922a); for examples of Webster's measurements of dosage see Webster (1930), pp. 909–29 and 931–48.
30. 'It likewise is true that dosage increases before the onset of an epidemic and that the value of this factor determines the slope of the epidemic curve. If, then, one wishes to believe that a change in dosage is essential of the epidemic, and there seems little doubt of it at present, the question remains as to what leads to the increase in dosage. We believe that changes in population susceptibility of one sort of another are responsible.' ('Report of the Director of the Laboratories to the Corporation for 1925–26,' October 15, 1926, *Scientific Reports*, Box 4, vol. 14, p. 392. Rockefeller University Archives. Rockefeller Archive Center, Tarrytown, N.Y.)
31. Report of the Director of the Laboratories to the Corporation for 1925–26,' October 15, 1926, *Scientific Reports*, Box 4, vol. 14, p. 451. Rockefeller University Archives. Rockefeller Archive Center, Tarrytown, N.Y.
32. 'Report of the Director of the Laboratories to the Corporation for 1924/25,' October 16, 1925 *Scientific Reports*, Box 4, vol. 13, p. 187, Rockefeller University Archives, Rockefeller Archive Center, Tarrytown, N.Y. See also Webster (1924), p. 139.
33. The geneticist, John Gowen, was recruited from the Maine Orono Agricultural Experiment Station (in preference to Sewall Wright!) on the recommendation of T.H. Morgan and W.H. Welch. By the early 1930s, when Gowen did not seem to have succeeded in establishing a coherent research program, attempts were made to send him off elsewhere, and in 1936 Gowen left to the Genetics Division at Iowa State University.
34. L. Webster to S. Flexner, March 1, 1929, Rockefeller University Archives, Faculty, Leslie T. Webster, Administrative Correspondence 1920–1943 (450 W 395(1)), Rockefeller Archive Center, Tarrytown, N.Y.
35. 'Report of the Director of the Laboratories to the Board of Scientific Directors,' April 19, 1930, *Scientific Reports*, Box 5, vol. 18, p. 81, Rockefeller University Archives, Rockefeller Archive Center, Tarrytown, N.Y.
36. 'Report of the Director of the Laboratories to the Board of Scientific Directors,' April 30, 1938, *Scientific Reports*, Box 6, vol. 26, p. 96, Rockefeller University Archives, Rockefeller Archive Center, Tarrytown, N.Y.
37. 'Report of the Director of Laboratories to the Board of Scientific Directors,' April 30, 1938, *Scientific Reports*, Box 6, vol. 26, p. 95, Rockefeller University Archives, Rockefeller Archive Center, Tarrytown, N.Y.
38. 'Report of the Director of Laboratories to the Board of Scientific Directors,' April 17, 1937, *Scientific Reports*, Box 6, vol. 25, p. 101, Rockefeller University Archives, Rockefeller Archive Center, Tarrytown, N.Y.
39. 'Report of the Director of Laboratories to the Board of Scientific Directors,' April 30, 1938, *Scientific Reports*, Box 6, vol. 26, p. 95, Rockefeller University Archives, Rockefeller Archive Center, Tarrytown, N.Y. See also Webster (1946).
40. In his various papers, Webster attempted to take some of Topley's and Greenwood's objections into account. For example, once the variously susceptible strains of mice were bred so as to optimize or minimize their susceptibility, he attempted to counter the objection that the relevant factor in the herd is not individual susceptibility but its distribution in the herd, which Webster interpreted not as a probability but as an average. Assembling mice populations with differing proportions of highly resistant and highly susceptible mice, and letting infection spread in the

herd, he observed that 'survivors are almost exclusively the individuals known at the outset to have been highly resistant.' Accordingly, he concluded that 'the presence of the individual with high susceptibility initiates the outbreak, no matter what the general level of susceptibility may be' (Webster and Hodes, 1939). Topley's and Greenwood's 'herd immunity' was for Webster nothing but the sum of individual resistances.

41. Greenwood and Topley, (1925), p. 97–98.
42. In fact, when the critique of experimental epidemiology came from one of the 'neo-Hippocratic' epidemiologists, it was far harsher than this: 'There is a real danger, at the present time, lest the Hippocratic Epidemiology, as developed by Baillou, by Sydenham, by Huxham, and—I will add—by Creighton and by Hamer, be swallowed up in Statistical Science. Statistics may swallow, but can never replace Epidemiology. There is no doubt, a statistical method of examining and appraising epidemiological data: that is admirable and praiseworthy enough. But epidemiology—the study of epidemics—has to do primarily, not with figures and numbers but with Peoples, Times, Seasons, Airs, Waters, and Places. Much the same may be said of the recent application, as it is called, of the Experimental Method to Epidemiology. This is, really, nothing but the practice of observation by analogy, and bears the same relation to true epidemiology as does Kriegspiel to Strategy or attendance at cinemas to the study of Human Life' (Crookshank, 1930).
43. 'Report of the Director of the Laboratories to the Board of Scientific Directors,' April 19, 1941, *Scientific Reports*, Box 7, vol. 29, p. 64–65, Rockefeller University Archives, Rockefeller Archive Center, Tarrytown, N.Y.
44. 'Report of the Director of the Laboratories to the Board of Scientific Directors,' April 15, 1944, *Scientific Reports*, Box 7, vol. 32, p. 61, Rockefeller University Archives, Rockefeller Archive Center, Tarrytown, N.Y.
45. 'Report of the Director of the Laboratories to the Board of Scientific Directors,' April 21, 1945, *Scientific Reports*, Box 7, vol. 33, pp. 104–107, Rockefeller University Archives, Rockefeller Archive Center, Tarrytown, N.Y.
46. 'Report of the Director of the Laboratories to the Board of Scientific Directors,' April 13, 1946, *Scientific Reports*, Box 7, vol. 34, p. 117, Rockefeller University Archives, Rockefeller Archive Center, Tarrytown, N.Y.
47. 'Report of the Director of the Laboratories to the Board of the Scientific Directors,' April 13, 1946, *Scientific Reports*, Box 7, vol. 34, p. 115, Rockefeller University Archives, Rockefeller Archive Center, Tarrytown, N.Y.
48. 'Report of the Director of the Laboratories to the Board of Scientific Directors,' April 21, 1945, *Scientific Reports*, Box 7, vol. 33, p. 107, Rockefeller University Archives, Rockefeller Archive Center, Tarrytown, N.Y.
49. When Schneider began using wild mice and mixed cultures of bacteria, he claimed—just like Webster and Topley did throughout their experiments—that 'such a view brings the significance of the infection model closer to the events in the real world,' since what the experimental studies were aiming at was a 'not merely a *tour de force* of the laboratory, but [a demonstration] that has meaning for the world of nature' ('Report of the Director of the Laboratories to the Board of Scientific Directors,' April 21, 1945, *Scientific Reports*, Box 7, vol. 33, pp. 104, 106, Rockefeller University Archives, Rockefeller Archive Center, Tarrytown, N.Y.).
50. Webster's memorandum was included in a letter commenting on it, sent by Alfred Cohn to L.T. Webster on November 25, 1930, Alfred Cohn Papers, Box 38, folder 5, Rockefeller University Archives, Rockefeller Archive Center, Tarrytown, N.Y.

REFERENCES

H.L. Amoss, 'Experimental Epidemiology. I. An Artificially Induced Epidemic of Mouse Typhoid' *Journal of Experimental Medicine* (1922), 36: 25–43.

H. Amoss, 'Experimental Epidemiology II. Effect of the Addition of Healthy Mice to a Population Suffering from Mouse Typhoid' *Journal of Experimental Medicine* (1922a), 36:45–69.

W. Anderson, 'Immunities of Empire: Race, Disease, and the New Tropical Medicine, 1900–1920' *Bulletin of the History of Medicine* (1996), 70: 94–118.

G.W. Corner, *History of the Rockefeller Institute. 1901–1953* (New York: Rockefeller Institute Press, 1964), p. 197.

F.G. Crookshank, 'Public Health Considerations Relating to Influenza, Pneumonia and Allied Epidemics—The Epidemiological Point of View' *Boston Medical and Surgical Journal* (1921), 184: 548–552.

F.G. Crookshank, *Epidemiological Essays* (London: Kegan Paul, Trench, Trubner & Co., 1930), p. vii.

H.R. Dean, and G.S. Wilson, 'William Whiteman Carlton Topley' *Journal of Pathology* (1944), 56: 451–467.

E. Fee, 'Public Health and the State: The United States' in D. Parker (ed.), *The History of Public Health and the Modern State* (Amsterdam: Rodopi, 1994), pp. 224–275.

S. Flexner, 'Experimental Epidemiology. Introductory.' *Journal of Experimental Medicine* (1922), 36: 9–14.

S. Flexner, 'The Developments of the First 20 Years of the Rockefeller Institute and Outlook for Further Growth' April 17, 1926, in *Scientific Reports*, Box 4 vol. 14. Rockefeller University Archives, Rockefeller Archive Center, Tarrytown, N.Y.

M. Greenwood, *Epidemiology: Historical and Experimental. The Herter Lectures for 1931* (Baltimore: The Johns Hopkins Press, 1932).

M. Greenwood, 'William Whiteman Carlton Topley, 1886–1944' *Obituary Notices of Fellows of the Royal Society* (1942–44) 4: 699–712.

M. Greenwood, and W.W.C. Topley, 'A Further Contribution to the Experimental Study of Epidemiology' *Journal of Hygiene* (1925), 24: 45–110.

M. Greenwood, and W.W.C. Topley, 'Experimental Epidemiology: Some General Considerations' *Proceedings of the Royal Society of Medicine. Section of Epidemiology and State Medicine* (1926), 19: 31–44.

M. Grimshaw, 'Scientific Specialization and the Poliovirus Controversy in the Years before World War II' *Bulletin of the History of Medicine* (1996), 69: 44–65.

I. Hacking, 'Self-Vindication of the Laboratory Sciences' in A. Pickering (ed.) *Science as Practice and Culture* (Chicago: Chicago University Press, 1992), pp. 29–64.

W. Hamer, *Epidemiology Old and New* (London: Kegan Paul, Trench, Trubner & Co, 1928), p. 16.

I. Löwy, 'From Guinea Pigs to Man: The Development of Haffkine's Anticholera Vaccine' *Journal of the History of Medicine and Allied Sciences* (1992), 47: 270–309.

I. Löwy, 'Whose Body? The Experimental Body and Twentieth Century Medicine' paper presented at a workshop on The Body in Twentieth Century Medicine, Manchester, September 1995.

C.J. Lynch, 'An Outbreak of Mouse Typhoid and Its Attempted Control by Vaccination' *Journal of Experimental Medicine* (1922), 36: 15–23.

A. Mendelsohn, 'The Romantic Reaction in Epidemiology after World War I' paper presented at the History of Science Society, Atlanta, November 1996.

J.R. Matthews, *Quantification and the Quest for Medical Certainty* (Princeton: Princeton University Press, 1995).

P.K. Olitsky, 'Obituary. Leslie Tillotson Webster, M.D., 1894–1943' *Archives of Pathology* (1943), 36:536–537.

N. Rogers, *Dirt and Disease: Polio Before FDR* (New Brunswick, NJ: Rutgers University Press, 1992).

C. Rosenberg, 'Explaining Epidemics' in his *Explaining Epidemics and Other Studies in the History of Medicine* (Cambridge: Cambridge University Press, 1992).

W.W.C. Topley, 'The Spread of Bacterial Infection' The Goulstonian Lectures. *The Lancet* (July 5, 12, and 19, 1919), 197, II: 1–5, 45–49, 91–96.

W.W.C. Topley, 'The First Milroy Lecture on Experimental Epidemiology' *The Lancet* (March 6, 1926), 210, I: 477–484.

W.W.C. Topley, 'The Second Milroy Lecture on Experimental Epidemiology' *The Lancet* (March 13, 1926a) 210, 1: 531–537.

W.W.C. Topley, 'The Third Milroy Lecture on Experimental Epidemiology' *The Lancet* (March 27, 1926b) 210, I: 645–51.

W.W.C. Topley, 'The Croonian Lectures: The Biology of Epidemics' *Proceedings of the Royal Society* (1941–42) B130: 337–59.

W.W.C. Topley, J. Ayrton, and E.R. Lewis, 'The Spread of Bacterial Infection. Further Studies on an Experimental Epidemic of Mouse Typhoid' *Journal of Hygiene* (1924), 23: 223–313.

W.W.C. Topley, and J. Wilson, 'Further Observations on the Role of Twort-d'Herelle Phenomenon in Epidemic Spread of Mouse Typhoid.' *Journal of Hygiene* (1925), 24: 295–300.

W.W.C. Topley, J. Wilson, and E.R. Lewis, 'The Role of Twort-d'Herelle Phenomenon in Epidemics of Mouse Typhoid' *Journal of Hygiene* (1925), 24:17–26.

W.W.C. Topley, and G.S. Wilson, *The Principles of Bacteriology and Immunity* (London: Edward Arnold, 1936), p. 967.

L.T. Webster, 'Experiments on Normal and Immune Mice with a Bacillus of Mouse Typhoid' *Journal of Experimental Medicine* (1922), 36: 71–95.

L.T. Webster, 'Contribution to the Manner of Spread of Mouse Typhoid Infection' *Journal of Experimental Medicine* (1923), 37: 269–274, p. 272.

L.T. Webster, 'Microbic Virulence and Host Susceptibility in Mouse Typhoid Infection' *Journal of Experimental Medicine* (1923a) 37: 231–267.

L.T. Webster, 'The Application of Experimental Methods to Epidemiology' *The American Journal of Hygiene* (1924), 4: 134–142.

L.T. Webster, 'Microbic Virulence and Host Susceptibility in Paratyphoid-enteritidis Infection of White Mice. IV. The Effect of Selective Breeding on Host Resistance' *Journal of Experimental Medicine* (1924a) 39: 879ff.

L.T. Webster, 'The Role of Microbic Virulence and Host Resistance in Determining the Spread of Bacterial Infections among Mice. I. Pasteurella Aviseptica Infections; II B. Friedlaender-Like Infection, and III. B. Enteritidis Infection' *Journal of Experimental Medicine* (1930), 52: 901–908, 909–929, 931–948.

L.T. Webster, 'Experimental Epidemiology' *Medicine* (1932), 11: 321–344.

L.T. Webster, 'Inherited and Acquired Factors in Resistance to Infection. I. Development of Resistant and Susceptible Lines of Mice through Selective Breeding' *Journal of Experimental Medicine* (1933), 47: 793–817.

L.T. Webster, 'Inherited and Acquired Factors in Resistance to Infection. A Comparison of Mice Inherently Resistant or Susceptible to Bacillus Enteritidis Infection with Respect to Fertility, Weight, and Susceptibility to Various Routes and Types of Infection' *Journal of Experimental Medicine* (1933a) 57: 819–43.

L.T. Webster, 'Experimental Epidemiology' *Medicine* (1946), 25: 77–111.

L.T. Webster, and I.W. Pritchett, 'Microbic Virulence and Host Susceptibility in Paratyphoid-Enteritidis Infection of White Mice. V. The Effect of Diet on Host Resistance' *Journal of Experimental Medicine* (1924), 40: 397–404.

L.T. Webster, and C. Burn, 'Studies on the Mode of Spread of *B. enteritidis* Mouse Typhoid Infection. II. Effects of External Conditions on the Occurrence of Smooth, Mucoid, and Rough Colony Types'; 'III. Studies of Bacterial Cells Taken from Smooth, Mucoid, and Rough Colonies'; 'IV. The Relative Virulence of Smooth, Mucoid and Rough Strains' *Journal of Experimental Medicine* (1927), 46: 855–870, 871–886, 887–907.

L.T. Webster, and H.L. Hodes, 'Role of Inborn Resistance Factors in Mouse Populations Infected with Bacillus Enteritidis' *Journal of Experimental Medicine* (1939), 70: 193–208, 205–206.

G.S. Wilson, 'Discontinuous Variation in the Virulence of Bact. Aertycke Mutton' *Journal of Hygiene* (1928–29) 28: 295–317, p. 313–16.

C.-E.A. Winslow, *The Conquest of Epidemic Disease. A Chapter in the History of Ideas* (orig. 1943; Madison: Wisconsin University Press, 1980).

M. Worboys, 'Treatments for pneumonia in Britain 1910–1940' in I. Löwy *et al.* (eds.), *Medicine and Change: Historical and Sociological Studies of Medical Innovation* (Paris: John Libbey, 1993), p. 317–336.

Chapter 5

MAKING HEREDITY IN MICE AND MEN: THE PRODUCTION AND USES OF ANIMAL MODELS IN POSTWAR HUMAN GENETICS.

Jean-Paul Gaudillière

INTRODUCTION

Historians as well as practionners often trace the origins of medical genetics to the aftermath of the Second World War. Although experimental geneticists moved away from eugenics in the 1920s and 1930s it is argued that two features contributed to change the study of human inheritance after 1945: on the one hand a more radical departure from eugenic methods and concerns was triggered by strong criticism of the uses of 'Rassenhygiene' by the Nazis; on the other hand, the inner dynamics of genetic research as exemplified by the biochemical study of hemoglobin or by the study of the mold *Neurospora* provided new means for investigating 'inborn errors of metabolism' and 'molecular pathologies' (Pichot, 1995; Lookard Conely, 1980; Mac Kusik, 1975; Childs 1973). The career of the British geneticist L. Penrose is considered emblematic of this displacement: Penrose was trained in the British tradition of biometrics and eugenics, but in the postwar era he was an influential translator of previous concerns with mental illnesses in the terms of biochemical genetics and chromosomal analysis (Kevles, 1985).

A different periodization however emerges in the historiography of American postwar biomedical research (Sloan, 2000; Kay, 1993; Patterson, 1987; Starr, 1982; Strickland, 1972). The Second World War changed the relationship between universities, research hospitals, health industries and government while the postwar era was characterized by new and more efficient therapies against bacterial infections. This in turn contributed to shift medical concerns from infectious to chronic diseases. A well analyzed example is the development of cancer research under the umbrella of the National Cancer Institute and of the American Cancer Society. Similarly, in the late 1950s, as the war against polio seemed to be won thanks to the Salk vaccine, the National Foundation for Infantile Paralysis reorganized its research effort, leaving aside the study of viruses to concentrate instead on 'birth defects'. Another seminal feature in the postwar rise of medical genetics was the establishment of the Atomic Energy Commission which supported the development of radiation studies in the 1950s with many projects dealing with the structure and changes of human chromosomes. Finally, the definition of human heredity was affected by the creation of a few academic centers for 'genetic counselling' (Kevles, 1985). The scientists working in these settings were concerned with disorders that could be linked to one or a very few number of genes through mendelian analysis. Building on estimates of frequencies and mendelian patterns of transmission, these 'human geneticists' produced risk figures, looked for 'carriers' of genetic defects and advised parents about reproduction strategies.

All these elements point to the decade between 1955 and 1965 as a critical period in the shaping of postwar human genetics. The first hypothesis underlying this paper is that rather than being a product of the research on 'genes and enzymes' conducted during the war, (human) biochemical genetics was a late product of postwar biomedicine. In other words, the notion of 'inborn errors of metabolism' became a key concept in clinical genetics in the late 1950s only when a new socio-cognitive configuration emerged, allying biochemists, geneticists, and clinicians. This is documented in the first part of the paper.

This realignment was facilitated by a series of technological innovations which included karyotyping techniques developed in the context of radiations studies, procedures for the biochemical diagnosis of genetic defects, as well as the selection of new animal models mimicking human pathologies. The second part of the paper focuses on the production and uses of the mouse mutants employed to study the etiology of disorders thought to be (partly) inherited. In the 1950s, the majority of these pathological mutants were found, maintained and produced at the Jackson Memorial Laboratory in the United States, then the only center mass-producing genetically homogenous mice for biomedical research.

Using mice as a model of human illnesses appears as a very problematic enterprise when one considers the moral and scientific difficulties of 'taking mice for men'. Firstly, animal models have to be similar enough to humans in order to enable researchers to say something about human pathologies on the basis of their handling of animals; yet animal models must differ from humans in such a way that experimentation on diseases with its legacy of suffering and death would not be viewed by practitioners as a critical moral and cultural issue. A second set of issues originates in the contradictory nature of the bio/medical enterprise. Both labor and negotiation are required for the stabilization of a 'device' mediating between the laboratory and the clinic which are worlds of their own characterized by heterogeneous aims, rules, languages, and material resources.

Medical scientists nonetheless view animal models as irreplaceable because mice and other organisms can be manipulated, have a high degree of plasticity and can more easily be standardized than humans. In a nutshell, animal models are valuable because time and resources can be invested to select systems fitting specific aims while reducing the variability of the living material in order to facilitate the replication of results. This will be discussed in the final section of the paper focusing on the uses of mice affected with a form of muscular distrophy. The case study shows that standardisation, replicability and commodification ensured the transposition of biochemical genetics into etiological knowledge but did not change the clinical practices.

INBORN ERRORS OF METABOLISM IN POSTWAR MEDICAL GENETICS

In 1946, the British geneticist L. Penrose introduced phenylketonuria (PKU) as an exemplar for 'the problems of eugenics as they appear in the light of recent knowledge' (Penrose, 1946). The topic also emphasizes the need for 'close cooperation between the clinician, the biochemist, the serologist, and the geneticist'. Penrose thought the biochemical investigation of the disease, i.e. the detection in the patient's urine of abnormal products of the amino-acid phenylalanine, to be decisive because of its medical applications. Detection of carriers rather than therapy was presented as a straightforward outcome of the knowledge of pedigrees and transmission paths:

> Another line of inquiry is the search for a functional test to differentiate carriers from the normal. At first it seemed as though experiments might demonstrate that the parents of phenylketonurics and other probable carriers had a low threshold of tolerance for doses of phenylalanine or of phenylpyruvic acid as compared with the normal average. Fölling thought that parents were liable to spontaneous excretion of phenylalanine. It now seems that, if there is any biochemical difference between carriers and the average person, it is every difficulty to detect. Perhaps the right test will yet be discovered. (Penrose, 1946, p. 952)

Penrose did not think of carrier detection as a means of eradicating phenylketonuria since 'we cannot reasonably sterilize 1% of the population' but as a means to 'reduce the incidence of phenylketonuria' by 'preventing consanguineous matings in affected families' and by 'discouraging the mating of partners who are both carriers'. To Penrose, phenylketonuria was not viewed as interesting because a causal link could be established between genes and metabolism. It became a reference case because the association of biochemical tools and mendelian studies of family trees could facilitate a form of 'eugenic counseling'.

This approach of phenylketonuria is a good illustration of the continuity of human eugen(et)ics throughout the war. It highlights the fact already noted by historians that postwar 'medical genetics' stressed a form of human genetics which emphasized statistical surveys, pedigree analysis, risk computation, and the control of reproductive units. For fifteen years—up to the late 1950s—the paradigm which dominated the analysis of pathological inheritance was not 'biochemical genetics' but 'mendelo-morganian genetics' as Maurice Lamy, the first professor of medical genetics in France, once put it.

A good illustration of this pattern is Lamy's textbook about 'the applications of genetics to medicine' published in 1943 and revised in 1952 (Lamy, 1943). The book was full of human pedigrees. It described mendelian patterns of inheritance through ideal exemplars such as albinism in the chapter on recessive inheritance or hemophilia in the chapter about sex-linked inheritance while a concluding chapter on inherited disease of medical importance deals with malformations, anemias, Tay-Sachs, diabetes, daltonism, Leber's disease, Huntington's chorea, hemophilia. However, phenylketonuria and 'inborn errors of metabolism' were not mentioned.

The concept of 'enzymatic blockade', Garrod's work, the idealization of Penrose's phenylketonuria, and a reference to Beadle and Tatum's one gene-one enzyme hypothesis actually appeared in medical genetic textbooks only in the late 1950s and early 1960s. In 1959, Lamy published a new synthesis entitled 'Hereditary errors of metabolism in infants'. The introduction stated:

> In his 1902 study of alcoptonuria, Garrod (with the collaboration of Bateson) has shown that an inherited metabolic deficiency may be caused by one single hereditary unit—a recessive gene as we call it today ... This was the first example of an inborn error of metabolism ... We now know many diseases of previously unknown origins and ill defined physiopathology which are in all probability caused by a similar inborn error of metabolism. Biochemical genetics expanded in parallel to these studies. One critical step were the experiments completed by Beadle and Tatum on the mold *Neurospora crassa*. (Lamy, 1959, p. 12).[1]

In spite of the reference to biochemical genetics, the aim seemed to be similar to Penrose's 1946 program, i.e. to define biochemical diseases associated with a familial pattern of transmission. Most descriptions of syndromes followed the same pattern: a summary of clinical symptoms was followed by the discussion of biochemical signs and pedigrees. With the exception of a handful of disorders in which specific enzymatic steps were targeted, there were few claims for a specific causal relationship between genes and diseases. This juxtaposition of some biochemistry (metabolites measurements) and some genetics (pedigrees) was nonetheless reinforced by the assumption that every single 'inborn defect of metabolism' showed deterministic links between genes, enzymes and clinical symptoms which would be identified sooner or later. This 'horizon d'attente' was especially visible in the chapter on diabetes as 'inborn errors of metabolism'. Lamy added speculations about the number and role of genes that might affect the production of insulin in the pancreas to evidence of a familial distribution of the disease.

That 'inborn errors of metabolism' actually entered medical genetics in the late 1950s has not gone unnoticed by historians of postwar genetics. Daniel Kevles, for instance, recalls that it was only in 1959 that Penrose opened a conference on human genetics by declaring 'At the present time the application of mathematical methods is no longer a dominating factor. Biochemical methods are now in the ascendant.' (Kevles, 1985, p. 235) Kevles attributes the new mood to changes in biochemistry and molecular biology, in particular to research on aminoacid metabolism and abnormal hemoglobin. Emphasis is put on the new understanding of 'molecular pathology' which was associated with Ingram's success in the identification of a peculiar amino-acid substitution responsible for the formation of abnormal hemoglobin molecules in patients affected with sickle-cell anemia (Chadarevian, 1998). However, one troubling feature is that Ingram's work and hemoglobin were of importance to molecular biologists but little discussed by medical geneticists before the mid 1960s.[2]

One alternative to this scheme may be to stress the role of medical and screening practices since phenylketonuria was the first biochemical defect which could be rapidly diagnosed in newborns. Yet, the relationship was probably the opposite: therapeutic developments and postnatal screening for phenylketonuria were greatly enhanced by the emblematic status of this disease for laboratory scientists (Paul and Edelson, 1998c; Paul, 1998b). Although the use of a special diet without phenylalanine was widely mentioned in the late 1950s, it was usually seen as a promising but still problematic perspective rather than a well defined therapy. The British geneticist Harry Harris for instance contrasted the rapid effect of the diet on the concentration of blood phenylalanine and the uncertainties regarding its efficiency on clinical parameters and mental illness.[3]

The basis for this chronology of *human* biochemical genetics (in contrast to the periodization of experimental biochemical genetics), is, I suggest, the transformation from a two-partner alliance between geneticists and clinicians which prevailed in the aftermath of World War II into a three partner-alliance including biochemists, geneticists and clinicians. This configuration may be traced back to the development of technologies which linked clinical examination, measurement of chemicals in body fluids, and establishment of medical pedigrees. These technologies included: a) a new generation of biochemical instruments which gradually entered laboratories in medical schools and hospitals in the aftermath of World War II, namely radioactive isotopes, electrophoresis or chromatography; b) new standardized procedures for the physio-chemical examination of patients, for instance the phenylalanine tolerance assay for detecting phenylketonuria carriers; c) animal models which mimicked human hereditary disorders.

PRODUCING MOUSE MUTANTS AT THE JACKSON LABORATORY.

In 1954, a brochure published by the US Jackson Memorial Laboratory stated:

> Just as the chemist has his pure and known chemicals with which he can do reproducible experiments, so at the Jackson Laboratory and in all the great laboratories of the world, thanks to Jackson Laboratory—the pure strains of inbred mice give the specialist an invaluable an essential pure living materials with which to work ... (the laboratory) has a function not duplicated at any other institution—to serve as the integrative center of mammalian genetics. It is the custodian of living, breeding, constantly renewing strains of standard mice. (Jackson Laboratory, 1958, p. 8)

By the time of its 25th anniversary, the setting established by C.C. Little in the 1930s had become the world leading center for both the study and the production of inbred—genetically homogenous—strains of mice (Rader, 1995). As recounted elsewhere, cancer research was then the most important market for mice 'with pedigrees' (Löwy and Gaudilliere, 1998; Gaudillière, 1998). This proved to have such long lasting consequences that the multiplication of mouse mutants showing pathological features during the 1950s and 1960s must be traced back to the changing organization of clinical research in the cancer field.

In the 1930s cancer chemotherapy studies were seen as a marginal, slightly disreputable subject (Patterson, 1987). By contrast, the search for anti-cancer drugs became a major research focus after the mid-1940s (Bud, 1978; Löwy, 1996). Two developments contributed to this change: the dramatic increase in funds available for cancer studies and the influence of war research. The war demonstrated the efficiency of large-scale, centrally-organized research such as the development of penicillin and the fight against malaria. In addition, the US. Office of Scientific Research and Development (OSRD) directly financed studies which led to the development of anti-tumor drugs. Research on war gases led to the application of nitrogen mustard to the therapy of lymphoma, while studies on vitamins led to the application of folic acid to the therapy of childhood leukemia (Einhorn, 1985; Farber, 1948).

In the early 1950s professionals and the lay public shared the feeling—based partly on the recent success of antibiotics—that the control of malignant diseases by drugs was imminent, and that it would be achieved through development of large-scale screening programs which would allow the discovery of the 'penicillin of cancer' (Gelhorn, 1953). This feeling was transformed into a direct political issue when in 1953 the US. Congress requested that the NCI develop an extramural (that is, sponsoring research beyond NCI's site) program of chemotherapy. The continuous pressure of Congress, together with growing demands on the part of non-NCI cancer specialists and of the chemical industry, led in 1955 to the development of the Cancer Chemotherapy National Service Center (CCNSC) (Löwy, 1996; Zubord, 1984; Zubord, 1977; Endicott, 1957). The CCNSC structure was a compromise: it was formally part of the NCI, but its decision-making power was delegated to panels which included numerous extramural scientists and physicians: a chemistry panel, a clinical studies panel, a pharmacology panel, an endocrinology panel and a screening panel. Congress rapidly allocated this organization 5.6 million dollars in 1956, 20 million in 1957, and 28 million in 1958.

The preclinical screening of drugs, unlike, for example, clinical research, can be relatively easily adapted to standardized, industry-like patterns of production (Löwy, 1996). The achievement of uniformity among experimental mice was one of the important elements in such standardization. The screening panel of the CCNSC decided that three transplantable tumors would be employed in all the screening tests. The decision to employ tumors transplanted in inbred strains of mice immediately created a need to enlarge the production of such mice. From 1956 on the CCNSC collaborated with the Jackson Memorial Laboratory to determine minimal standards for laboratory animals and develop a mouse production infrastructure. Mass production was first conducted in Bar Harbor, and later was extended to commercial laboratories. All the animals used in screening tests had to be supplied by producers accredited by the CCNSC, and all the demands of supply of mice were processed through CCNSC's Mammalian Genetics and Animal Production Section. The control of genetic purity was achieved by the Jackson workers who supplied certified breeding pairs of mice to other commercial breeders. The CCNSC thus opened a huge specialized market which enabled the large scale production of 'more uniformly healthy, well fed mice, with known genetic background and variability.'[4]

A few years later, in 1959, Gelhorn, one of the leading U.S. specialists in the field of cancer drug therapy, strongly criticized CCNSC methods of screening for anti-tumor drugs. In his testimony before a meeting of the National Advisory Cancer Program, Gelhorn contended that the expanded screening program did not lead to the discovery of new, clinically important classes of agents. The method employed by the CCNSC, he added, was inefficient: 'the mass and mechanized type of screening now employed is less likely to be productive than the observations of the individual investigators.' (Gelhorn, 1959, p. 8). Scientists associated with the CCNSC program observed in the meantime poor correlations between screening results obtained in the three screen system and the results of their first clinical trials. Therefore, they proposed to enlarge CCNSC screening to include other tumors and other laboratory animals.

These changes in the organization of CCNSC services did not, however, end the controversy between scientists and doctors who advocated an industrial-type preclinical selection of cancer-inhibiting compounds and the supporters of a more traditional style of investigation, based on individual expertise. Facing the criticism that too large an amount of public money was spent on a program which was only efficient at eliminating drugs and failed to uncover new anti-cancer drugs, CCNSC directors argued that beside tangible achievements in the organization of efficient testing for anti-cancer drugs, their program brought important benefits to the scientific community as a whole. One of the most important results of the chemotherapy screening program, they explained, was 'the development of enough high-quality animal resources to meet the needs of the program and the entire scientific community. The program was a key factor in anticipating and providing such resources for the major expansion of biomedical research in the past decade.' (CCNSC, 1966, p. 398).

The impact of this industrial research operation on the community of 'mousers' was actually tremendous. The CCNSC-based mass-production did make some inbred strains available to biomedical scientists but it also significantly altered the nature of the Jackson products. In 1955, the Jackson Laboratory sold 200,000 mice. Five years later, the production of inbred animals or first generation hybrids for cancer research reached one million. Scaling-up required more production rooms, more technicians and caretakers, cheaper and simpler means for bedding, feeding, housing the mice.

Growing was not only a problem of quantity but a problem of quality as well. In lieu of guaranteeing the uniformity of dozens of thousands of animals, the 'laboratory' was to operate on a plant size and standardize dozens of hundreds of thousands of mice. Scaling up the production of mice by two orders of magnitude meant increased risks of contamination, diseases, and unnoticed variations in the living material. Scaling up also required increased control of the living conditions, increased control of the personel, and a well-organized and routinized control of quality. In other words, mass-production changed the definition of tasks, the organization of the laboratory, and the division of labor.

A reformed production scheme operated from 1956–57 onward. It relied on the coordination of three departments: a) the 'foundation stocks' where scientists and technicians maintained pedigreed and systematically controlled animals from every inbred strain; b) the 'research expansion stocks' and 'production expansion stocks' where technicians produced hundreds of breeding pairs a year; c) the 'production stocks' where caretakers and fore(wo)men put these pairs to their reproductive work.

In the newly built mouse production rooms, not only new mice but new careers could be made. Care-taking was not 'taylorized' in the sense of time-motion studies but task distribution became much more specific. Local recommendations completed the general rules set up through the CCSNC. Critical to the success of the organization was keeping records of the mouse circulation, as when moving pairs from the production expansion stocks to the production rooms.[v] Breeding cards showing the 'complete record of an animal's life' were kept on permanent file, coded and transferred on IBM card for statistical analysis.

Another seminal issue was the surveillance of health and the disposal of infected, cancerous and otherwise ill adults. Breeding pens had to be 'taken down from the rack and checked for condition once a week' until 'retirement.' Increased control of the mice both generated and required a form of knowledge closer to that of veterinarians and agricultural engineers involved in livestock production. A few engineers were employed to conduct studies of breeding behavior, longevity, nutrition, causes of deaths, and growth rates. They soon joined colleagues in the other production centers to develop a laboratory animal care society.

In contrast to what might have been expected from the concerns of a homogenous, large-scale production, the development of the 'research expansion stocks' actually opened a new space between the laboratory and the plant. In this space, a few researchers were responsible for systematic quality control based on grafting assays and for the selection of scientifically interesting stocks:

> PES and RES are both research and service departments. The need for this arise from the fact that it is necessary to have the cooperative attention of several staff members and the trained service of research assistants to make sure that our inbred lines are what they purport to be . . . At present (1959) sixteen inbred strains, widely used in the laboratory and elsewhere are maintained in relatively large colonies. The other thirteen strains and the muscular dystrophy stock are maintained in smaller numbers . . . (Jackson Laboratory, Annual Report 1958–1959, p. 91)

The 'muscular dystrophy stock' was a typical result of the changes triggered by scaling up. Following the mid-1950s reorganization, the Jackson workers entered a new world inhabited by dozens of mutant strains of mice. In contrast to *Drosophila* studies, the chances of observing morphological or physiological mouse variants remained low until the 1950s. With the changing scale of production and increased attention to abnormalities in large stocks a threshold was passed: the 'gray zone' between the laboratory and the plant started to produce numerous good mutants for chromosome mapping and for biomedical research. This pattern may be illustrated with the origins of the mouse mutants described after 1945.

As shown in Table 5.1, the number of mutants found by the Jax workers increased from less than 25% of the total in the early 1950s to 40% of the total in the early 1960s. Among the medically interesting

TABLE 5.1 MOUSE MUTANTS DESCRIBED AFTER 1945. SOURCE: *BIOLOGY OF THE LABORATORY MOUSE*. E. GREEN (ED), JACKSON MEMORIAL LABORATORY, 1966.

	JACKSON	OTHER CENTER	PATHOLOGICAL MUTANTS	NEUROLOGICAL MUTANTS	SKELETON	TOTAL
1946–1950	0	12	6	0	2	12
1951–1955	11	37	24	9	9	48
1956–1960	15	49	34	12	10	64
1961–1965	48	68	52	19	11	116

mutants, the classical morphological variants (merely mice affected with skeleton abnormalities) and neurological mutants which had been described before World War II were supplemented in the 1950s with physiological mutants such as *jaundice, hemolytic anemia, dystrophia muscularis, obese*, etc. Finally, in the 1960s, biochemical variants such as *serum esterase, immunoglobulin-1*, or *amylase* were described.

The isolation of these mutants not only opened new research venues, but it also opened new markets. By the early 1960s, special production schemes had been established for three mutants: *dystrophia muscularis, obese*, and *hairless*. The Jax dystrophic mice were the most widely supplied. One may therefore wonder how the debates about the etiology of human muscular atrophies were affected by the availability of these animals.

HOW HUMAN MUSCULAR DYSTROPHY BECAME AN 'INBORN DEFECT OF METABOLISM': THE CIRCULATION OF *DYSTROPHIA MUSCULARIS*

In 1963, the American specialist of neuromuscular diseases Carl Pearson commented on recent advances in the study of human muscular dystrophy:

> Until several years ago there was almost complete lack of agreement about classification of the dystrophies and innumerable subgroupings of minor clinical disease variants were included in some series. The studies of Walton and Nattrass which were published in 1954 were unique in that they were based both upon clinical observations and careful genetic evaluations. These studies have provided the basis for the currently most acceptable classification. The inclusion of the genetic approach into the study of a disorder such as muscular dystrophy is valid because nearly all cases have shown a genetic basis. This is of more than academic interest since the genetic variants ultimately mirror fundamental biochemical differences in etiology. (Pearson, 1963, p. 633)

Pearson's chronology points to the fact that by the late 1950s, human muscular dystrophy did not only become a series of truly genetic diseases which could be distinguished on the basis of transmission mode rather than clinical course, but it also became an 'inborn error of metabolism'. As such it opposed other etiologies of muscular atrophies: nutritional deficiency or enervation.

Etiology is often presented as the medical search for the causes of diseases which gradually gained a scientific status by conforming to theoretical rules provided by Koch's postulates in microbiology or Bradford Hill's postulates in statistical epidemiology. Medical researchers tend to view the endless debates about the contribution of genetic, biochemical, physiological, immunological, bacteriological, psychological, and social causes involved in most human diseases as a 'theory-practice' problem: the theoretical ideals summarized by these postulates are difficult to translate into practice. Following the debates triggered by the circulation of the Jackson mice suggests that it may be useful to turn this approach upside down and to look at etiologic debates in terms of situated practices which consist of peculiar sets of laboratory, clinical and statistical operations. Rather than seeing the use of experimental devices like the Jax mice as an attempt to test postulates, this section takes etiologic claims as reified forms of the use of mice.

In 1965, the New York Academy of Science organized a large symposium on muscular dystrophies in animals and humans. Homburger, who had recently developed a strain of hamster affected with muscular atrophy, concluded his contribution with a word of caution about the multiplicity of animal models of human muscular dystrophy which did not only include dystrophies caused in rats and rabbits by avitaminosis E or by experimental enervation, but also genetic models:

> Because of this known similarity of anatomical lesions of muscles affected by different deficiencies, caution is indicated in concluding that an experimental myopathy may be related or unrelated to human dystrophy . . .
>
> The researcher is caught between the scylla of disregarding clinically important findings because they occurred in animals, and the charybdis of mistakenly accepting as clinically meaningful, purely species-specific phenomena because they may resemble human material. The avian, murine, and hamster hereditary dystrophy-like myopathies occupy a special place in comparative pathology because their genetic etiology resembles that of many human dystrophies. (Homburger, 1966)

Homburger did not only capture what was new in the mouse models introduced in the 1950s but accurately described the problem of using and juxtaposing several animal models. The use of Jax inbred mice in cancer research was dominated by the work of institutions such as the CCNSC. By contrast, the circulation of Jax mutants lacked a similar regulatory agency. Developments were less constrained. As illustrated by the fate of dystrophic mice in Bar Harbor and Paris: the manipulation of *dystrophia muscularis* did displaced etiologic theories but multiple meanings and variable uses prevailed.

In the Spring of 1956, the newsletter of the Jackson Memorial Laboratory announced that 'Jackson Mice Give New Hope in Muscular Dystrophy'.[6] The article recounted the discovery by E.S. Russell of weak-looking and dragging mice in a colony of the '129' strain. These mice were shown to be affected with a defect of skeletal muscles caused by a mutation. Hope was associated with the linkage between production and research: with the support of the Muscular Dystrophy Association plans were made for producing large numbers of dystrophic mice while a special team investigating muscular dystrophy was created.

P.J. Harman, a pathologist from New York University Medical School, temporarily joined the Jax workers for a description of the dystrophic mice. The original report stressed three points: the mutation was transmitted as an autosomal recessive gene, the dystrophic mice displayed a 'characteristic' microscopic picture of degenerating muscles, there was alteration of the central nervous system. Thus, the disease could be viewed as 'the first experimental animal of known genetic background to present a myopathy similar to that found in the human muscular dystrophies.' (Michelsohn, 1956, p. 1079)

Further research concentrated on the nature of inheritance, the histological symptoms, and the biochemical changes associated with the disease. Biochemical studies focused on what the Jackson workers viewed as the 'most investigated biochemical change that accompanies muscular dystrophy in man—an alteration in the normal pattern of creatine storage and excretion.'.[7]

The resuscitation of this ancient target of biochemical analysis was typical of the venues opened by the availability of the mouse model. On the one hand, as the Jax biochemist A. Kandtush put it 'interpretation of the results obtained so far may be complicated by the fact that creatine metabolism in the normal mouse appears to differ from that in other animals' including man. Discrepancies between the accepted pattern of creatine/creatinin in the human disease and observation in dystrophic mice could then be used both ways: either as evidence that the mouse was not a good model, or as evidence that a defect in creatine metabolism was not a good hypothesis. On the other hand, husbandry provided opportunities for manipulation. Jax work focused on the nutritional approach. In the following years, variability in the creatine metabolism was refereed back to the quality of the protein fed, and more particularly to the amount of 'amino acid glycine'. Attempts to use this relationship to control the disease however gave mixed results: 'mice maintained on diets supplemented with glycine grew more rapidly during the first two or three months of life ... life-span however was not prolonged.'[8]

From late 1950s onwards, such mixed results and the multiplication of mutants increasingly shifted the Jackson management of dystrophic mice from the research side toward the production side. Production was initially based on the standardization of 'ovarian transplantation' techniques. These techniques had been introduced on a small scale in the 1950s for producing obese mice. They were employed to circumvent the sterility of dystrophic female and to increase production yield from 1/4 (obtained when crossing heterozygotes) to 1/2 (theoretically) obtained when ovaries from a dystrophic mice were implanted in normal females which were later fertilized by heterozygous males. In the 1960s fertilization procedures were improved with the use of artificial insemination. Survival time and resistance were increased by changing the 'genetic background', namely by transferring the dystrophic gene into one of the standardized strains mass-produced for cancer research. Sales, however, did not expand beyond a few thousands animals a year since most centers started their own reproduction programs while ordering certified breeding pairs from the Jackson.

So what was done with all these dystrophic animals? In 1965, the Jackson Laboratory circulated a general inventory of all the papers based on the use of the Jax model published in a decade. As shown in Table 5.2, patterns inside and outside the Jackson Laboratory were similar: biochemical papers dominated while preclinical studies and drug testing amounted to a little 11%.

TABLE 5.2 THEMES OF THE ARTICLES BASED ON THE USE OF JAX DYSTROPHIC MICE (1957–1964). SOURCE: 'DYSTROPHIA MUSCULARIS IN THE HOUSE MOUSE: A BIBLIOGRAPHY' BY JOAN STAATS, JACKSON MEMORIAL LABORATORY, 1965.

PATHOLOGICAL ANALYSIS	13
HISTOPATHOLOGY AND MICROSCOPY	21
ELECTROPHYSIOLOGY	16
ENDOCRINOLOGY AND CIRCULATION	9
COMPARISON WITH AVITAMINOSIS E	10
ENZYMATIC MEASUREMENTS	78
THERAPEUTIC TRIALS	19

In France, the Jax mice were imported in one single setting: the laboratory of biochemistry at the Hospital for Sick Children in Paris where scientists were already interested in human muscular dystrophy. This research unit and service laboratory was created in the aftermath of World War II under the patronage of Robert Debré, a prominent French pediatrician. Benefiting from the 'reconstruction' of biomedical research after 1945, Debré obtained from the French medical authorities the creation of genetic and biochemical services associated with his department of pediatrics at the Hôpital des Enfants Malades (Gaudillière, 2000; Gaudillière, 1992). The former was headed by Maurice Lamy, the latter by George Schapira. As a departure from prewar practices, the young physicians within the biochemical unit were sent in the United States for postdoctoral training. As they respectively returned, J.C. Dreyfus and J. Krüh launched projects based on their command of new tools for deciphering metabolic pathways such as the use of radioactive isotopes.

Local studies of muscular dystrophy began within the medical genetics unit with a large collection of pedigrees organized by Jean de Grouchy, a student of Maurice Lamy. The collection started with patients referred to the nearby consultation for polio patients and later extended to specialized institutions for crippled children. Analyzing familial transmission, de Grouchy distinguished two patterns of transmission: sex-linked recessive on the one hand, and autosomal recessive on the other. These patterns characterized two different forms of one single clinical entity (la myopathie) to be distinguished from muscular atrophies of neural origins (de Grouchy, 1953).

On the biochemical side, the basic approach was to look for features specific to this disease by assessing enzymatic activities in biopsies of skeletal muscles obtained from patients affected by 'the' disease. Early observations lead to the conclusion that clinical muscular dystrophy was associated with a decrease in the activities of a few enzymes involved in the carbohydrate metabolism and with a marked increase of one among these: aldolase. Surveying a few dozens cases, Schapira and Dreyfus were rapidly convinced that the rise in serum aldolase was grossly specific to muscular dystrophy and could be employed for biological diagnosis (Schapira, 1953). Although it was tempting to suggest that 'these enzymatic defects are the cause of the dystrophy', Dreyfus carefully concluded that these enzymatic changes were a 'sequel of a more primary biochemical flaw' since other investigators had documented similar alterations in dystrophy of a different etiology, namely experimental muscular atrophies caused by nerve section (Dreyfus, 1954).

What did the Jax mice change in this work?

Breeding pairs from Bar Harbor crossed the threshold of the Hospital des Enfants Malades early in 1957. A few measurements made the case. Dystrophic mice also displayed the same increase in serum aldolase (Schapira, 1957). The convergence of results worked both ways. It contributed to make the human hyperaldolasemia more robust while reinforcing interests in the murine model. The use of mice thus established a new hierarchy of symptoms:

> Hyperaldolasemia is found in progressive muscular dystrophy (PMD), in experimental dystrophy due to avitaminosis E and in hereditary dystrophy in mice. However while in muscle from dystrophic patients or mice enzyme level is decreased, it is increased in nutritional muscular dystrophy. From a biochemical point of view, the hereditary dystrophy of mice and the clinical disease display many similarities while nutritional dystrophy behaves in a very different way. (Dreyfus, 1958, p. 237)

One year earlier, hyperaldolasemia had been observed in both rat muscular atrophy caused by avitaminosis E and human muscular dystrophy. The feature then opposed hereditary and nutritional degeneration on the one hand, atrophies from nervous origin on the other hand. By echoing the clinical classification, the use of the rat model had stabilized the biochemical unity of the disease. The domestication of Jax mice now resulted in a further divide between nutritional and hereditary dystrophies. This distribution of models which reinforced the hereditary etiology of the human disease was soon echoed in claims that attempts to cure dystrophic mice with vitamin E failed. Manipulation of the Jax mice thus led to a promising correlation between altered levels of aldolase and hereditary dystrophy.

This did not mean that alterations of glycolytic enzymes came to be viewed as the primary defect in the disease. Building on their interests in the structure of muscle proteins, Schapira and his colleagues considered rather than the main event was either 'a trouble of the rate of synthesis or degradation of muscle proteins' or 'an increase of permeability' which may explain the release of proteins into the blood stream (Dreyfus, 1958, p. 247). From the biochemist viewpoint, one great advantage of mice was that one could inoculate C14 labeled amino-acids and trace the synthesis of enzymes. Experiments conducted in 1959 with the dystrophic mice strongly reinforced the idea that the first biochemical defect was an increase in the degradation of proteins. The new biochemical targets were the cellular mechanisms controlling the turnover of proteins.

The arrival of Jax mice nonetheless favored the local conjunction between heredity and biochemistry in two ways. Firstly, different forms of muscular atrophies were granted different biochemical features. This individualized progressive muscular dystrophy as a hereditary disease characterized by biochemical changes. Consequently, the human disease could be compared with a typical 'inborn error of metabolism'. Writing for patients' families and general practitioners, Schapira discussed the prospects of a rational therapy by making an analogy with galactosemia:

> What is left is another approach: the search for a rational therapy based on the knowledge of the pathological process. Think of a genetic disorder. Cells lack one enzyme involved in a peculiar metabolic pathway ... A good example is galactosemia in babies ... Avoiding dairy products is enough to have all the symptoms banish and cure the baby. This the case of an enzymatic deficiency which may be cured. Our job is to find out whether muscular dystrophy patients lack one enzyme (Schapira, 1963, p. 399).[9]

Secondly, the increasing emphasis on the metabolism of proteins facilitated an internal reshuffling which linked muscular dystrophy to molecular biology rather than physiopathology. One theory defeated in the years 1958–1960 was that of vascular etiology. At the Enfants Malades, this theory had been substantiated by Jean Demos a young clinician working in the biochemical laboratory. In addition to caring for the muscular dystrophy patients, Demos participated in the surveys of enzymatic activities. His main commitment was however to a more physiological approach. Circulation parameters were first investigated for diagnostic purposes but rapidly became a sign. From the late 1950s onward, Demos viewed muscular degeneration to be caused by a defect in blood circulation which might be corrected by drugs dilating blood vessels (Bach, 1998). By contrast, Schapira and Dreyfus wrote in 1962: 'The vascular lesions of PMD are probably important in aggravating the situation. It is not possible at present to admit that they are the primary event' (Dreyfus et Schapira, 1962).

In 1963, when reviewing recent observations of muscular dystrophy, Pearson echoed these changes and noted that from a practical viewpoint the main achievement of research was the enzymatic detection of carrier state: 'The biochemical identification of the carrier state in sex-linked recessive Duchenne dystrophy is now possible by the routine measurement of serum creatine-phosphate kinase (CPK) and aldolase in females from families in this form of dystrophy is known to exist.' (Pearson, 1963, p. 639) By the early 1960s, measuring CPK, and to lesser degree aldolase, made possible early diagnosis of human muscular dystrophy of the Duchenne type and a form of genetic counseling.[10]

The measurement of CPK in human muscular dystrophy was a by-product of technical changes in enzymology. This development was based on small-scale surveys in humans and on statistical evidence. Mice played no direct role in the innovation. Moreover, the irony of the case is that the comparison between mice and men could have blocked the process. When CPK blood levels were measured in the dystrophic mice conflicting result surfaced. In most instances, the increase observed in humans was not observed in the Jax model. Clinical developments were already on their way. So the mice results were simply dismissed (Hosenfeld, 1962; F. Schapira, 1963).

This biochemistry-based genetic counseling was, in the early 1960s, the one significant change in the medical handling of muscular dystrophy. Therapy—or better the lack of therapy—was barely affected by the redefinition of etiology. This points to the contrasted fate of the Jax dystrophic animals in biochemical and clinical settings. By contrast to the biochemists, many specialists of neuromuscular disorders remained highly skeptical of the utility of the murine model. J.N. Walton, author of the most used clinical and genetic classification of muscular dystrophies did use the Jax in his department of neurology at the University of Newcastle upon Tyne. Yet, it was to denounce the claims of analogous histopathological features in mice and humans. In 1965, Walton warned the participants at the New York Academy of Science workshop on muscular dystrophy in animals and humans:

> Although it is plain that investigation in each of these types of myopathy occurring naturally or produced experimentally in animals in likely to throw light upon the fundamental processes of muscle degeneration and repair, it seems likely that the solution to the problem of muscular dystrophy will come only in the final analysis as a result of painstaking investigations in the nature of the pathological process that occurs within the muscle cell in the human disease. (Walton, 1965, p. 315)

This discrepancy ellicits two remarks. Firstly, the use of the Jax mutants was far from always being synonymous of a move toward genetic explanations as it was the case with muscular dystrophy or diabetes.

This may be illustrated with the fate of the Jackson obese mice. Although these mutants were widely employed in the 1960s for biochemical and pathological studies, their circulation did not contribute to displace research practices toward the notion that obesity has a genetic etiology. Secondly, the case of muscular dystrophy suggests that mutant mice made a more significant impact on etiologic debates than they did on clinical practices. The influence of therapeutic assays with mice on the clinical handling of muscular dystrophy patients has been fairly limited if existing at all. This contrasts the fate of numerous anti-cancer drugs tested in inbred mice which entered preliminary clinical trials although they did not remain in use for a long time. This dissimilarity suggests that heavy investments and direct regulation were often necessary to make the mouse metaphor operational in the clinic.

CONCLUSION

This paper has followed the fate of the dystrophic mice produced at the Jackson Laboratory as a means to investigate the uses of the mouse models of human pathologies produced after 1945 and to address the impact of these mediating devices on the development of human genetics in the postwar era, in particular on the generalization of the notion of 'inborn error of metabolism'.

The logic of production of genetically homogenous mice in part explains the postwar chronology of human biochemical genetics. Pathological mutants eventually multiplied in the late 1950s and 1960s as a boost in the demand for inbred animals to be employed in cancer chemotherapy led to a major reorganization of the Jackson Laboratory. This reorganization of research and production tasks enabled the selection of an increasing number of mouse mutants which could be employed either in mapping studies or as models of human pathologies.

The fate of the dystrophic animals produced at the Jackson Laboratory in the 1960s shows that the uses and meanings of these pathological mutants was a matter of local domestication. The manipulation of Jax mice was usually juxtaposed to existing research practices and did not radically modify the established division of labor. Thus, in Schapira's biochemical laboratory *dystrophia muscularis* was employed in etiologic studies based on enzymatic measurements directly related to the service function of the laboratory. Mouse mutants were nonetheless instrumental in stabilizing results and displacing etiologic hypotheses. Comparison of mice and men narrowed down the uncertainties of evaluating the causes of putative hereditary disorder. The analogy worked both ways: genetic causation was reinforced because mice were taken to be closer to men than other laboratory animals while inbred mice were legitimated because they visualized putative hereditary transmission. Thus they provided a point of departure for chains of analogies which resulted in the ordering of usually conflicting results.

Mouse models of hereditary pathologies such as *dystrophia muscularis* favored new arrangements of clinical diagnosis, mendelian analysis of family trees, and biochemical measurements. Since the Jackson inbred mutants were tools showing a certified homogenous genetic background that could easily be employed for collecting blood, urine or organs in order to make chemical analysis, the Jax products *embodied* a juncture between medical genetics and medical biochemistry and favored the acceptance of 'inborn error of metabolism' as a concept emblematic of this boundary.

Our case study finally opposes the use of inbred strains in cancer chemotherapy and the fate of the Jax mutants. The latter's impact on therapeutic practices has been fairly limited if it exists at all. Such a contrast suggests that heavy investments, political visibility, and direct regulation were often necessary to cross a technological threshold, stabilize meanings, and make the mouse models legitimate in clinical settings.

NOTES

1. 'Dans un travail qui date de 1902, peu de temps après la redécouverte de Mendel et qui porte sur l'alcaptonurie, il (Garrod) montre (avec Bateson) comment un trouble métabolique constitutionnel peut être causé par l'action d'une unité héréditaire, un gène dans notre langage d'aujourd'hui, en l'espèce un gène récessif à l'état homozygote ... C'était là un premier exemple de ces 'inborn errors of metabolism' ... Nous connaissons maintenant nombre de maladies dont la cause était obscure et le mécanisme physiopathologique mal déterminé et qui traduisent à n'en pas douter une erreur innée du métabolisme. Parallèlement à ces études, se développait la génétique biochmique dont les étapes essentielles sont marquées par ... les remarquables expériences de Beadle et Tatum sur une espèce particulière de moissure: *Neurospora crassa* ...'
2. The most significant exception is H. Harris who was closely associated with Haldane and Penrose. His *Human Biochemical Genetics* already includes an important chapter on abnormal hemoglobins and 'molecular pathologies' of the blood proteins in 1959. On haemoglobin, medicine and molecular biology see L. Kay, *op. cit.* and S. de Chadarevian 1998.
3. H. Harris, 1959, chapter III.
4. The National Program of Cancer Chemotherapy Research: Information statement,' *Cancer Chemotherapy Reports*, (1959), 1: 99–104, quote on p. 100.
5. 'Each complete litter not used to make new breeding pairs within the PES should be put into a pen marked with small cards noting strain, pedigree number of parents (female number is always written first), birthdate of offspring, number of males and females ... Arrange these animals in

trios and in brother-sister pairs for the production stocks supply assigning to each mating an identifying number, which is recorded in the strain ledger in a completely different serial numbering system, related to the PES individual animal pedigree number only through the parents data recorded in this ledger. These identifying numbers follow these animals throughout their existence in the production colony.' Note for the new personel, undated typescript, Jackson Laboratory Archives.

6. 'Jackson Mice Give New Hope in Muscular Distrophy', *The Roscoe B. Jackson Memorial Laboratory Quarterly*, (1956), pp. 1–2.
7. Jackson Memorial Laboratory, Annual Report, 1956–1957.
8. Jackson Memorial Laboratory, Annual Report, 1959–1960.
9. 'Il nous reste par conséquent la dernière méthode de travail, celle d'une thérapeutique rationnelle fondée sur la compréhension de la maladie. Considérons un example de maladie héréditaire: il manque un enzyme nécessaire à la suite des évènements métaboliques ... Nous en avons un example dans un autre domaine qui est celui de la galactosémie du nourrisson ... Il suffit de supprimer l'apport lacté pour que tous les troubles disparaissent et que l'enfant guérisse si on le traite à temps. Vous avez donc ici une maladie héréditaire par déficit enzymatique et curable. A nous de trouver dans la myopathie s'il y a une enzyme qui manque.'
10. Dans la mesure où la génétique nous permet de dire que, dans un certain pourcentage, un couple est susceptible de donner naissance à des enfants atteints, nous nous trouvons devant le problème délicat de l'eugénique médicale et, par conséquent, il faudrait pouvoir soumettre les frères et les soeurs à une consulmtation de génétique. Ainsi, on pourrait éventuellement rassurer les sujcts normaux ou bien préciser dans quel pourcentage ils sont susceptibles de donner naissance à des enfants atteints de la maladie. (Schapira, 1963, p. 398)

REFERENCES

M.A. Bach, 'Building the French Muscular Dystrophy Association: The Role of Doctor/Patient Interactions' *Social History of Medicine* (1998) 11: 233–253.

R.F. Bud, 'Strategy in American cancer research after World War II: A case study' *Social Studies of Science* (1978) 8: 425–459.

Cancer Chemotherapy National Service Center, 'The Cancer Chemotherapy Program, 1965' *Cancer Chemotherapy Reports* (1966) 50: 397–401.

S. de Chadarevian, 'Following Molecules: Hemoglobin between the Clinic and the Laboratory' in *The Molecularization of Biology and Medicine* H. Kamminga, S. de Chandarevian (eds), (London: Harwood Academic Publishers, 1998), pp. 171–201.

B. Childs, 'Garrod, Galton, and Clinical Medicine', *Yale Journal of Biology and Medicine* (1973) 46: 307–308.

J.C. Dreyfus *et al.*, 'Biochemical Study of Muscle in Progressive Muscular Dystrophy', *Journal of Clinical Investigation* (1954) 33: 794–797.

J.C. Dreyfus *et al.*, 'Serum enzymes in the physiopathology of the muscle' *Annals of the New York Academy of Science* (1958) 134: 235–249.

J.C. Dreyfus et G. Schapira, *Biochemistry of Hereditary Myopathies* (London: Charles Thomas Publisher, 1962).

J. Einhorn, 'Nitrogen mustard: the origins of chemotherapy for cancer' *International Journal of Radiology and Oncology* (1985) 11: 1375–1378.

K.M. Endicott, 'The chemotherapy program' *Journal of the National Cancer Institute* (1957), 19: 275–293

S. Farber, L.K. Diamond, R.D. Mercer *et al.*, 'Temporary remission in acute leukemia in children by folic acid antagonist, 4-aminoptenyl-glutamic acid (aminopterin)' *New England Journal of Medicine* (1948), 238: 787–793.

J.P. Gaudillière, 'Entre laboratoire et hôpital: Biochimistes et biomédecine après la guerre' *Sciences Sociales et Santé* (1992), 4: 106–147.

J.P. Gaudillière, 'Circulating Mice and Viruses: The Jackson Memorial Laboratory, the National Cancer Institute, and the Genetics of Breast Cancer' in *The Practices of Human Genetics* M. Fortun, E. Mendelsohn (eds), (Dordrecht: Kluwer Academic Publishers, 1998), pp. 89–124.

J.P. Gaudillière, 'Whose work shall we trust? Genetics, pediatrics and hereditary diseases in postwar France' in Controlling our Destinies: Historical, Philosophical and Ethical Perspectives on the Human Genome Project (Notre-Dame: University of Notre Dame Press, 1999).

A. Gelhorn, 'A critical evaluation of the current status of clinical cancer chemotherapy' *Cancer Research* (1953) 13: 202–215.

A. Gelhorn, 'Invited remarks on the current status of research in clinical cancer chemotherapy' *Cancer Chemotherapy Reports* (1959) 5: 1–12.

J. de Grouchy, *Contribution à l'étude de la myopathie*, Thèse de doctorat en médecine, Paris, 1953.

D.J. Hosenfeld, U. Wiesmann, R. Richterich, 'Plasma Creatine Kinase Activity in Mice with Hereditary Muscular Distrophy' *Enzymologia, Biologia et Clinica* (1962–63) 2: 246–249.

Jackson Laboratory, *Genes, Mice and Men, A quarter century of Progress at the Roscoe B. Jackson Memorial Laboratory* (Bar Harbor: Jackson Laboratory, 1954).

L. Kay, *The Molecular Vision of Life* (Oxford: Oxford University Press, 1993).

D. Kevles, *In the Name of Eugenics, Genetics and the Uses of Human Heredity* (Berkeley: University of California Press, 1985).

M. Lamy, *Applications médicales de la génétique* (Paris: Doin, 1943).

M. Lamy, *Maladies héréditaires du métabolisme chez l'enfant* (Paris: Masson, 1959).

I. Löwy, *Between Bench and Bedside* (Cambridge: Harvard University Press, 1996).

C. Lookard Conley, 'Sickle Cell Anemia—the First Molecular Disease' in *Blood, Pure and Eloquent*, M. Wintrobe (ed) (New York: Mac Graw Hill, 1980) p. 380.

I. Löwy, and J.P. Gaudillière, 'Disciplining cancer: Mice and the practice of genetic purity' in *The Invisible Industralist: Manufactures and the Production of Scientific Knowledge*, (London: Macmillan, 1998), pp. 209–249.

V. Mac Kusik, 'The Growth and Development of Human Genetics as a Clinical Discipline', *American Journal of Human Genetics* (1975) 27: 264–269.

A.M. Michelsohn, E.S. Russell, P.J. Harman, 'Dystrophia muscularis: A hereditary primary myopathy in the house mouse.' *PNAS* (1956) 41: 1079–1084.

C.M. Parson, 'Muscular Distrophy' *American Journal of Medicine* (1963) 35: 632–645.

J.T. Patterson, *The Dread Disease: Cancer and Modern American Culture* (Cambridge: Harvard University Press, 1987).

D. Paul and PJ Edelson, 'The Struggle in Metabolic Screening' in *The Molecularization* pp. 203–220.

D. Paul, 'Phenylketonuria' in *The Practices of Human Genetics*, M. Fortun, E. Mendelsohn (eds), (Dordrecht: Kluwer Academic Publishers, 1998).

L.S. Penrose, 'Phenylketonuria: A problem in eugenics', *Lancet* (June 29, 1946) i: 949–953.

A. Pichot, *L'eugénisme ou les généticiens saisis par la philanthropie* (Paris: Hatier, 1995).

K. Rader, *Making Mice: C.C. Little, the Jackson Laboratory, and the Standardization of Mus musculus for Research*, PhD thesis, University of Indiana, 1995.

G. Schapira *et al.*, 'L'élévation du taux de l'aldolase sérique, test biochimique des myopathies', *Semaine des Hopitaux de Paris* (10 juin 1953) i: 1917–1920.

G. Schapira, 'Recherche et myopathie' *La médecine infantile* (Aout-Septembre 1963) i: 397–400.

F. Schapira, *et al.*, 'Hyperalodolasémie chez la souris myopathique' *Comptes Rendus de l'Académie des Sciences* (5 aout 1957) 243: 753–755.

F. Schapira, J.C. Dreyfus, 'La créatine-kinase du sérum chez la souris myopathique' *Enzymologia, Biologia et Clinica* (1963) 3: 53–57.

P. Sloan (ed), *Controlling our Destinies: Philosophical, Historical and Ethical Studies of the Human Genome Project* (Notre-Dame: University of Notre-Dame Press, 2000).

P. Starr, *The Social Transformation of American Medicine* (New York: Basic Books, 1982).

S.P. Strickland, *Politics, Science and the Dread Disease* (Cambridge: Harvard University Press, 1972).

J.N. Walton and R.J.T. Pennington, 'Studies on Human Muscular Dystrophy with Particular Reference to Method of Carrier Detection' *Annals of the New York Academy of Science* (1965) 134: 315.

C.G. Zubrod, S.A. Schepartz and S.C. Carter, 'Historical background of the National Cancer Institute's drug development trust' *National Cancer Institute Monographs* (1977), 45: 7–11.

C.G. Zubord, 'Origins and development of chemotherapy at the National Cancer Institute' *Cancer Treatment Reports* (1984) 68: 9–19.

Chapter 6

EXPERIMENTAL ARRANGEMENTS AND TECHNOLOGIES OF VISUALIZATION: CANCER AS A VIRAL EPIDEMIC, 1930–1960

Angela N.H. Creager and Jean-Paul Gaudillière

INTRODUCTION

In 1943, the *Journal of the American Medical Association* published a long article on cancer written by a well-known pathologist at the Rockefeller Institute for Medical Research. Peyton Rous's article proposed a new viral explanation of cancer:

> To account for the worldwide distribution of tumors ... it has been necessary to suppose that the body carries resident viruses just as it does resident bacteria, indigenous viruses, as they have been termed [.] ... They would give no sign of their presence in most instances and would in due course be passed on to the young. But if a provocative carcinogen happened to work on the cells with which such a virus was associated, thus altering its milieu, it might undergo variation and, taking a hand in cell affairs for the first time, give rise to a tumor. (Rous, 1943, p. 580)

The fact that Rous endorsed tumor viruses hidden in cells as a cancer cause was not unprecedented; thirty years earlier Rous had made his scientific if not his medical reputation with a similar claim based on the study of transmissible tumors in fowls. In the meantime, however, 'Rous sarcoma agent' had been transformed into a non-infectious, intracellular entity, and interest in tumor viruses among cancer specialists had waned (Löwy, 1990; Rheinberger, 1995; Gaudillière, 1998). Why was Rous returning to viruses and again taking the risk of being ridiculed by cancer clinicians and pathologists, who were well aware of the fact that cancer is not contagious? And what sort of virus was it that could exist as a hidden element in the host, indistinguishable for generations from normal cellular constituents?

Rous's proposition illustrates one among many of the strange twists taken by debates about the etiology of cancer during the years that saw the emergence of molecular biology. During the 1920s and 1930s the most popular theories of cancer causation had focused on the role of chemicals. Such carcinogens could be either irritants acting on the body tissue, such as coal tar, which incited skin outgrowths upon application, or analogs of hormones and metabolities that induced cancer by disturbing the normal regulation of growth. Early interest in cancer transmission by means of either inheritance or infection had receded and transmissible agents played very little role in medical oncology, although these conceptions were kept alive in a few experimental settings. This consensus changed rapidly in the aftermath of the Second World War. It is not that chemical carcinogenesis disappeared but rather that the etiology of cancer was enriched by a whole new series of cancer viruses that could be passed from one individual to another, were hidden in cells and sometimes became part of the material transmitted to offspring. In order to shed some light on the origins of this intriguing combination of infection and heredity—viruses and genes—which put the etiology of cancer into an increasingly molecular disguise, we will follow the production and use of a new series of models and machines in tumor virus research. Our selection of examples is informed by the notion that etiological theories can be understood as reified practices.

Etiology is usually understood as the medical search for the cause of disease, which eventually attains a scientific status by conforming to theoretical rules, such as Koch's postulates in microbiology or Hill's postulates in statistical epidemiology. Medical researchers tend to view the endless debates about the contribution of genetic, biochemical, physiological, immunological, bacteriological, psychological, and social causes involved in most human diseases as a 'theory-practice' problem: the theoretical ideals summarized by these postulates are difficult to translate into practice because the bodily experience of disease is so complex. Recent social and cultural studies of medical knowledge suggest that it may be useful to turn this approach upside down and to look at etiological debates in terms of situated practices which consist of peculiar sets of laboratory, clinical, and statistical operations. Rather than seeing the discovery of tumor viruses as an attempt to test postulates, this paper takes theories of cancer causation as reified forms of the manipulation of new pieces of equipment and new inscriptions.

Such a perspective facilitates historical recognition of the commonalities between laboratory practice in cancer research and in experimental biology more generally during the middle decades of the twentieth century. Historians of the life sciences have often emphasized the importance of new instrumentation, particularly the impact of the electron microscope, the ultracentrifuge, electrophoresis apparatus, and X-ray diffraction techniques upon biological research; the prevalence of these new technologies helped to distinguish postwar molecular biology from its predecessors. Recent scholarship has demonstrated that the successful incorporation of these new machines into the laboratory relied upon the establishment of collectively agreed-upon skills, technical procedures, and modes of representations (e.g., see Rasmussen, 1996). In other words, a new form of biological experimentation was at stake. The work associated with the conceptual resurrection of the tumor viruses in the 1940s was integrally related to the development of this new material culture. The legitimating strategies worked both ways: tumor viruses made possible the integration of 'molecular' means of investigation into reorganized practices, while the new technologies made the tumor viruses visible and acceptable.

In analyzing these developments, we have found it useful to speak of an *experimental arrangement* to designate the infrastructure through which viruses, including tumor viruses, were made visible. We use experimental arrangement to refer to a particular combination of techniques and tools employed by skilled operators to generate sets of images and inscriptions which—for all practical purpose—are evidence for the existence of a specific class of biological agents.[1] During the middle decades of the twentieth century, cancer viruses may be said to have been constituted on three distinct experimental arrangements: the pathological, the macromolecular, and the molecular genetic. It may be worth stressing that, for our cases, these modes cannot be reduced to disciplines or scientific styles (not that they are inherently cross-disciplinary); the arrangements are shared with other practitioners who may be working on different problems and systems entirely. Such experimental arrangements therefore represent units of historical analysis which are tangential to disciplines as well as to theoretical paradigms.

We will examine three model systems represented upon these different experimental constructs to follow the changing etiology of cancer: the rabbit papilloma system investigated by F. Peyton Rous, Joseph Beard and Richard Shope in the 1930s at the Rockefeller Institute; the chicken leukemia system employed by Joseph Beard in the 1950s at Duke University; and finally the chicken sarcoma system mobilized by Harry Rubin in Wendell Stanley's Virus Laboratory at Berkeley in the 1960s. These various experimental systems were not simply determined by the putative underlying viruses, but also relied upon technologies of visualization that constituted cancer viruses in various ways—as histological lesions, images of macromolecules, or plaques on cell cultures. The development of new experimental arrangements by interwar and postwar biomedical scientists provided the material basis for (re)placing cancer between infection and heredity.

THE PATHOLOGICAL ARRANGEMENT: INOCULATION, IMMUNOLOGY, AND CANCER VIRUSES AS MASKED AGENTS

The Rockefeller Institute for Medical Research was established in 1901 to support a wide range of research in medical science. Under the directorship of Simon Flexner, laboratories were organized around divisions of pathology and bacteriology, experimental surgery, physiology and pharmacology, and chemistry, with units of cancer research, biophysics, and animal and plant pathology being added over the years (Corner, 1964). During the early years of the Rockefeller Institute's development, filterable viruses became central to medical research and etiology. Flexner made their study a priority among pathologists at the institute, encouraging investigations of human, animal, and even plant viruses.[2]

The emphasis on virus research had ramifications in unexpected areas. In 1911, institute pathologist F. Peyton Rous found that chicken sarcoma virus could be transmitted by a filterable virus, an observation which attracted worldwide attention even as it met the skepticism of cancer specialists (van Helvoort, in press).[3] Rous's assistant (and subsequently colleague) James B. Murphy continued these research efforts at the institute (as Rous turned his own attention to blood preservation), but over the course of two decades of continuing experimentation on other systems, Murphy became less and less convinced of the role of viruses in cancer. As Ilana Löwy and Hans-Jörg Rheinberger have recounted, when Murphy picked up the chicken tumor agent again in 1928, he employed new biochemical and biophysical methods to characterize it (Löwy, 1990; Rheinberger, 1995). The resulting observations, including the agent's resistance to inactivation by ultraviolet light, led Murphy to differentiate the tumor factor from the 'virus class of disease-producing agents' and present it as an endogenous cellular substance of 'enzyme-like nature' (Murphy *et al.*, 1928, as quoted in Rheinberger, 1995, p. 55). After unsuccessful attempts to isolate the agent by chemical precipitation, the ultracentrifuge was used to isolate a component from the tumor fraction by the late 1930s; over the course of the next ten years this was classified as a mitochondrial particle, and then christened a microsome (Rheinberger, 1995, pp. 60–65).

A few months before the publication of Rous's 1943 review in the *Journal of the American Medical Association*, Murphy summarized his own views on cancer causation in the same journal:

> There is insufficient indication at present that viruses play any important role in the general picture; therefore no attempt will be made to discuss at length the possible relation of this group of agents to cancer. The only evidence of importance in this field is derived from the study of a group of fowl tumors which may be transmissible by filtrable agents. As no mammalian malignant tumor has yet been transmitted except by living tumor cells, the relationship of the fowl tumors to mammalian tumors cannot be considered as established. Furthermore, there is no agreement as to the nature of the transmitting agents and their relationship to the viruses (Murphy, 1942, p. 110).

In Murphy's mind, the association of microsomes with cancer had become gratuitous, and Rous's agent was no longer presented as infectious. However, Rous's claims on the viral nature of cancer was gaining support from another model system.

Shope's Passages and Masked Viruses

In 1932, research results obtained by an animal pathologist at the Princeton outpost of the Rockefeller Institute reawakened early interest in tumor viruses. During the past year, Richard Shope, one of the young fellows of the institute, collected some rabbits showing skin outgrowths. Back in the laboratory, the connective tissue masses were cut, sliced, stained, and examined under the microscope. Shope found that the disease could be transmitted to domestic rabbits by inoculating with filtered extracts of the growths, suggesting the presence of a tumor virus akin to Rous' chicken agent (Andrewes, 1971, p. 648).

Discussing the findings with friends from his home state of Iowa, Shope learned that midwestern wild rabbits, referred to as 'horned' or 'warty,' often presented horn-like protuberances over various parts of the body. Shope recruited locals to ship excised 'warts' as soon as they could find a case of the disease. Upon receiving his first sample, Shope grafted fragments of the wart under the skin of domestic rabbits. Microscopic examination confirmed his suspicion that the warts were small non-spreading masses of skin tissue. At this stage, the phenomenon was simply one more example of a natural mammalian abnormal outgrowth, something that a medical pathologist might describe as a neoplastic but not a cancerous growth. What made it interesting to Shope was that these 'warts' manifested a contagious disease, not quite an infectious cancer but at least a transmissible growth (Shope, 1933).

At the time of these initial findings, Shope was immersed in epidemiological studies of swine influenza, and he was inclined to investigate the 'warts' infection in like manner—by following experimental transfer of the disease. Passage was Shope's key asset. It could be used to refer to the successful, artificial transmission of a disease from one laboratory animal to another by means of inoculation; it could also refer to the natural transfer of an infectious agent from one type of host to another. Passaging bacteria and viruses in the laboratory was a means to make visible unseen infectious agents as well as the invisible steps in the parasitic vectors' reproductive cycles. This strategy had been most effective in the case of swine influenza. Through inoculation studies in the laboratory in concert with epidemiological studies of the disease in the field, Shope elucidated the cycle of swine influenza virus: sick pigs infect their natural parasites, the lungworms; the worms cast off eggs that carry the virus into the pig's feces; these eggs are eaten by earthworms, which in turn are ingested by new pigs; then any physical shock—exposure to hard rain or cold—give the pigs influenza.[4]

In the case of the rabbit warts, however, the promises of passage did not last long. Shope had been successful in inducing skin growths by inoculating domestic rabbits with his wild material, but he could not manage to pass the tumor to other rabbits following this first transfer. For all practical purposes, the system was a dead-end. Even when Shope arranged to receive regular live wild rabbits with warts from Kansas, he was unable to pass the growths serially between domestic rabbits by means of filtered extracts.[5] Because passage from wild to domestic rabbits rendered the papilloma virus uninfectious, Shope became convinced that this case demonstrated specificity in the host-virus relationship. The tumor agent was transparent—beyond the reach of the filtration-inoculation techniques of visualization (Shope, 1933).

Believing that papilloma virus was still lurking in the nodules of the inoculated rabbits even through they failed to transmit the disease, Shope kept tinkering with the system. In 1937, Shope provided evidence that 'second-passage' domestic rabbits—those inoculated with non-infectious papilloma extracts of domestic rabbits previously infected with cotton-tail rabbit outgrowths—were protected against the same virulent virus from cotton-tails. Immunological specificity was taken as evidence of completed passage; resistance was a proof of the continuity of the virus. According to Shope, the virus was continuously transmitted, but its presence was 'masked' for an unknown reason:

> The outcome of these immunization experiments makes it clear enough that the usual non-transmissibility of papillomatosis serially in domestic rabbits is due to no lack of virus in the domestic rabbit papillomas. It is referable instead, to the efficient masking in some unknown fashion of the virus they contain. It seems possible that this masking of virus is the end-result of the host-parasite antagonism and may represent a defence utilized by the domestic rabbit against unrestrained parasitism by the agent of papillomatosis (Shope, 1937, pp. 230–231).

Thus, masking accounted for the invisibility of the virus. This putative phenomenon, drawn from Shope's epidemiological reasoning, became important also for pathological experimentation in Rous's laboratory.

Rous's Pathological Images

Soon after his discovery, Shope offered Rous his tumor material: 'I have been wondering for some little time whether you would care to have this filterable rabbit "tumor" that I have been working with lately ... [T]his letter is sent to you merely as a "feeler" in case you really would like to have the virus and had not felt like requesting it of me.'[6] Rous was indeed interested in the virus, which Shope passed on to him, but not on account of the puzzling failure of passage between rabbits. What appealed to Rous in the papilloma system was the possibility of turning Shope's material into a reservoir of viruses that could be made into an acceptable cause of mammalian cancer. Addressing cancer pathologists in the first place, Rous needed pathological visibility, in the form of typical lesions and good microscopical slides that would be convincing to physicians.

The first round of morphological studies Rous conducted, with his assistant Joseph Beard, were disappointing. There was no malignancy: the papillomas did not metastasize. The growths remained localized and sometimes actually dried up and fell off without much adverse effect on the rabbit's health. Histological studies provided a means for rationalizing this lack of spreading. The papillomas did not show undifferentiated cancer cells but rather an irregular mass of epithelium cells that followed their normal pattern of differentiation. As a thick outer circle of keratinized cells developed, growth stopped (Beard and Rous, 1934). However, with the hope that the more active papillomas 'would eventually become carcinomatous,' Rous selected rabbits for long-term observation. As he recounted, 'the change occurred soon, within less than six months, in seven of ten domestic rabbits. The cancers have been multiple in the majority of instances.'[7] Photographs of metastasizing papillomas and micrographs of irregular mass of flat 'carcinoma' cells underlying the classical papilloma were arranged according to an increasing gradient of 'cancerous' cells in order to argue that a virus-induced growth could progress into a real cancer in a manner analogous to 'instances of a graded alteration from papilloma to cancer in human pathology.' (Rous and Beard, 1935) (See Figure 6.1.)

Photographic images notwithstanding, Rous faced a dilemma in advancing the system as a model for viral-induced cancer. The link between inoculation of the virus and appearance of an actual cancer seemed incidental rather than direct. Extracts prepared from papillomas only induced non-cancerous growths upon inoculation into new hosts. Once the papilloma growth had progressed to a weak cancer (a step that appeared to be unrelated to inoculation), ground and filtered extracts no longer induced new growths at all. Even grafting of the cancerous material did not produce new tumors in a reliable fashion. Thus the progression to cancer appeared to eliminate the virus, threatening the postulated link between a new virus and a mammalian cancer.

One possible solution was to keep looking for a virus in established tumors and try to visualize it by other means. The simplest available techniques for this approach in a pathology laboratory with rabbits were immunological, and John Kidd, another of Rous's assistants, embarked on a series of immunological tests. He demonstrated that rabbits with transplanted cells from a papilloma-derived cancer produced sera capable of protecting unimmunized rabbits against the Shope papilloma virus. Obtaining new strains of virus from Shope and inbreeding the rabbits 'to get a better strain for transplantation' helped improve passage and immunological detection.[8]

Because the virus was now visualized by the very same technique that Shope used, Rous and Kidd began also borrowing the 'masking' interpretation. In 1939, Rous commented on their most malignant cancer transplant: 'Though masked it [the virus] persists and increases in transplanted cancer deriving from the papillomas of domestic rabbits, as tests of the new host have shown. One such cancer is now growing rapidly in the thirteenth successive group of animals to which it has been transferred.'[9] At the same time, the notion of masking was adapted to the local situation. The concept was employed to address the visible pathological event, namely the transition from growth to cancer. Rous was now explaining the transition as a change of the virus: 'Yet under the ordinary circumstances of its action ... the virus produces only papillomas as standardized as so many Ford engines ... [T]here is a

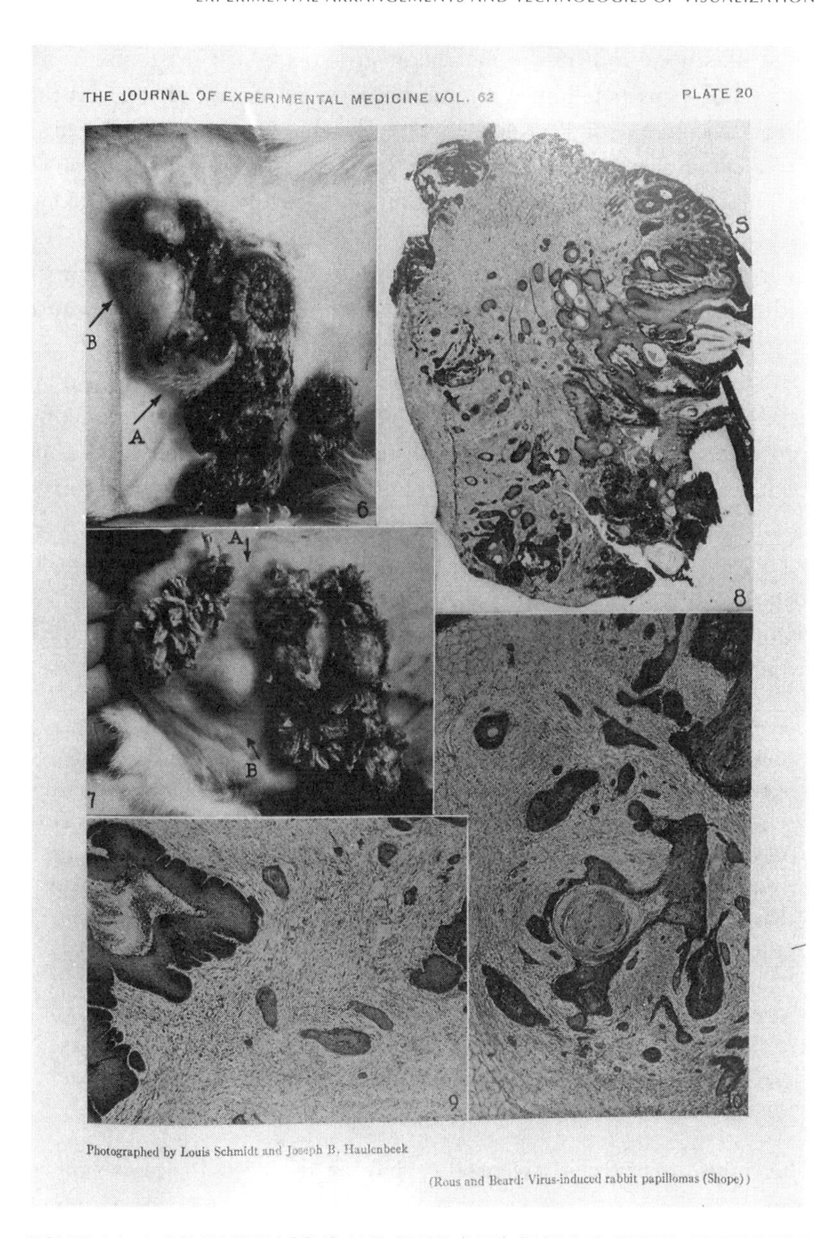

FIGURE 6.1 A FIGURE FROM ROUS AND BEARD (1935) SHOWING SEVERAL HISTOLOGICAL SECTIONS FROM RABBIT PAPILLOMAS THAT HAVE BECOME MALIGNANT. ALL SECTIONS WERE STAINED IN EOSIN AND METHYLENE BLUE; 'S' INDICATES THE SIDE OF THE SKIN SURFACE. PHOTOGRAPH 6 SHOWS AN ULCERATING CANCER THAT HAS REPLACED A PAPILLOMATOUS MASS; PHOTOGRAPH 7 SHOWS A FURTHER BIOPSY OF CANCERS A AND B IN PHOTOGRAPH 6. PHOTOGRAPH 8 SHOWS A SECTION OF A, A MALIGNANT PAPILLOMA THAT HAS BROKEN UP INTO SQUAMOUS CELL CARCINOMA AND INVADED MUSCLE TISSUE. PHOTOGRAPHS 9 AND 10 SHOW FURTHER SECTIONS OF THE SAME TUMOR, THE SECOND OF WHICH SHOWS SPREADING AFTER IMPLANTATION INTO OTHER MUSCLE TISSUE. REPRODUCED FROM *THE JOURNAL OF EXPERIMENTAL MEDICINE* BY COPYRIGHT PERMISSION OF THE ROCKEFELLER UNIVERSITY PRESS.

great likelihood that in the malignant growths which originated from the cell of virus papillomas are the outcome of variant changes in the virus associated with these elements.'[10] The change that turned the virus into a carcinogenic agent was not, in Rous's view, a simple alteration of the virus but rather a transformation akin to the previous masking because it was taking place in an abnormal and changing milieu: 'The environment of the virus tends to become strange, ... the papillomas with which it is associated becoming more and more abnormal because of the local disturbances they set up.'[11]

By the time that Rous adopted Shope's masking perspective for rabbit papilloma, he was promoting a three-way explanation of cancer, taking into account a causative virus, a disturbance of the cellular environment, and what others called carcinogens, like coal tar and other chemicals, which he viewed as mere triggering agents. The mouse milk agent exhibited this kind of complex etiological relationship to cancer, inspiring Rous to develop this model system in concert with that of rabbit papilloma.[12] He reported his current thinking in the 1941 scientific report of the institute director:

> For experimental purposes Dr. Rous, Dr. Kidd and Dr. Friedewald consider tumors in general as due to viruses. This hypothesis has led them to a broad scrutiny of neoplastic phenomena ... The eligibility to mammary cancer in mice of some strains is known to be conferred on the suckling young by way of the milk, and there are facts indicating that the animal body contains resident viruses, just as it contains resident bacteria, which ordinarily do no harm. These viruses gaining entrance to the body, perhaps shortly before or after birth, may persist in association with the cells which they happen to reach, and produce no injury unless subjected to abnormal cytological conditions such as may be brought by the action of 'carcinogenic agents' so called. Then at one spot or another the virus may undergo variation, cease to be a mere commensal, and so influence the cell with which it is associated that it become a neoplastic cell.[13]

The mouse experiments were also shaping the pathological vision of epidemics: Ludwig Gross, working on the mouse mammary tumor system, used it to define a new class of infectious diseases characterized by 'vertical epidemics,' the transmission of latent infectious agents from one generation to the next (Gross, 1946, 1949).

By the early 1940s, the resurrection of the viral theory of cancer was associated with the development or the transformation of three experimental systems: the rabbit papilloma virus, the mouse milk tumor factor, and the Rous chicken sarcoma agent. Work on all three tumor virus systems relied on similar forms of experimentation, which may be described in terms of pathological work. Rous, Shope, and Gross employed similar sets of instruments and procedures; this 'pathological arrangement' visualized 'filterable' tumorigenic agents transmitted from body to body in the form of irregular tissue masses and microscopical slides of spreading cells (e.g., Figure 6.1). The core practices originated in medical bacteriology and consisted of the maceration of tumor tissues, the preparation of filtered extracts, and the inoculation of test animals which—when successful—was followed by the formation of growths that could be cut, sliced, stained and mounted to be displayed as histological evidence of cancer. These basic components were associated with immunological procedures which provided alternative traces of viruses when inoculation failed to induce disease. On the pathological scene, these procedures were taken to reveal infectious agents that were passed among experimental bodies in essentially the same way they circulated in the 'field.' One key component of the pathological arrangement was the existence of reference collections against which new findings could be calibrated. These reference collections included slides of cancerous tissues, and animals that hosted particular virus strains or were immunized against them, as well as inoculation records.

Researchers in this pathological mode of cancer research shared many techniques—for instance, the filtration and inoculation procedure—with medical scientists performing studies of other animal viruses. The pathological setups for tumor viruses also had commonalities with other research practices of cancer specialists, such as reliance on histological staining. The combination of these methods and tools grounded a new regime of explanation for cancer that is best illustrated by the genesis of the 'masked' virus hypothesis. In the case of the rabbit papilloma virus, the discrepancies between pathological and immunological techniques of visualization were superseded with the 'masking' theory, which granted the tumor virus a lack of virulence due to its significant immunizing power. Masking had another advantage: it could be linked to 'latency' and to the low predictability of cancer 'passage,' which were, at that time, critical features relevant to the transmission of the scourge outside the laboratory.

THE MACROMOLECULAR VIRUSES

> The new whirligig of science, the ultracentrifuge, . . . extracts into plain view disease viruses which heretofore could not be seen even with microscopes. (*The New York Times*, 29 June 1937)

Ultracentrifuges and Viruses at the Rockfeller Institute

During the same period that Rous, Beard, and Kidd sought to track the hidden papilloma virus, new machines for visualizing pathological agents as macromolecules were being built and employed at the Rockefeller Institute. In the wake of the Svedberg's invention of the oil-turbine analytical ultracentrifuge as a precision instrument for observing proteins, other physicists had begun to develop ultracentrifuges with an eye towards biomedical application. Jesse Beams and his student Edward Pickels, at the University of Virginia, adapted Henriot and Huguenard's 'spinning top' design to make a centrifuge rotor supported and driven by air (Beams and Weed, 1931). Pickels was most interested in the prospects of the air-driven ultracentrifuge in medical research, and concluded his dissertation: 'For the study of biological materials, particularly filterable viruses and bacteria, the contributions may be a definite step in opening an entirely new field of research.'[14]

Beams and Pickels's efforts attracted the attention of Johannes Bauer, on the research staff of the Rockefeller Foundation's International Health Division investigating yellow fever virus. Pickels began working with Bauer in 1934 and was formally appointed to the laboratory staff as soon as he completed his Ph.D.[15] There Pickels began change the specifications of his ultracentrifuge to suit research on viruses (Bauer and Pickels, 1936).[16] Because the Virus Laboratory of the International Health Division was housed at the Rockefeller Institute, the presence of Pickels brought a diverse group of medical researchers into contact with the ultracentrifuge. Virus researchers were strongly represented among the laboratories of the institute, and Pickels began collaborating as well with Thomas Rivers and Joseph Smadel of the institute's hospital (Smadel *et al.*, I. and II., 1938; Smadel *et al.*, 1939; Smadel *et al.*, 1940).

The head of biophysics at the Rockefeller Institute, Ralph W.G. Wyckoff, was also eager to develop the air-driven ultracentrifuge for medical research.[17] Following the widely cited publication of institute researcher Wendell Stanley's isolation of crystalline tobacco mosaic virus (TMV) in the summer of 1935 (Stanley, 1935; Kay, 1986), Wyckoff and Stanley began to collaborate on ultracentrifugal studies of TMV.[18] Through the spring of 1936, Stanley was sending Wyckoff samples of TMV, including different strains of the disease and preparations from hosts other than tobacco. Despite recurrent equipment problems,[19] Wyckoff's runs showed TMV to be a homogeneous protein with a molecular weight between 15 and 20 million daltons

(Wyckoff *et al.*, 1937).[20] Partly in response to this promising collaboration, Wyckoff moved his centrifuges to Princeton.[21]

In Princeton Wyckoff met with another institute associate who had transferred from New York. In the fall of 1935, Joseph Beard, Rous's assistant, moved to the Department of Animal Pathology at Princeton to continue his work on the Shope papilloma virus and to benefit from a collaboration with Shope himself. Since Stanley's crystallized plant virus was still in the headlines, animal pathologists were racing to achieve the first pure animal virus. As Shope wrote Rous that summer, 'Have you read Stanley's announcement that he has crystallized tobacco mosaic virus[?]. It sounds like the real thing. The papilloma virus, with its stability towards heat and prolonged storage, seems to me one of the best animal viruses to try to put through the same paces and I think Beard and I will take a fling at that next fall (or Beard alone).'[22] Rous, who was entirely committed to values of pathological experimentation, was initially unenthusiastic about the macromolecular approach: 'For Beard's career the attempt to purify a virus would, in my opinion, be nothing less than suicidal.'[23] Carl TenBroeck, head of the Princeton department, disagreed with Rous's pessimism; he encouraged Beard's efforts at purifying the rabbit papilloma virus even when his initial attempt at using chemical precipitation with John Northrop failed.[24]

Around this same time an unexpected finding by Wyckoff and his coworker Robert Corey opened up a further possible use of the ultracentrifuge for virus work. They found that when they centrifuged clear juice from Stanley's mosaic-infected tobacco plants at 25,000 r.p.m., a pellet of fibrous material sedimented to the bottom of the cell, that, when analyzed in a microscope, was seen to be composed of crystals (Wyckoff and Corey, 1936). Thus the ultracentrifuge could be used to purify, even crystallize, viruses from serum or sap. This method was especially promising for plant viruses that were too dilute or fragile to withstand chemical extraction,[25] but its utility for animal viruses was also quickly proven. On February 15, 1936, Rous sent his approval for Beard's attempt to prepare papilloma virus using the ultracentrifuge, an experiment that was in fact already being performed. His letter stressed the symbolic value of crystallization: 'Just now I have [read] the Stanley-Wyckoff article ... It's good stuff and reading it has told me somewhat more about the concentration of the papilloma virus that it was possible to learn in our hurried talk ... Best luck to you and Wyckoff. I hope the virus soon crystallizes. Otherwise people will ask whether you have done more than merely concentrate it in association with a pathological protein.'[26]

Shortly thereafter, Beard and Wyckoff announced in *Science* that they had purified 'a homogeneous heavy protein from virus-induced rabbit papillomas' (Beard and Wyckoff, 1937). Because of the small amount of material they had isolated, the conclusions were provisional. As Beard wrote Rous, 'After six ultracentrifugations there was just enough for one attempt for the X-ray [analysis of the sedimented material]. Dr. W. interpreted the photograph as showing definite evidence of crystallization but the findings were not conclusive. Material is scarce, and we shall do no further work of this sort this year. The best we can do is to study the chemical properties of the virus to prepare the way for large scale operations next year.'[27] Further studies of purified papilloma in the ultracentrifuge showed that it sedimented as a homogeneous species; papilloma had successfully followed TMV onto the macromolecular arrangement. Wyckoff viewed the work with papilloma as illustrative of the general applicability of the ultracentrifuge, and Stanley became interested in the virus-cancer relation (Wyckoff, 1937).[28]

Beard at Duke: Chicken Leukemia and the Quantifying Spirit of Postwar Virology

The subsequent turmoil and organizational innovation occasioned by the mobilization of science for World War II did not leave the emerging macromolecular infrastructure untouched.[29] Over the course of the war, the nature and uses of the ultracentrifuge were altered in at least three different ways, which in turn affected the postwar fate of the ultracentrifuge-based tumor viruses. First, the preparative value of the apparatus was strongly reinforced by the numerous war projects in which the ultracentrifuge was used as a tool for purifying biologicals, including vaccines. These projects in turn enhanced the serial production of the instrument and its dissemination (Creager, in press). Second, as indicated by the considerable research efforts towards development of an influenza vaccine, immunological means of disease prevention were central to the war effort (Gaudillière, 1998). Within this context, the ultracentrifuge was not only closely associated with immunological reactions and methods of assessing the specificity of viruses, but also became a widely accepted base for the purification of macromolecules: Photographs of ultracentrifuge sedimentation boundaries and computations of molecular size and weight became key signs of both macromolecular behavior and vaccine purity. Finally, by the end of the war, as biological electron microscopy was expanded by virtue of RCA support and instruments, the ultracentrifuge was combined with the new imaging technology as part of an enriched macromolecular set of machines (Rasmussen, 1997).[30]

Before the outbreak of the war, in his new laboratory at Duke University Medical School, Beard continued research on the rabbit papilloma virus, visualizing the virus protein in the ultracentrifuge, the diffusion chamber, and the electrophoresis apparatus (Beard *et al.*, 1939; Sharp *et al.*, 1942, 1946). The protein chemist Hans Neurath, also at Duke, provided Beard with local expertise for interpreting the hydrodynamic behavior of the protein in the ultracentrifuge and diffusion chamber. Rabbit papilloma appeared as a homogeneous macromolecule, with a molecular weight of 47 million daltons calculated from these two techniques (Neurath *et al.*, 1941). Even though rabbit papilloma gave a sharp boundary in the ultracentrifuge, its sedimentation constant varied from run to run. But as the researchers noted, the variations were no more 'than those encountered in sedimentation studies of plant viruses,' which provided the standard of macromolecular purity for the field of virus research. Beard had access to the whole panoply of equipment for physicochemical characterizations of macromolecules, but the ultracentrifuge, as the instrument which made the virus molecular, was given some epistemological privilege. As he observed in an review on purified animal viruses, 'ultracentrifugation has been the principal procedure for the purification and physical study of viruses.' (Beard, 1948, p. 52) The same equipment and approaches were central to his work on influenza virus during the war, as part of the Army's vaccine development effort (Board for the Investigation and Control of Influenza and Other Epidemic Diseases, 1941).

After the war, Beard resumed work on animal tumor viruses as macromolecules. Following a short-term postwar collaboration with W. Ray Bryan from the National Cancer Institute, who was investigating the chicken sarcoma virus, Beard started to consider alternative models. There were few candidates. An opportunity to develop a new system arose in the late 1940s when Edward Eckert, a NCI scientist, moved to Duke University with a project concerning chicken leukemia. Early on, Eckert and Beard developed a cell-free technique for transmitting the disease by means of filtered plasma inoculation (Eckert *et al.*, 1951). A relatively simple ultracentrifuge preparation of the infective fraction showed that a homogeneous species was sedimenting, and electron micrographs displayed a simple distribution of particle sizes. Based on this good fit with expected macromolecular patterns, the chicken leukosis agent became a tumor virus.

After promising beginnings, the pursuit of a macromolecular virus slowed when Eckert and Beard recognized that they could not standardize the pathological assay for the virus in any straightforward way. During the war, because of its relationship to vaccine production and standardization, biological metrology, in particular the measurement of infectivity, had become a major concern. Linear dose-response relationships—between the amount of viral inoculum and the number of animals developing the disease—were increasingly viewed as important criteria in drawing etiological conclusions. With the chicken leukosis virus, however, consistent dose-response correlations could not be obtained. Attempts to better standardize inoculation procedures did not prove of much help, so Eckert and Beard turned to other elements of the system. In 1954, they adopted a new inbred line of chickens that had been selected for high susceptibility to another blood disease. They produced few eggs and failed to hatch those they laid (as the authors commented 'As chickens, the birds were relatively poor specimens.'), but they were highly susceptible to the leukosis agent and the infectivity counts became more consistent (Eckert *et al.*, 1954a). Beard and Eckert also altered the way in which infectivity was represented to generate a more linear relationship between virus dosage and the lag-time that separated inoculation from the appearance of the disease. From 1954 onward, a new set of standard curves was used to check the yield and efficiency of virus preparations (Eckert *et al.*, 1954a and 1954b; Eckert *et al.*, 1955a).

A more quantitative bioassay for the chicken leukosis virus did not secure the purity of the agent. Eckert complemented his immunological characterization of the chicken leukosis agent with the same sort of biochemical analysis Stanley and other researchers on macromolecular viruses were using, but a new and puzzling problem emerged by 1954.[31] Infectious fractions from the ultracentrifuge seemed to contain an enzyme reacting with ATP—one of the building blocks of nucleic acids and central to bioenergetics. Since no activity of this sort had ever been reported for a virus, there was great suspicion that the enzyme was a contaminant and that the virus preparation was impure even if it appeared homogeneous in the electron microscope and the ultracentrifuge. The researchers blamed the problem on capricious protein aggregation (Eckert *et al.*, 1955b). The enzymatically active fraction and the infective agent remained associated through multiple ultracentrifuge runs aimed at disengaging them. When macromolecular techniques for resolving the problem seemed to be exhausted, Eckert focused on immunological testing in order to distinguish the contaminating enzyme and the virus particles. The use of immune sera, however, produced another unexpected association. The ultracentrifuged extracts reacted with sera prepared with known antigenic materials from normal chicken cells. These results introduced a second series of putative contaminants that neither the ultracentrifuge nor precipitation reactions could dissociate from the virus. Triangulating between different representational devices, Eckert and Beard finally employed electron microscopic pictures to show that the spherical viral components were included in large aggregates formed during precipitation reactions with the various sera.[32] Thus they concluded that the associated particles were not contaminants at all, that 'material indistinguishable antigenically from normal chick (host) tissue ... are [sic] vital, intrinsic constitutional elements of the virus.' (Eckert *et al.*, 1955b, p. 635)

For this particular line of research, such a redefinition of the chicken leukosis virus was a hazard to the clear distinction between infective agent and host cell that physico-chemical instruments had operationalized. But more generally, the combination of machines, immunological reactions, chemical analysis, and purity which characterized Beard's research continued to serve as an experimental alternative to a pathological style of work. The potential conflict between this vision of the macromolecular cancer virus with the representation of tumor viruses through the pathological arrangement is illustrated by Beard's attack on Shope's masking concept.

In a talk Shope delivered at a Caltech conference on viruses in 1950, he stressed the inherent differences between animal viruses and plant and bacterial viruses. Not only were animal viruses less simple and tractable than the 'macromolecular' plant and 'genetic' bacterial viruses (see below), he argued, but host variability, complex patterns of interference between animal viruses, and host-virus interactions meant that research on animal virus systems would always be more complicated. Analogies between systems should be consequently qualified; it was 'questionable how much light phage research can throw specifically on the events of other virus infections[.]' (Shope, 1950, p. 79) Recapitulating his definition of masking based on the rabbit papilloma case, Shope insisted on the peculiar combination of immunological reactivity and non-transmissibility that made the 'masked' virus a non-infective but carcinogenic agent and efficient antigen. Since conventional inoculation procedures were far from efficient for visualizing such latent and hidden viruses, it might well be that 'agents more elusive than the conventional tumor viruses may exist and may even be quite common so far as anything we know.' (Shope, 1950, p. 82) In other words, a general masking process typical of animal viruses passing through different hosts—a process not too different from Gross's vision of vertical transmission—might be responsible for the variable incidence of cancer in animals.

In Beard's estimation, Shope's claim was a wild generalization:

> [T]he early concept of masking has been expanded inordinately, without benefit of continued investigation or explanation, to include widely and entirely unrelated problems ... The masking theorist would imply that virus may be present in human cancer in only masked and, thus, hidden form; that it may be a variant virus from another host species; or, indeed, that the relationship between virus and host may be any one or all of those which have been conjured up for the papilloma and other animal tumor viruses. (Beard, 1956, p. 281)

Beard challenged masking by reinterpreting the observations as an issue of mere quantification. He argued that natural host variability could attenuate virus infectivity and replication (without necessitating specific immunological responses), resulting in widely varying numbers of virus particles in cotton-tail and domestic rabbits. As evidence he cited the variable amounts of papilloma virus isolated through the ultracentrifuge from different rabbits. In this view, domestic rabbits were not immunological resistant, but rather their low-level virus infection was not detected by researchers: 'The relations described here show that wholly infectious virus could have been present in concentration of approximately 560 million or less/gm of wart tissue and still have failed to induce growths in the most susceptible rabbits. *Any amount of virus which might have existed in domestic rabbit warts below this threshold of detection was that which was supposed to be masked.*' (Beard, 1956, p. 285, emphases in original) The link between particle number and infectivity was visualized by a figure that represented wart virus content as mass or volume, a schematic easily misread as a picture of relative particle size (Figure 6.2).

Beard attributed such variable viral replication not to immunological activity but to genetics: 'The influence of the genetic status of the host, with respect to the individual, to the strain, and to the species, on tumor virus infection and multiplication can scarcely be overemphasized.' (Beard, 1956, p. 287) This account left out the immunological explanations and the attention to variability in pathological processes

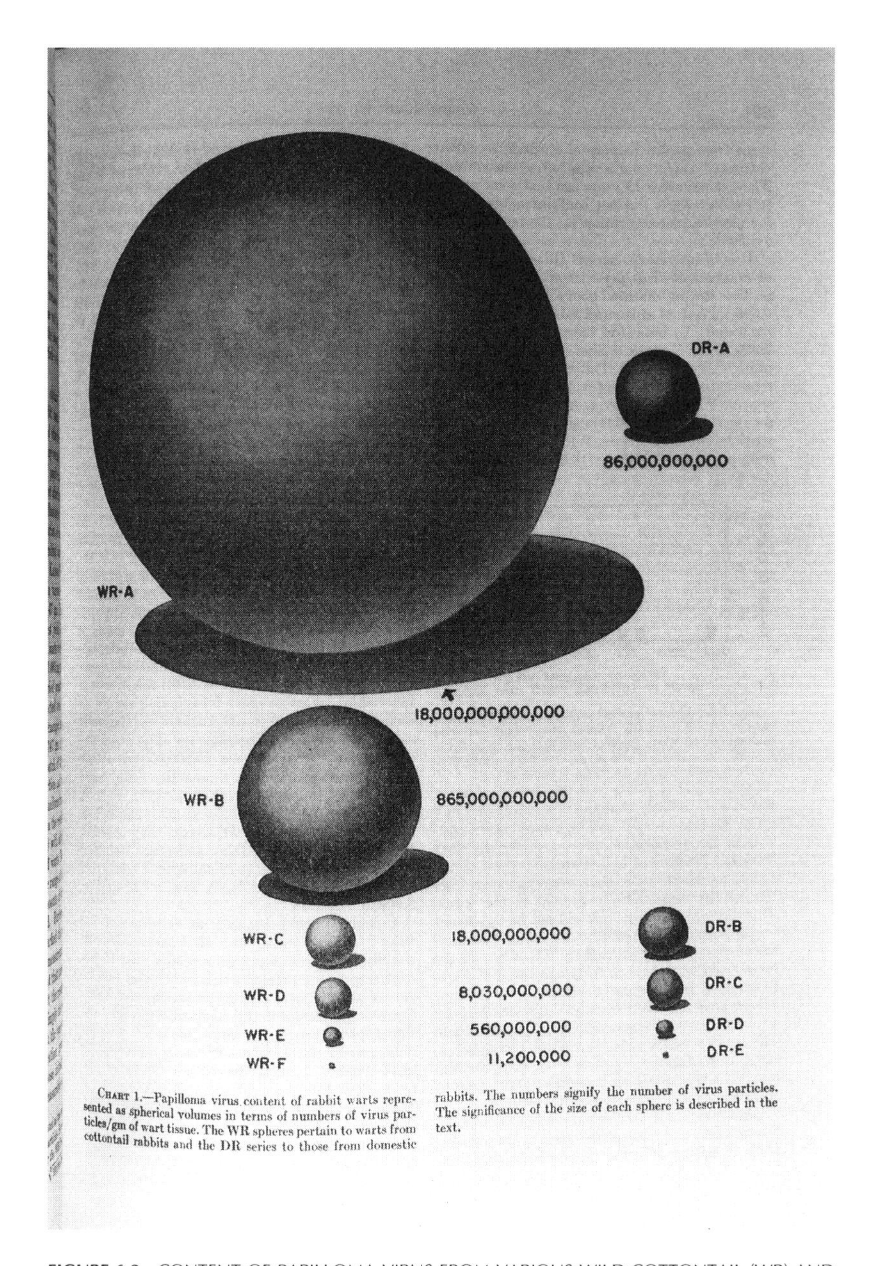

FIGURE 6.2 CONTENT OF PAPILLOMA VIRUS FROM VARIOUS WILD COTTONTAIL (WR) AND DOMESTIC (DR) RABBIT WARTS. THE NUMBERS SIGNIFY THE NUMBER OF VIRUS PARTICLES, ALSO REPRESENTED BY THE RELATIVE VOLUME OF EACH SPHERE. BEARD (1956), P. 283. REPRODUCED FROM *CANCER RESEARCH* BY COPYRIGHT PERMISSION OF THE AMERICAN ASSOCIATION FOR CANCER RESEARCH, INC.

that were characteristic of the bacteriological style. Generalizing from his own experience with chicken strains, Beard used the variation in virus infectivity to argue that genetic differences between individual hosts provided the key to the cancer problem.

Beyond its etiological significance, Beard's redefinition of 'masked' tumor viruses was fundamentally instrumental and quantitative. It culminated in a strong plea for the sort of metrology that had expanded under wartime virus production:

> From the foregoing considerations it is clear that the clarification of many of the most fundamental problems of the virus tumors is dependent on the application of quantitative methods of study ... Much of significance of bioassay is lost without correlation of biological activity with physical means or particle number of the etiological agents. With access to modern tools it is by no means impossible, though in many instances it is tedious and costly, to establish the relationships of mass and activity. (Beard, 1956, pp. 289–290)

In comparison to the complex relationship between transmission, infectivity, and tumorigenesis that characterized the vision of Rous and Shope, Beard was proposing a scheme in which tumor viruses differed in only one respect from other epidemic viruses: their low level of cellular replication and, as a consequence, the need for more systematic and powerful instruments for visualizing them. Beard's approach was a by-product of the World War II macromolecular production system, which linked biophysical machines, purification, and immunological prevention. It could also be viewed as a drastic simplification of the hidden tumor virus problem, blazing the trail toward an intensified technological approach to finding the viruses responsible for human cancer.

Unlike the earlier pathological arrangement for virus experimentation, macromolecular approaches did not precede work on papilloma virus, mouse mammary tumor virus, and chicken sarcoma virus. Investigations on these systems were gaining momentum in the 1940s, and the macromolecular arrangement offered an alternative means to materialize such viruses. Central to this approach was the ultracentrifuge as an instrument for both producing and visualizing viruses, as developed at the Rockefeller Institute around Stanley's exemplary plant viruses. Wyckoff and Beard extended this reliance on biophysical measurement in their identification of 'pure' animal viruses in the form of sharp sedimentation boundaries (e.g., Figure 6.3). Ultracentrifugation results were calibrated against images obtained with the electron

(A)

(B)

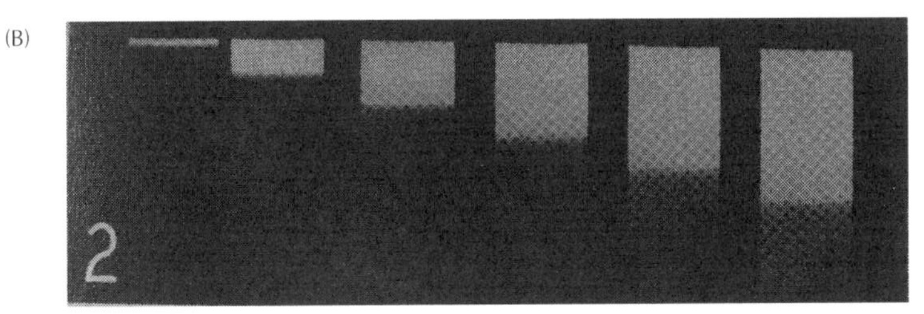

FIGURE 6.3 (A) VIEW OF R.W.G. WYCKOFF'S AIR-DRIVEN ULTRACENTRIFUGE AND ITS MOUNTINGS. BISCOE, PICKELS, AND WYCKOFF (1936a). REPRODUCED BY *THE JOURNAL OF EXPERIMENTAL MEDICINE* BY COPYRIGHT PERMISSION OF THE ROCKEFELLER UNIVERSITY PRESS. (B) THE SEDIMENTATION DIAGRAM OF A SALINE SOLUTION OF HOMOGENEOUS PROTEIN FROM HIGHLY INFECTIOUS RABBIT PAPILLOMA WARTS AT SEVERAL TIME POINTS OVER A SEDIMENTATION RUN. THE SHARPNESS OF THE BOUNDARIES AND THE FACT THAT NO CHANGE COULD BE SEEN IN THE SEDIMENTATION PROFILE AFTER STORING THE PREPARATION FOR 36 HOURS IN THE ICE BOX ATTEST TO THE PURITY OF THE AGENT. BEARD AND WYCKOFF (1938). REPRODUCED FROM *THE JOURNAL OF BIOLOGICAL CHEMISTRY* BY COPYRIGHT PERMISSION OF THE AMERICAN SOCIETY FOR BIOCHEMISTRY & MOLECULAR BIOLOGY.

microscope and complemented with standardized immunological assays.

Thus in contrast to the pathologist's reliance on organisms, the macromolecular arrangement focused on instruments developed by physicists and on standardized quantitative practices.[33] These new modes of visualizing tumor viruses contributed to the postwar displacement of questions about the existence and etiological role of the tumor viruses (How do the tumor viruses relate to other cancer causes? Are they peculiar to a few animal species? Why is transmission rarely visible?) by technological questions (What are the best means to identify and prepare isolated and pure tumor viruses? What is the amount of viruses in the tissues? How can immunological prevention be achieved?). Once redefined in engineering terms, the search for animal (and human) tumor viruses was not only feasible, but it also matched the rapid development of American biomedical research and the culture of technological mobilization, culminating in the massive virus screening programs launched by the National Cancer Institute (NCI) in the 1950s and 1960s (Gaudillière, 1998).

THE GENETIC VIRUSES: THE (SECOND) JUNCTURE WITH MOLECULAR BIOLOGY

Plaquing the Tumor Viruses: The Genetic Visualization

If the macromolecular vision of viruses represented the original point of convergence between the 'molecular biology' emerging from the Second World War and cancer research, it was not the last. The genetic research of bacteriophage became an equally important model for investigating cancer viruses. By the early 1950s, once lysogeny was no longer an artifact but a real phenomenon, the integration of phage into bacterial chromosomes and subsequent replication within the host came to be used as an illustration of the possible fate of mammalian tumor viruses.[34] Most historical accounts of this convergence present phage as a theoretical model for conceptualizing cancer. By contrast, we will indicate how the practice of plaquing was integral to a new experimental arrangement that could be used to visualize tumor viruses.

The plaquing of bacteriophage on a lawn of bacterial cells in a Petri dish was devised by Emory Ellis at Caltech using a traditional bacteriological implements and methods (Ellis, 1966, p. 57).[35] Plaque counting was further developed through Ellis's collaboration with Max Delbrück into an assay that visualized the hypothetical phage life cycle (Ellis and Delbrück, 1939). A plaque, or hole in a layer of bacterial cells, represented the progeny of a single virus particle which had subsequently replicated, killing the infected bacteria. Plotting the number of plaques against the dilution of the phage sample yielded linear 'dose/response' curves. Moreover, this simple test could be employed to identify and study genetic strains of bacteriophage. Following the exemplar of *Drosophila* genetics, Caltech workers attempted to produce (and map) phage mutants showing, for instance, unusual plaques.[36]

It was the genetic rather than bacteriological foundation of phage research that Delbrück stressed when presenting the system. Nonetheless, through postwar patronage the practice of phage plaquing was extended to include study of human pathogens. Beginning in January 1947, the National Foundation for Infantile Paralysis (NFIP) began sponsoring virus research at Caltech.[37] One argument for bringing Delbrück back to Caltech at that time (after several years at Vanderbilt University) was precisely that he 'could take over an important segment of the program to be financed by the $300,000 grant just made to the Institute by the NFIP.'[38] In addition, the James G. Boswell Foundation had given Caltech substantial funding for building up a research program on animal viruses (Kevles, 1993).

In 1950–51, Delbrück and Renato Dulbecco visited virology labs on the East Coast to consider which animal viruses were worthy of their study.[39] As Daniel Kevles has pointed out, Dulbecco's invention of a plaque assay for the equine encephalitis virus built on the methods of animal virologists like George Gey, John Enders, and Wilton Earle (Kevles, 1993). Dulbecco's innovation was directly inspired by the bacterial assay. He managed to cultivate embryo cells in monolayers reminiscent of a bacterial cell lawn, to keep newly produced viruses in a discrete cellular vicinity, and to visualize the foci of cell destruction by staining. He established a linear relationship between the number of plaques and the dilution of the virus inoculum. Early results supported the notion that animal virus infection cycles were similar to phage multiplication: 'This result [the production of infectious spots in cell culture] on the one hand indicates that the plaque count is an efficient assay technique; on the other hand, it establishes a basic concept concerning animal virus action, namely, that infection of an embryo is produced by one virus particle.' (Dulbecco, 1952, pp. 751–52).

The NFIP, which had been supporting this work on animal viruses at Caltech, quickly provided assistance to extend the technique to the polio virus.[40] By 1954, an assay for the polio virus had been devised by

Marguerite Vogt and Dulbecco along the same lines: plaque formation, a linear relationship, and a single-particle infection model (Dulbecco and Vogt, 1954). Moreover, Dulbecco and Vogt suggested that the plaque assay could enable the selection of 'morphological' variants from cells infected with irradiated viruses, on the model of bacteriophage plaque mutants (Dulbecco and Vogt, 1955).

Polio was not the only medically relevant target. By 1958, the Rous Sarcoma Virus (RSV) had been 'plaqued' in Dulbecco's lab by Harry Rubin, a veterinarian and postdoctoral fellow, and Howard Temin, a graduate student (Temin and Rubin, 1958). In addition to the usual plaquing techniques, the assay relied on a special high-titer strain of RSV developed from chicken tumors provided by Bryan of the National Cancer Institute, a former collaborator of Beard. The main difference with previous work in plaquing was that the foci did not consist of an area of cell destruction but rather of a pile of growing cells no longer inhibited by cell contact inhibition. Each spot made visible both the replication of a virus and the multiplication power of 'transformed' cancer cells. (See Figure 6.4) The plaque procedure was therefore turned into a transformation assay that built-in the genetic vision of the 'enemy within': as refracted by the transformation-plaque procedure, cancer was caused by an infective virus that could enter cells, integrate into chromosomes, and trigger continuous growth.[41] One important aspect of the new system of visualization was that it relied on skills and procedures rather than on biophysical instrumentation, allowing it to travel faster than the macromolecular arrangement.

Conflicts and Common Ground about the Viral Etiology of Cancer

In 1958, Harry Weaver, the research director of the American Cancer Society, was inspired by the new work on animal tumor viruses to organize a conference on 'the possible role of viruses in cancer,' which took place in 1959. Stanley became the driving force behind the fifteen scientists serving on the organizing committee, a diverse group including the three historical figures in the domain (Rous, Shope, and Beard), bacteriologists working on infectious viruses (John Enders, Hilary Koprowski, Thomas Puck), molecular biologists working on phage (Salvador Luria, André Lwoff, Renato Dulbecco), and a specialist of cancer chemotherapy (Cornelius Rhoads).[42]

At the meeting, the phage delegation (Dulbecco and Luria) advanced their molecular genetic interpretation of vertical transmission that treated viruses as 'agents of infective heredity' (Luria, 1960, p. 679).

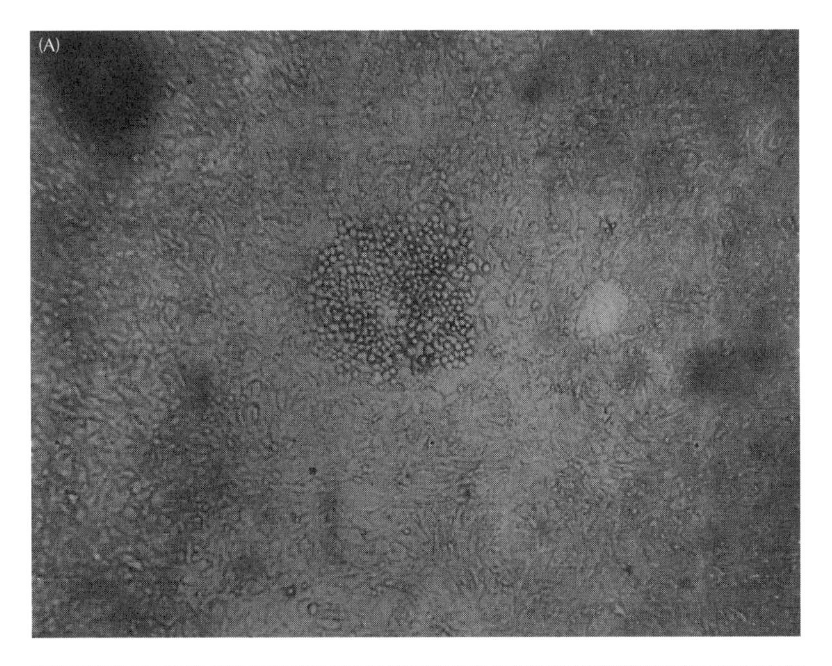

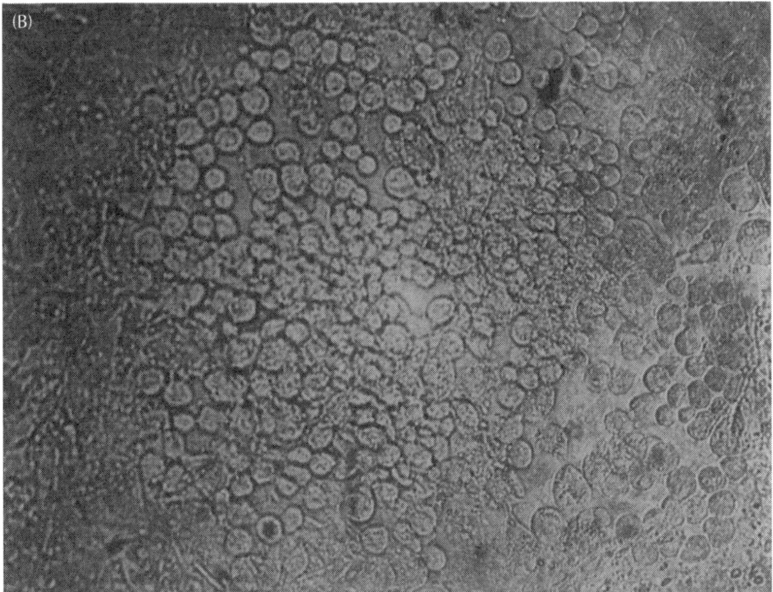

FIGURE 6.4 A SINGLE FOCUS OF ROUS SARCOMA CELLS ON A CONFLUENT BACKGROUND OF NORMAL CULTURED FIBROBLASTS SIX DAYS AFTER INFECTION BY A SINGLE RSV PARTICLE. (A) IS MAGNIFIED 25X; (B) IS MAGNIFIED 100X. RUBIN AND TEMIN (1959), P. 366, FROM *RADIATION BIOLOGY AND CANCER* EDITED BY THE STAFF OF THE UNIVERSITY OF TEXAS M.D. ANDERSON HOSPITAL & TUMOR INSTITUTE, COPYRIGHT © 1959. BY PERMISSION OF THE UNIVERSITY OF TEXAS PRESS.

> In the evolution of virus-induced cancers, . . . the question arises . . . whether exogenous infection, or 'vertical transmission' through the germ cell or the genital tract, or even *de novo* origin from noninfective proviruses or other cellular constituents, may be implicated for the presence of these viruses in the animals in which they are found. Especially with agents such as Gross's mouse leukemia factor, the narrow host-range specificity and the difficulty of transmission by extract suggest a remarkably inefficient adaptation to natural spread. (Luria, 1960, p. 678)

Central to the vision of tumor viruses as genetic particles interacting with or even integrated within the cell's DNA was the exemplar of bacteriophages. Luria conceptualized cancer agents as 'genetic elements . . . akin to the genetic materials of all cells' that interacted with cellular genes, modified the cell metabolism, and remained quiescent for a long time before reproducing and infecting other cells (Luria, 1960, p. 679). According to Luria and Dulbecco, this approach provided a vantage point from which it was no longer necessary to see the 'somatic mutation hypothesis and the virus hypothesis of cancer origin as alternative and mutually exclusive.' (Luria, 1960, p. 679) Rather, they would be reconciled in a general vision of cancer focusing on the role of endogenous cancer genes, small pieces of occasionally transferable DNA encoding the information for inducing an endless multiplication of cells.

This genetic theory of cancer causation was opposed by a well-known specialist in influenza viruses, the British bacteriologist C.H. Andrewes, who argued for the viral—exogenous, infective—nature of tumor viruses:

> I may not know very much about cancer—I know I don't—but I have been in the virus field for 36 years, and I can confidently dispute the suggestion that tumor viruses are in a class of their own . . . When [Luria] says that 'virus infection is a cellular mutation' I am personally at a total loss. I don't think it even occurred to me during the 1957 epidemic of Asian influenza that we were suffering in our hundreds of thousands from cellular mutations! (Andrewes, 1960, p. 690).

Echoing Beard and his emphasis on the variability of infectivity, Andrewes also contended that the 'genetic' hypothesis could do little to account for numerous features cherished by animal virologists and cancer pathologists, beginning with the variable effects of the 'same' tumor viruses in different hosts or the changes of virulence.

As director-at-large of the ACS since 1957 and frequent supporter of NCI's virus efforts before Congress, Stanley was probably the most prominent scientific spokesperson arguing for the viral etiology of cancer. By virtue of his role as scientist-advisor and of his diversified research commitments—purification and crystallization of the polio virus, biophysical instrumentation, nucleic acid studies—Stanley was in a good position at the meeting to mediate the virologists' disagreement. However, in his attempt to stress common ground, he failed to reconcile the differing methods and criteria. On the one hand, he defined the viral theory of cancer as a problem of alternatively 'masked' and 'unmasked' viruses, the pathological legacy. On the other hand, he stressed the importance of the biochemical and genetic perspectives for understanding the control of viral replication and the enzymatic changes determined by the virus's nucleic acids. In conclusion, Stanley pointed to the practical developments that linked phage and tumor viruses, infection and tumorigenesis:

> Certainly the mammalian cells and the viruses are now available for extensive research in areas comparable to transduction, lysogeny and lysogenic conversion in the bacteria-bacterial system, and fortunately some investigators are already at work ... [Dr. Rubin] has developed a precise and efficient method for infecting and altering cells *in vitro* and is using this technique for studying the relationships between the Rous virus and host cells in a manner similar to that exploited so successfully in the study of bacterial viruses. (Stanley, 1960, pp. 801 and 803).

In Stanley's view, the reconfiguring of older pathological tumor viruses in the manner of bacterial phages provided a nexus for consensus between branches of animal virology. The Caltech plaque assay for virus studies served to organize a new consortium of molecular biologists, animal virologists, and cancer researchers; this experimental arrangement was displacing the centrality of macromolecular instrumentation in Stanley's own laboratory.

Rubin and the Deficiencies of RSV

Since moving his laboratory to the University of California at Berkeley in 1948, Stanley had extended his program on the biophysics and biochemistry of viruses to include work on cancer viruses. In addition to macromolecular characterizations of papilloma virus (Knight, 1950; Schachman, 1951), workers in his Virus Laboratory began growing a number of human malignant cell cultures to study the effects on them of known viruses and to try to detect suspected tumor viruses (Stanley, 1958). The studies, however, were turning up chromosome changes in the malignant cell lines, pointing to genetic rather than to viral associations with cancer.

In 1960 Harry Rubin joined Stanley's laboratory to expand tumor virus studies. He began to use RSV to study the relationship between infection, replication, and focus formation (representing malignant transformation), using the assay he had developed at Caltech with Temin. The advantage of RSV over other model systems for virus-induced cancer was the rapidity of transformation: inoculation with the virus produced malignant growths promptly and directly, whereas other tumor viruses 'transform cells only occasionally and after a long delay in the animals they infect, as if the development of the malignancy were just an accidental by-product of their multiplication.'[43] (Rubin, 1964, p. 46) Even the limitations of the culture assay could be turned into meaningful phenomena: limits to the cancer-producing power of the virus might shed light on the biological mechanisms of tumor resistance. Although one could obtain a linear dose response of foci to virus concentration, there was an upper limit to the number of chick cells (10%) in a culture that would produce foci after exposure to RSV. One explanation for this result derived from cell culture asynchrony and the attendant inefficiency of infection (Rubin, 1960a).

Resistance to RSV transformation in cell cultures also occurred spontaneously in the laboratory. When Rubin attempted to determine the growth curve of RSV in chick cells in comparison with that exhibited by chicken leukosis factor, he found that many batches of cells would neither transform nor reproduce when infected with RSV. The use of growth curves drew on analogous experimental work by phage geneticists that mapped the multiplication of various bacteriophages. Now, it seemed that avian viruses showed patterns of interference resembling those of the T bacteriophages—Rubin was able to find in the uncooperative cells a 'resistance-inducing' factor (RIF), which turned out to be another virus, a naturally-occurring leukosis virus often endemic in chickens (Rubin, 1960a and 1960b; Friesen and Rubin, 1961). In an effort to isolate this factor, Rubin and his coworkers found yet another virus that was indistinguishable from RIF both in its appearance in the electron microscope and in its biochemical behavior but that was immunologically different. It was, however, immunologically identical to RSV, indicating that both viruses were composed of the same coat protein. This virus was named Rous-Associated Virus, or RAV. Attempts to separate RSV from RAV-infected strains were fruitless. In fact, transformed cells that were rid of RAV were also conspicuously lacking in RSV. RSV, it seemed, was defective in reproducing in the absence of RAV, although it could transform cells (Rubin, 1964, p. 51; Hanafusa *et al.*, 1963). These cells passed on the nucleic acid of RSV when they divided, and thus maintained the virus-induced transformed state, but they did not produce infectious virus (which required RAV coat protein). The seeming proliferation of related avian viruses continued, with an RAV2 being isolated in 1964.[44]

Two remarks on Rubin's growing set of avian viruses are in order. First, Rubin had adapted the bacteriophage model in specific ways. Once the plaque assay was working, he sought to understand unusual growth and infection patterns by looking for interactions between different viruses whose relationships could be established through immunological, biochemical, and biophysical techniques. In this sense his work on the plaque assay related to the macromolecular research dominant in the Virus Laboratory. Second, through this set of experiments Rubin and his coworkers translated the old 'masking' problem into new genetic terms according to the analogy with lysogeny in bacteriophage: a tumor virus could be latent and uninfectious, transmitted genetically as a resident of the chromosome. What was required to make this gene infectious, to allow it to be constituted as a virus, was co-infection with a helper virus. Thus the problem of masking could be understood in terms of DNA, and of the divisibility of the nucleic acid and protein parts of the virus (Rubin and Hanafusa, 1963; Rubin, 1965). A viral nucleic acid without a coat protein was simply a gene that could be a cancer-producing hitchhiker in the host genome. But if virus-induced cancer could be transmitted genetically in the absence of 'infection,' then 'masking' was just another name for the transformation of viral nucleic acids into endogenous genetic informat-ion, and the viral explanation of cancer was indeed a problem of circulating genes.

Rubin's vision of the defects of RSV as defining its malignant potential was soon challenged. Rous argued that the 'defectiveness' of RSV was both strain-specific and recent in origin, and Rubin, unable to isolate active virus from Rous's old stocks, had no means by which to demonstrate the generality of his findings.[45] More distressingly, Rubin's previous coworker on the project, Hanafusa, began reporting experi-mental evidence that 'transformation of cells and replication of the RSV genome are under the control of the RSV genome and do not require the intervention of helper virus.'[46] At the same time, the striking variability of RSV had unexpected benefits; in the early 1960s, 'nearly every strain of RSV has become capable within the last few years of causing tumors in mammals as they never could before.'[47] Also, as Rubin's specific interpretation of the role of helper viruses began to fade, the theoretical scaffolding based upon the plaquing assay was extended upward by many other workers. Hanafusa continued his studies of RSV at the Rockefeller Institute (renamed Rockefeller University) and used protein chemistry in conjunction with mutant selection (as enabled by the plaque formation assay) to find transforming proteins. Several groups adopted the mutant strategy and searched for defective viruses inducing impaired cellular growth or no growth at all. Thus the rearrangement of tumor virus research in terms of plaquing eventually opened the path to (viral) oncogenes, in the work of Howard Temin, Stephen Bishop, and Harold Varmus (Kevles, 1995; Fujimura, 1996).

In contrast to the first encounter between tumor viruses and molecular biology, which focused on macromolecules and machines, the second encounter was to a large extent the making of a new etiological theory. The infrastructure of this change was the plaque assay, which became a new procedure for visualizing tumor viruses in the late 1950s (Figure 6.4). Like the pathological arrangement, the transformation-plaquing assay rested principally on skills, such as the sterile manipulation of Petri dishes, cell cultures, and virus strains, rather than upon machines. Affiliated techniques from nucleic acid biochemistry, tissue culture, and phage studies (such as the use of mutant strains) completed the plaque assay as the basis for a molecular genetic mode of experimentation. As visualized by this procedure, the cancer-causing viruses became circulating nucleic acids that could enter chromosomes and control growth processes.

The molecular genetic arrangement was often used in parallel with the macromolecular arrangement and occasionally in the same laboratory. This however did not preclude the fact that these modes of visualization prompted divergent medical strategies. To combat an efficient attack on an enemy that was usually 'within'—in other words, 'masked' or 'latent' viruses—one could hardly rely on classical immunological tools, i.e. vaccines. From the mid-1960s onward, the new molecular dream was to beat cancer genes of viral origins on their own terms, namely by finding and neutralizing the cancer proteins that these genes putatively encode. This was to remain a powerful motivation until the rise of genetic engineering in the 1970s was accompanied by major reshaping of the genetic arrangement.

CONCLUSION

The problem of 'masking,' or more generally the unpredictable occurrence of viral diseases without retrievable etiologic agents, has been endemic in twentieth-century laboratories where researchers have sought to stabilize the relations of invisible agents to diseases. The converse, the issue of asymptomatic carriers of pathogenic microorganisms, has haunted medicine since the turn of the century, occasioning tense debates about the detection, natural history, and etiology of transmissible diseases (Mendelsohn, this volume). The history of medical bacteriology may accordingly be viewed as a history of 'unmasking' procedures beginning with serological analysis. However, the attribution of cancer to latent viruses strained the etiological credibility of germ theory even more than did the existence of asymptomatic carriers of typhoid or tuberculosis. Because human cancer was never a contagious scourge, the nature of its transmission pattern and the very existence of cancer viruses depended upon the ability of biomedical researchers to make sense of 'masking,' i.e. of the relationship between the absence of clinical signs or typical cancer lesions and the presence of virus.

The resurrection of tumor viruses in the 1930s was therefore intimately related to the invention of new modes of visualization of *hidden* cancer viruses and, for decades, discrepancies between differently visualized viruses were at the core of the debates about cancer causation. Conflicts caused by divergent images of cancer existed at several strata of biomedical activity: (a) between biological work and clinical activities—for instance between postmortem examination and serological tests; (b) between different techniques and instruments—such as immunological reactions versus electron microscopy; and (c) between different styles of theorizing—for example the opposing bacteriological and the genetic conceptions of viruses. Analyzing some of these conflicts, this paper has documented the displacements from a pathological approach to the cancer transmission problem towards machine-based visualization of viruses, and then to a postwar gene-based approach that placed a series of cancer viruses between infection and heredity.

We have interpreted these shifts in terms of three coherent modes of visualization—the pathological, the macromolecular, and the molecular genetic—whose infrastructures of tools, skills, and operations we have designated as experimental arrangements. On these three technical stages different material condensations of viruses appeared—in the guise of histological lesions, macromolecules, or cell culture plaques.

While we stress the distinctions between the three—papilloma as a pathological specimen differed from papilloma the macromolecule—they were not mutually exclusive. For example, researchers using macromolecular tools to retrieve cancer agents and those seeking them in cell culture plaques shared an instrumental commitment to quantification, despite frequent problems with reproducibility that were encountered with tumor viruses. Moreover, the experimental arrangements we have described sometimes occupied the same institutional locations: the pathological and macromolecular modes of visualization co-existed at the Rockefeller Institute in the 1930s, as did the macromolecular and molecular genetic modes in Berkeley in the 1960s. Nonetheless, conflicts between the approaches based on these arrangements emerged, and debates over the etiological significance and material reality of cancer viruses hinged on the divergent practices and epistemological commitments associated with these three experimental infrastructures. For example, the meaning and attractiveness of Shope's notion of masking varied among researchers in these three schools of practice. Whereas for pathologists masking handily explained why transmitted papilloma virus could neither induce growths nor be detected after cross-strain passages, for postwar biochemists it evaded the central issue of virus concentration in various hosts. For molecular geneticists, masking could be translated into the terms of viral DNA, which could integrate into the host genome and so avoid detection as an exogenous agent.

More broadly, thinking in terms of *experimental arrangements* is less a means to emphasize the trivial fact that medical findings are instrument-based than a way to stress three features of *modeling practices* in twentieth-century biomedical research. Firstly, instruments and technical procedures are never used in isolation but are calibrated against one another. In cases where the representations being developed from different instruments are in conflict, a hierarchy of tools is established, so that core instruments (e.g. the ultracentrifuge for macromolecules) supply the key reference data and others (e.g. the electron microscope) provide confirmatory information. Secondly, an experimental arrangement in our sense can support many different experimental systems; the effective visualization of one virus using one arrangement of tools and skills serves as an inspiration to image other viruses, such as cancer viruses, using the same combination. Finally, these different modes of visualization can be used to stage etiological agents for broader audiences, including physicians and patients. In our case, experimental arrangements not only constituted the material infrastructure of etiological claims but made the animal tumor viruses highly visible and so available to support (public) health enterprises.

ACKNOWLEDGEMENTS

We thank Ilana Löwy for her comments on an earlier version of this essay and Gail Schmitt for help in editing the manuscript. Thomas Rosenbaum and Lee Hiltzik of the Rockefeller Archive Center and Scott de Haven of the American Philosophical Society offered generous assistance to us in locating archival sources. The research was enabled, in part, by NSF Grant SBE 94-12291 and a Princeton University faculty research grant.

NOTES

1. Our emphasis in *experimental arrangements* is on the integration of machines, skills, and interpretation associated with the use of a particular set of instruments. See the conclusion below for a discussion of the relationship between experimental arrangements and experimental systems as analyzed by Hans-Jörg Rheinberger (Rheinberger, 1997, chapter 2).
2. See Simon Flexner, 'The Developments of the First Twenty Years of the Rockefeller Institute and An Outlook for Further Growth,' Scientific Reports, Rockefeller Archive Center (hereafter RAC) RU 439, Vol. 14, 1925–26, p. 28.
3. Flexner's pre-existing interest in the pathological study of tumors played a role in Rous' coming to the Institute (Corner, 1964, p. 59 and p. 109). On the reception of Rous' initial work on chicken sarcoma virus, see Corner, 1964, pp. 110–111 and Rous, 1967.
4. See Shope's reports in the Report of the Scientific Directors, RAC RU 439, Vol. 19, 20, and 21, 1930–33.
5. P. Rous lecture for the award of the Kobler Medal to R.E. Shope in 1957, RAC 450, SH 77, folder 'Biographical 1935–1963'; Shope, 1935, p. 830.
6. Shope to Rous, 10 May 1932, Shope folder #7, Francis Peyton Rous papers, American Philosophical Society (hereafter APS) BR77. For Rous' viewpoint on this episode see his presentation address for the Kobler Medal *op. cit.* n. 5.
7. Report of Rous's laboratory in the Report of the Scientific Directors, RAC RU 439, Vol. 23, 1934–35, pp. 32–37.
8. Report of Rous's laboratory in the Report of the Scientific Directors, RAC RU 439, Vol. 27, 1938–1939, pp. 4–5.
9. Report of Rous' laboratory in the Report of the Scientific Directors, RAC RU 439, Vol. 28, 1939–1940, p. 40.
10. *Ibid.*, p. 41.
11. *Ibid.*, p. 41.
12. *Ibid* and Rous papers, folder Bittner, APS. For more on the history of research on the mouse mammary tumor factor, see Löwy and Gaudillière, 1998 and Gaudillière, 1999.
13. Report of Rous's laboratory in the Report of the Scientific Directors, RAC RU 439, Vol. 30, 1941–42, p. 39.
14. Elzen, 1988, p. 173, quoting Pickels, *The Air-Driven Ultra-centrifuge*, unpublished dissertation, 1935, University of Virginia, Charlottesville, VA, p. 56.
15. 'Mr. Pickels—Appointment,' 19 Sep 1936, RAC RU 5, series 4, box 22, folder 260; Elzen, 1988, p. 161.

16. Early on, Pickels added on optical system (to view the sedimenting material) to the apparatus, and emulated other features of Svedberg's oil-turbine machine such that the air-driven instrument could also be used to measure molecular weights. He also designed a rotor that could accommodate larger volumes of liquid, so that the ultracentrifuge could be used to *prepare* as well as *analyze* materials. For more on air-driven ultracentrifuges, see Elzen, 1986 and 1988; for more on the relationship between virus research and ultracentrifuge design, see Creager (in press).
17. Wyckoff collaborated with Pickels, publishing two papers with him in 1936 (Biscoe *et al.*, 1936a, 1936b), but also competed with him for the attention and trust of medical researchers at the institute. See Memorandum to Dr. Arthur F. Coca, May 2, 1939, Rous papers, folder Ralph Wyckoff #2, APS.
18. Wyckoff and Stanley first collaborated on X-ray diffraction studies of crystalline TMV, but these were not as fruitful as ultracentifugal studies. Ralph Wyckoff to W.M. Stanley, 22 Jan 1936, Stanley papers, carton 14, folder Wyckoff, University of California, Berkeley, Bancroft Library (hereafter UCB) 78/18c.
19. Wyckoff to Stanley, 6 May 1936, Stanley papers, carton 14, folder Wyckoff, UCB.
20. Stanley had also collaborated with Svedberg, who determined a molecular weight in the same range but found that the virus was not chemically homogeneous (Eriksson-Quensel and Svedberg, 1936). See Kay, 1986, and Creager, in press.
21. Wyckoff was also sensing that he would not be kept on the institute staff beyond 1937, when his contract as a research associate expired. Memorandum to Dr. Arthur F. Coca, 2 May 1939, Rous papers, folder Ralph Wyckoff #2, APS.
22. Shope to Rous, 12 Jul 1935, folder Shope #8, Rous papers, APS.
23. Rous to Shope, 21 Jul 1935, folder Shope #8, Rous papers, APS.
24. See Joseph W. Beard, Memorandum to Arthur F. Coca, 18 May 1939, Rous papers, folder Joseph Beard #2, BR77, APS. It seems that the efforts to isolate papilloma virus via salt precipitation methods did not advance far enough to be reported in the laboratory's annual reports. On the significance of Northrop's enzyme and bacteriophage research to studies of purified viruses, see Kay, 1986.
25. With Stanley, Wyckoff used the ultracentrifuge to isolate potato X virus and tobacco ring spot disease (Stanley and Wyckoff, 1937); with another Princeton plant pathologist, Wyckoff isolated two strains of cucumber mosaic disease and tobacco necrosis virus (Price and Wyckoff, 1938, 1939).
26. Rous to Beard, 15 Feb 1937, folder Beard, Rous paper, APS.
27. Beard to Rous, 3 Mar 1937, folder Beard, Rous papers, APS.
28. Stanley gave an address on purified viruses in relation to cancer at the Third International Cancer Congress in September 1939; abstract enclosed with letter to William H. Woglom, 2 May 1939, Stanley papers, carton 1, folder May 1939, UCB. See also Stanley's letter to Svedberg, 26 Mar 1937, on plant viruses as a model for cancer agents, Stanley papers, carton 13, folder Svedberg, UCB.
29. On the reorganization of biomedical sciences during and after the war see Starr, 1982 and Strickland, 1972.
30. For an account of the co-calibration of the ultracentrifuge and the electron microscope, see Gaudillière, 1998.
31. On this linkage of ultracentrifugation and biochemical methods, see the analysis of Claude's work at the Rockefeller Institute in Rheinberger, 1995.
32. On triangulation, see Star, 1986.
33. On the role of these biophysical machines and approaches in defining 'molecular biology,' see, Kay, 1993, Creager, 1996, and Rasmussen, 1997.
34. Evidence for some continuity of this line of thinking is easy to find in repeated comparison of lysogenic bacteria with tumor cells discussed in the 1950s and 1960s. For a preliminary analysis see Galperin, 1994.
35. Delbrück to H.W. Davenport, 28 Jul 1980, Delbrück papers, folder 7.8, Caltech Archives. For a detailed analysis of the early uses of phage, see Summers, 1993. Ellis asserted that step-growth curves were already plotted when Delbrück made his entry in phage studies (Ellis, 1966).
36. For early analysis of phage work, see Olby, 1994 (2nd edition of 1974 book).

37. As analyzed by Lily Kay, in 1945, Pauling and Beadle designed a plan for molecular biology at Caltech that was initially submitted to the Rockefeller Foundation. The plan focused on protein physical chemistry as a unifying approach for both the Chemistry and the Biology divisions. Negotiations achieved nothing. Thus Beadle and Pauling approached other sources: governmental agencies like the Public Health Service, the Atomic Energy Commission, and the Office of Naval Research, and philanthropic sponsors such as the American Cancer Society to the National Foundation for Infantile Paralysis. (Kay, 1993, p. 225 ff.)
38. G. Beadle to Dean Watson, 4 Dec 1946, Beadle papers, folder 3.4, Caltech Archives.
39. M. Delbrück to G. Beadle, 11 Aug 1950, Beadle papers, folder 3.4, Caltech Archives; R. Dulbecco to M. Delbrück, 4 Feb 1951, Delbrück papers, folder 6.22, Caltech Archives.
40. R. Dulbecco, Grant application to the NFIP, May 1953, Biology Division papers, box 49, folder 23, Caltech Archives.
41. On the uncertainties and further development of the transformation procedure, see Gaudillière, 1994.
42. D.E. Johnson to Stanley, 10 Apr 1958, Stanley papers, carton 13, folder American Cancer Society Research Committee, UCB.
43. In his *Scientific American* article, Rubin makes a strong case for the utility of RSV for studying cancer causation, comparing the rapid carcinogenic action of RSV and the delayed action of mammary cancer and leukemia viruses in mice. 'Often these viruses are transmitted from parent mice to offspring for many generations without producing any disease.' The mouse model systems thus presented a complex of infectious and hereditary aspects (Gross's vertical and horizontal transmission). Rubin also points out that polyoma virus in mice, rabbit papilloma virus, and simian SV-40 produce cancer in hosts long after inoculation. (Rubin, 1964, pp. 46–47)
44. Rubin to Rous, 10 Dec 1964, Stanley papers, carton 12, folder Rous, UCB.
45. Stanley to Rous, 8 Feb 1965, Stanley papers, carton 12, folder Rous, UCB.
46. Hanafusa's conclusion in galley proofs from proceedings of Beard's symposium on chicken tumors, held ~1963, as quoted in Rous to Stanley, 21 Jan 1965, Stanley papers, carton 12, folder Rous, UCB.
47. Rous to Rubin, 30 Nov 1964, Stanley papers, carton 12, folder Rous, UCB. This occurrence allowed the creation of a new experimental system: mouse sarcoma, in which viruses could be plaqued from mammalian cells.

REFERENCES

C.H. Andrewes, 'Discussion of Dr. Luria's Paper,' *Cancer Research* (1960), 20: 689–694.

C.H. Andrewes, 'Francis Peyton Rous, 1879–1970,' *Biographical Memoirs of Fellows of the Royal Society* (1971), 17: 642–661.

J.H. Bauer, and E.G. Pickels, 'A High Speed Vacuum Centrifuge Suitable for the Study of Filterable Viruses,' *Journal of Experimental Medicine* (1936), 64: 503–528.

J.W. Beams, and A.J. Weed, 'A Simple Ultra-Centrifuge,' *Science* (1931), 74: 44–46.

J. Beard, and P. Rous, 'A Virus-Induced Mammalian Growth with the Character of a Tumor (The Shope Rabbit Papilloma), II. Experimental Alterations of the Growth on the Skin: Morphological Considerations: The Phenomena of Regression,' *Journal of Experimental Medicine* (1934), 60: 723–740.

J.W. Beard, and R.W.G. Wyckoff, 'The Isolation of a Homogeneous Heavy Protein from Virus-Induced Rabbit Papillomas,' *Science* (1937), 85: 201–202.

J.W. Beard, and R.W.G. Wyckoff, 'The pH Stability of the Papilloma Virus Protein,' *Journal of Biological Chemistry* (1938), 123: 461–470.

J.W. Beard, W. Ray Bryan, and R.W.G. Wyckoff, 'The Isolation of the Rabbit Papilloma Virus Protein,' *Journal of Infectious Diseases* (1939), 65: 43–52.

J.W. Beard, 'Review: Purified Animal Viruses,' *Journal of Immunology* (1948), 58: 49–108.

J. Beard, 'The Fallacy of the Concept of Virus "Masking": A Review,' *Cancer Research* (1956), 16: 279–291.

J. Biscoe, E.G. Pickels, and R.W.G. Wyckoff, 'An Air-driven Ultracentrifuge for Molecular Sedimentation,' *Journal of Experimental Medicine* (1936a) 64: 39–45.

J. Biscoe, E.G. Pickels, and R.W.G. Wyckoff, 'Light Metal Rotors for the Molecular Ultracentrifuge,' *Review of Scientific Instruments* (1936b) 7: 246–250.

Board for the Investigation and Control of Influenza and Other Epidemic Diseases, *Army Medical Bulletin* (1941), 64: 1–22.

A.N.H. Creager, 'Wendell Stanley's Dream of a Free-Standing Biochemistry Department at the University of California, Berkeley,' *Journal of the History of Biology* (1996), 29: 331–360.

A.N.H. Creager, 'That "Whirligig of Science," The Ultracentrifuge in Virus Research,' Chapter 4 of *The Life of a Virus: TMV, as an Experimental Model*, 1930–1935 (Chicago: University of Chicago, in press).

G.W. Corner, *A History of the Rockefeller Institute: 1901–1953, Origins and Growth* (New York City: The Rockefeller Institute Press, 1964).

R. Dulbecco, 'Production of Plaques in Monolayer Tissue Cultures by Single Particles of an Animal Virus,' *Proceedings of the National Academy of Sciences, USA* (1952), 38: 747–752.

R. Dulbecco, and M. Vogt, 'Plaque Formation and Isolation of Pure Lines with Poliomyelitis Viruses,' *Journal of Experimental Medicine* (1954), 99: 176–177.

R. Dulbecco, and M. Vogt. 'Biological Properties of Poliomyelitis Viruses as Studied by the Plaque Technique,' *Annals of the New York Academy of Science* (1955), 61: 790–800.

E. Eckert, D. Beard, and J. Beard, 'Dose-Response Relations in Experimental Transmission of Avian Erythromyeloblastic Leukosis. I. Host-Response to the Virus,' *Journal of the National Cancer Institute* (1951), 12: 447–463.

E. Eckert, D. Beard, and J. Beard, 'Dose-Response Relations in Experimental Transmission of Avian Erythromyeloblastic Leukosis. III. Titration of the Virus,' *Journal of the National Cancer Institute* (1954a) 14: 1055–1066.

E. Eckert, D. Beard, and J. Beard, 'Dose-Response Relations in Experimental Transmission of Avian Erythromyeloblastosis Leukosis. IV. Strain Differences in Host Response to the Virus,' *Journal of the National Cancer Institute* (1954b) 14: 1067–1079.

E. Eckert, D. Beard, and J. Beard, 'Dose-Response Relations in Experimental Transmission of Avian Erythromyeloblastosis Leukosis. V. Influence of Host Age and Route of Virus Inoculation,' *Journal of the National Cancer Institute* (1955a) 15: 1195–1207.

E. Eckert, D. Sharp, D. Beard, I. Green, and J. Beard, 'Virus of Avian Erythromyeloblastic Leukosis. IX. Antigenic Constitution and Immunologic Characterization,' *Journal of the National Cancer Institute* (1955b) 16: 593–637.

E. Ellis, and M. Delbrück, 'The Growth of Bacteriophage,' *Journal of General Physiology* (1939), 22: 365–384.

E.L. Ellis, 'Bacteriophage: One-step Growth,' in J. Cairns, G. Stent and J.D. Watson (eds.) *Phage and the Origins of Molecular Biology* (Cold Spring Harbor, New York: Cold Spring Harbor Laboratory of Quantitative Biology, 1966), pp. 53–62.

B. Elzen, 'Two Ultracentrifuges: A Comparative Study of the Social Construction of Artefacts,' *Social Studies of Science* (1986), 16: 621–662.

B. Elzen, *Scientists and Rotors*, Ph.D. dissertation, University of Twente, 1988.

I. Eriksson-Quensel, and T. Svedberg, 'Sedimentation and Electrophoresis of the Tobacco-Mosaic Virus Protein,' *Journal of the American Chemical Society* (1936), 58: 1863–1867.

B. Friesen, and H. Rubin, 'Some Physicochemical and Immunological Properties of an Avian Leukosis Virus (RIF),' *Virology* (1961), 15: 387–396.

J. Fujimura, *Crafting Science: A Sociohistory of the Quest for the Genetics of Cancer* (Cambridge: Harvard University Press, 1996).

C. Galperin, 'Virus, provirus et cancer,' *Revue d'Histoire des Sciences* (1994), XLVII: 7–56.

J.-P. Gaudillière, 'Wie man Modelle für Krebsentstehung konstruiert. Viren und Transfektion am (US) National Cancer Institute.' in M. Hagner, H.-J. Rheinberger, B. Wahrig-Schmidt (eds.) *Objekte, Differenzen und Konjonkturen* (Berlin: Akademie Verlag, 1994), pp. 233–257.

J.-P. Gaudillière, 'The Molecularization of Cancer Etiology in the Postwar United States: Instruments, Politics and Management,' in H. Kamminga and S. de Chadarevian (eds.) *Molecularizing Biology and Medicine: New Practices and Alliances, 1910s–1970s* (Amsterdam: Harwood Academic Publishers, 1998), pp. 139–170.

J.-P. Gaudillière, 'Circulating Mice and Viruses: The Jackson Memorial Laboratory, the National Cancer Institute, and the Genetics of Breast Cancer, 1930–1965,' in E. Mendelsohn and M. Fortun (eds.) *Sociology of Science Yearbook* Vol. 28, Special Issue: The Practice of Human Genetics (Amsterdam: Kluwer Academic Publisher, 1999), pp. 89–124.

L. Gross, 'The Possibility of Exterminating Mammary Carcinoma in Mice by a Simple Preventive Measure,' *New York State Journal of Medicine* (1946), 46: 327–331.

L. Gross, 'The Vertical Epidemic of Mammary Carcinoma in Mice,' *Surgery, Gynecology, and Obstetrics* (1949), 88: 295–308.

H. Hanafusa, T. Hanafusa, and H. Rubin, 'The Defectiveness of Rous Sarcoma Virus,' *Proceedings of the National Academy of Science, USA* (1963), 49: 572–580.

T. van Helvoort, 'A Century of Research into the Cause of Cancer: Is the New Oncogene Paradigm Revolutionary?,' *History and Philosophy of the Life Sciences* (1999), 21: 293–330.

L.E. Kay, 'W.M. Stanley's Crystallization of the Tobacco Mosaic Virus, 1930–1940,' *Isis* (1986), 77: 450–472.

L.E. Kay, *The Molecular Vision of Life: Caltech, The Rockefeller Foundation, and the Rise of the New Biology* (Oxford/New York: Oxford University Press, 1993).

D.J. Kevles, 'Renato Dulbecco and the New Animal Virology: Medicine, Methods, and Molecules,' *Journal of the History of Biology* (1993), 26: 409–442.

D.J. Kevles, 'Pursuing the Unpopular: A History of Courage, Viruses and Cancer' in R.B. Silvers (ed.) *Hidden Histories of Science* (New York: New York Review, 1995), pp. 69–112.

C.A. Knight, 'Amino Acids of the Shope Papilloma Virus,' *Proceedings of the Society for Experimental Biology and Medicine* (1950), 75: 843–846.

I. Löwy, 'Variances in Meaning in Discovery Accounts: The Case of Contemporary Biology,' *Historical Studies in the Physical and Biological Sciences* (1990), 21: 87–121.

I. Löwy, and J.-P. Gaudillière, 'Disciplining Cancer: Mice and the Practice of Genetic Purity,' in J.-P. Gaudillière and I. Löwy (eds.) *The Invisible Industrialist: Manufactures and the Production of Scientific Knowledge* (London: Macmillan, 1998), pp. 209–249.

S. Luria, 'Genetic Concept of Virus Infection,' *Cancer Research* (1960), 20: 677–688.

J.B. Murphy, O.M. Helmer, and E. Sturm, 'Association of the Causative Agent of a Chicken Tumor with a Protein Fraction of the Tumor Filtrate' *Science* (1928), 68: 18–19.

J.B. Murphy, 'An Analysis of the Trends in Cancer Research,' *Journal of the American Medical Association* (1942), 120: 107–111.

H. Neurath, G.R. Cooper, D.G. Sharp, A.R. Taylor, D. Beard, and J.W. Beard, 'Molecular Size, Shape, and Homogeneity of the Rabbit Papilloma Protein,' *Journal of Biological Chemistry* (1941), 140: 293–306.

R.C. Olby, *The Path to the Double Helix: The Discovery of DNA* (London: Dover, 2nd edition, 1994).

W.C. Price, and R.W.G. Wyckoff, 'The Ultracentrifugation of the Proteins of Cucumber Viruses 3 and 4,' *Nature* (1938), 141: 685–686.

W.C. Price, and R.W.G. Wykoff, 'Ultracentrifugation of Juices from Plants Affected by Tobacco Necrosis,' *Phytopathology* (1939), 29: 83–94.

N. Rasmussen, 'Making a Machine Instrumental: RCA and the Wartime Origins of Biological Electron Microscopy in America, 1940–1945,' *Studies in History and Philosophy of Science* (1996), 27: 311–349.

N. Rasmussen, *Picture Control: The Electron Microscope and the Transformation of Biology in America, 1940–1960* (Stanford: Stanford University Press, 1997).

H.-J. Rheinberger, 'From Microsomes to Ribosomes: "Strategies" of "Representation",' *Journal of the History of Biology* (1995), 28: 49–89.

H.-J. Rheinberger, *Toward a History of Epistemic Things: Synthesizing Proteins in the Test Tube* (Stanford: Stanford University Press, 1997).

F.P. Rous, and J. Beard, 'The Progression to Carcinoma of Virus-Induced Rabbit Papillomas (Shope),' *Journal of Experimental Medicine* (1935), 62: 523–548.

F.P. Rous, 'The Nearer Causes of Cancer,' *Journal of the American Medical Association* (1943), 122: 573–581.

F.P. Rous, 'The Challenge to Man of the Neoplastic Cell,' *Science* (1967), 157: 24–28.

H. Rubin, and H.M. Temin, 'Radiation Studies on Lysogeny and Tumor Viruses,' *Radiation Biology and Cancer*, symposium papers published for the University of Texas M.D. Anderson Hospital and Tumor Institute (Austin, Texas: University of Texas Press, 1959), pp. 359–381.

H. Rubin, 'An Analysis of the Assay of Rous Sarcoma Cells *in vitro* by the Infective Center Technique,' *Virology* (1960a) 10: 29–49.

H. Rubin, 'A Virus in Chick Embryos Which Induces Resistance *in vitro* to Infection with Rous Sarcoma Virus,' *Proceedings of the National Academy of Science, USA* (1960b) 46: 1105–1119.

H. Rubin, and H. Hanafusa, 'Significance of the Absence of Infectious Virus in Virus-Induced Tumors,' *Viruses, Nucleic Acids, and Cancer*, Proceedings from the Seventeenth Annual Symposium on Fundamental Cancer Research, The University of Texas M.D. Anderson Hospital and Tumor Institute (Baltimore: Williams and Wilkins, 1963), pp. 508–517.

H. Rubin, 'A Defective Cancer Virus,' *Scientific American* (1964), 210: 46–52.

H. Rubin, 'Virus without Symptoms and Symptoms without Virus: Complementary Aspects of Latency as Exemplified by the Avian Tumor Viruses,' in Morris Pollard (ed.) *Perspectives in Virology*, Vol. 4 (New York: Harper & Row, 1965), pp. 164–174.

H.K. Schachman, 'Physical Chemical Studies on Rabbit Papilloma Virus,' *Journal of the American Chemical Society* (1951), 73: 4453–4455.

D.G. Sharp, A.R. Taylor, D. Beard, and J.W. Beard, 'Electrophoresis of the Rabbit Papilloma Virus Protein,' *Journal of Biological Chemistry* (1942), 42: 193–202.

D.G. Sharp, A.R. Taylor, A.E. Hook, and J.W. Beard, 'Rabbit Papilloma and Vaccinia Viruses and T_2 Bacteriophage of *E. coli* in "Shadow" Electron Micrographs,' *Proceedings of the Society for Experimental Biology and Medicine* (1946), 61: 259–65.

R.E. Shope, 'Infectious Papillomatosis of Rabbits,' *Journal of Experimental Medicine* (1933), 58: 607–624.

R.E. Shope, 'Serial Transmission of Virus of Infectious Papillomatosis in Domestic Rabbits,' *Proceedings of the Society for Experimental Biology and Medicine* (1935), 32: 830–832.

R.E. Shope, 'Immunization of Rabbits to Infectious Papillomatosis,' *Journal of Experimental Medicine* (1937), 65: 219–231.

R.E. Shope, '"Masking," Transformation, and Interepidemic Survival of Animal Viruses,' in M. Delbrück (ed.) *Viruses 1950* (Pasadena: California Institute of Technology, 1950), pp. 79–92.

J.E. Smadel, E.G. Pickels, and T. Shedlovsky (II. only), 'Ultracentrifugation Studies on the Elementary Bodies of Vaccine Virus. I. General Methods and Determination of Particle Size; II. The Influence of Sucrose, Glycerol, and Urea Solutions on the Physical Nature of Vaccine Virus,' *Journal of Experimental Medicine* (1938), 68: 583–606 (I.); 607–627 (II.).

J.E. Smadel, T.M. Rivers, and E.G. Pickels, 'Estimation of the Purity of Preparations of Elementary Bodies of Vaccinia,' *Journal of Experimental Medicine* (1939), 70: 379–385.

J.E. Smadel, E.G. Pickels, T. Shedlovsky, and T.M. Rivers, 'Observations on Mixtures of Elementary Bodies of Vaccinia and Coated Collodion Particles by Means of Ultracentrifugation and Electrophoresis,' *Journal of Experimental Medicine* (1940), 72: 523–529.

W.M. Stanley, 'Isolation of a Crystalline Protein Possessing the Properties of Tobacco-Mosaic Virus,' *Science* (1935), 81: 644–645.

W.M. Stanley and R.W.G. Wyckoff, 'The Isolation of Tobacco Ring Spot and Other Virus Proteins by Ultracentrifugation,' *Science* (1937), 85: 181–185.

W.M. Stanley, 'Relationships, Established and Prospective, Between Viruses and Cancer,' *Annals of the New York Academy of Sciences* (1958), 71: 1100–1113.

W.M. Stanley, 'Virus-Induced Neoplasia—Outlook for the Future,' *Cancer Research* (1960), 20: 798–804.

S.L. Star, 'Triangulating Clinical and Basic Research: British Localizationists,' *History of Science* (1986), 24: 29–48.

P. Starr, *The Social Transformation of American Medicine* (New York: Basic Books, 1982).

S.P. Strickland, *Politics, Science & Dread Disease: A Short History of United States Medical Research Policy* (Cambridge, MA: Harvard University Press, 1972).

W.C. Summers, 'How Bacteriophage Came To Be Used by the Phage Group,' *Journal of the History of Biology* (1993), 26: 255–267.

H.M. Temin, and H. Rubin, 'Characteristics of an Assay for Rous Sarcoma Virus and Rous Sarcoma Cells in Tissue Culture' *Virology* (1958), 6: 669–688.

R.W.G. Wyckoff, and R.B. Corey, 'The Ultracentrifugal Crystallization of Tobacco Mosaic Virus Protein,' *Science* (1936), 84: 513.

R.W.G. Wyckoff, 'The Ultracentrifugal Purification and Study of Macromolecular Proteins,' *Science* (1937), 86: 92–95.

R.W.G. Wyckoff, J. Biscoe, and W.M. Stanley, 'An Ultracentrifugal Analysis of the Crystalline Virus Proteins Isolated from Plants Diseased with Different Strains of Tobacco Mosaic Virus,' *Journal of Biological Chemistry* (1937), 117: 57–71.

Part 3
HEREDITY, MEDICINE, HEALTH POLICIES

Chapter 7

DEGENERATION THEORY AND HEREDITY PATTERNS BETWEEN 1850 AND 1900

Patrice Pinell

While the various approaches to heredity in terms of degeneration have led to an abundant literature from researchers dealing with the history of mentalities, culture or politics, they have rarely attracted the interest of historians of science and medicine even though they dominated medical thought about hereditary transmission for over fifty years.

The purpose of this paper is to shed light into a dark area of medical genetics by analysing how the preexisting theoretical and practical framework developed, and by identifying how Mendelian genetics was introduced into medicine. Two different phases may be identified in the evolution of this framework. The first period (1850 to the mid-1880's) saw the evolution of a degeneration theory 'explaining' pathological heredity; it grew alongside, but unconnected to, various conceptions of 'normal' *vs.* 'pathological' heredity. The second period (1880 to the early 20th century) started with the first attempts at structuring theories on pathological *vs.* normal heredity. The evolution in the approach to 'hereditary/familial' diseases will be considered here as a vital lead since it determined how Mendelian genetics was integrated.

FROM AUGUSTIN-BÉNÉDICT MOREL TO CHARLES FÉRÉ: DEGENERATION AS AN ETIOLOGICAL, UNIFYING BASIS FOR MEDICAL PATHOLOGY

Morel, an alienist and fundamentalist Catholic, hostile to any evolutionistic idea, was openly 'fixist' (Bing, 1994; Pick, 1989; Castel, 1979). His theory of degeneration derived meaning from a religious perspective, but it gained such internal consistency that it was possible to render it autonomous of its premises. Insanity, a distinct feature of fallen creatures, was the consequence of a physical, social or moral disorder which somehow embodied itself in a disorder of the nervous system, and was the result of a process developing over several generations. On this point, Morel subscribed to transformistic ideas: once the nervous disorder has occurred in an individual and has been transmitted to his/her progeny as a morbid predisposition, it becomes, under the action of all kinds of 'determining causes,' a pathological condition symptomatologically different from what it was in the ancestor. The hereditary morbid predisposition is thus transmitted from generation to generation, but in such a way that descendants become psychologically frailer and frailer. The defect worsens with each transmission and manifests itself through increasing mental disorders until it produces completely degenerate individuals, incapable of breeding (the bedridden idiot is a typical example). Once the process is completed, the lineage eventually dies out. This conception of insanity, as one and multiform, an indefinite set of variations from one matrix through the dual action of determining causes and the shift from predisposition to patency over successive generations, led Morel to reject his predecessors' symptomatological classifications and the desaggregation of mental disorders into countless entities.

Degeneration theory proposed to go beyond descriptive approaches by using a type of pathogenesis and etiology capable of integrating all components of the diversity found in a clinical, unifying environment. Morel accepted all causes, from intoxication to the effects of revolutionary crises, including the complete range of vices, and indifferently conceived them as either 'causes of the hereditary predisposition' or 'determining causes' activating the predisposition and giving insanity its symptomatic form. Degenerative heredity was different from normal physiological heredity through its polymorphic manifestations. As Féré put it, in a commentary, 'transmittable diseases transform' (Féré, 1884, p. 8).

Within one family and over several generations, the theory allowed one to establish a relationship among the most diverse mental disorders and gave meaning to their occurrence. This meaning, which can instantly be translated into an ideological discourse linking social living conditions and mental disorders, was not irrelevant to Morel's success. But the adjustment of degeneration theory to the socio-political concerns of the ruling classes was not enough to explain why this theory spread outside of France and served as the major frame of reference until right after World War I (Bing, 1994). As he theorized insanity from a pathogenic viewpoint which rooted it in the nervous system (against the supporters of psychogenic theories on mental disorders) and hereditary etiology, Morel started bridging the gap between psychiatric and general medicine (Castel, 1979), a movement of which *La Famille névropathique*, published in 1884 in *Archives de neurologie*, is part. A former intern of Charcot and a lab assistant at the recently created clinical chair for diseases of the nervous system, its author Chalres Féré was one of those physicians of the 'Salpêtrière

School' who had decided to dedicate themselves to mental disorders. That same year, Féré was appointed head of a department for the mentally impaired at the Bicêtre hospital. 'Based on clinical experience,' *La Famille névropathique* aimed at demonstrating that 'morbid psychological conditions' and the 'diseases of the nervous system affecting motion and sensitivity' were in fact two psychopathic and neuropathological branches 'of one single family inextricably bound by the laws of heredity' (Féré, 1884). Féré subscribed to most of Morel's approach, removed its metaphysical references and extended its field of explanatory relevance. Now unburdened by 'fixism' and the Original Sin, the new degeneration theory was part of an eclectic evolutionism borrowing the Darwinistic 'struggle for life' concept and the Neo-Lamarckian notion of hereditarily acquired characteristics. The reference to Darwin accounted for the dynamics leading to families of degenerates dying out because their defects put them at a disadvantage in the survival of the fittest, while hereditarily acquired characteristics explained the onset of the degeneration process, *i.e.* the transformation of a nervous disorder into hereditary predisposition.

While Féré, modeling his approach on Morel's, distinguished between the causes creating the predisposition and the determining causes shaping the nervous disease, his model differed in that it analyzed the mechanics of these causes. Morel put moral and physical causes on the same level, arguing on the consubstantiality of body and soul, whereas Féré, resolutely materialistic in his approach, suggested theories which aimed to define a middle ground between social and/or moral disorders and biological disorders. The first crucial notion to fulfill this mediating function was nervous exhaustion. The end result of an early environment-based process, nervous exhaustion is an intermediate step leading to a second process: its transformation into hereditarily transmittable functional disorders. Any kind of overtaxing is capable of producing nervous exhaustion: whether the result of emotional, sexual or intellectual stress, including and most frequently this specific mental overtaxing that is neurasthenia, the most common neurosis.

The second 'mediating' notion was intoxication. What nervous exhaustion accomplishes can be reproduced by many types of intoxication such as lead poisoning, whose influence Roques wanted to demonstrate in the degeneration process (Roques, 1872), and especially alcoholism with its more serious impact. Féré approached alcoholism as a metabolic disorder affecting the parent(s) and transmitted to their descendants as morbid nervous predispositions (which the high rate of epilepsy and psychosis among children of alcoholic parents could account for).

When describing the psychopathic branch of the family, Féré did not break much new ground. He simply took up theories formulated by Morel and other alienists like his followers Legrand du Saulle and Magnan. He was innovating in the section devoted to the different diseases of the neuropathological branch. In each instance, it mentioned clinical cases where the usual set of symptoms specific to the entity (chorea, paralysis agitans, Parkinson's disease, Basedow's disease, locomotor ataxia, multiple sclerosis, spinal amyotrophia, etc.) was accompanied by mental disorders which could be found either in the patient or in his/her relatives. These clinical cases came from the observations made at the Salpêtrière hospital and were sometimes completed by case studies or statistics from the literature.

Once this branch-by-branch development was made, Féré wrote that, thus formed into a 'natural group,' his 'neuropathological family' was in no way separate from other major pathologies and had frequent connections with nutritional diseases, malformations and development impairment (Féré, 1884, p. 45–63). And to demonstrate its kinship with rheumatism and other major causes of degeneration (arthritis/neuropathology, rheumatism/hysteria or chorea, gout/cephalalgia or delirium or epilepsy, hypocondriasis or even general paralysis), Féré concluded that 'arthritis and neuropathological diathesis are two co-generic conditions resulting from a diversely specialized nutritional disorder. As degenerative conditions, neuropathology, scrofula, tuberculosis and arthritis are diversely combined within families and, under certain conditions, their manifestations are transformed or enhanced by one another' (Féré, 1884, p. 63).

Féré's book was enthusiastically received. Charcot even used it as his main reference when dealing with heredity (Charcot, 1889); Déjerine considered that it was 'the most significant dissertation ever published since Morel's in this field' (Déjerine, 1886, p. 29), and it was used by all those in France who were writing about morbid heredity. The success of *La Famille névropathique* can be explained by the programmatic direction legitimized by its central assumption and the explanatory capacity of its theoretical framework.

In 'programmatic' terms, by extending Morel's logic to motion/sensitivity-affecting diseases, Féré suggested a nosographic unification of the 'mental' and 'nervous' fields. He developed the direction suggested by Morel, albeit within a context deeply changed by the emergence of the École neurologique de la Salpêtrière and in such a way that, if the reunion of mental and clinical medicine was to be made by a link to neurology, this would happen under the aegis of neurologists.

Cut off from the general hospital by asylum institutions and also having a specific professional status, the medical study and treatment of insanity had been shaped by two bipolar trends from the outset. One trend tended to unite this field with anatomico-clinical medicine whereas the other claimed that these diseases without lesions were specific, had a psychogenic etiology and came within the framework of moral treatment (Castel, 1979). The Paris Faculty of Medicine strongly supported the first trend, pushing for the clinical chair on mental diseases opened at Saint-Anne to be entrusted to a 'generalist' clinician (Ball) as opposed to an asylum physician. The 'Salpêtrière School' was clearly involved in the conflict and later acted as an institutional fulcrum for the emergence of non-asylum 'neuropsychiatry.' Bourneville, who became an alienist at the Bicêtre hospital, and Jouffroy, who succeeded Ball at Sainte-Anne, were, like Féré, Charcot's students. But *La Famille névropathique* did not only have a programmatic value for the development of psychiatry based on neurology it also suggested, through degenerative heredity, a link between neurology, general medicine and the fight against social scourges. During the late 19th century, social-health concerns, crystallized around tuberculosis, syphilis, nervous/mental diseases, and childhood diseases, tended to converge and blend into a common obsession-decadence. The fear of decadence for France and the 'French' race found its objective bases in a stagnating birth rate, its alleged causes were found in the various social curses and its unity was found in the degeneration process. And works such as Féré's suggested that the degeneration process was multiform, the various nervous, tuberculous or syphilitic heredities combining to give a far-reaching scope to the phenomenon. A synthetical formalization of ideas that were already part of common opinion within the medical elite, *La Famille névropathique* could not but receive general acceptance; in addition, the hereditarily morbid pattern it suggested, because it was so simple, had remarkable explanatory power in 'ordinary' clinical studies.[1]

The degeneration theory as developed by Féré allowed for connections to be made, within a single family, between any two pathological, nervous manifestations (one just had to find two individuals affected). In other words, it produced meaning. With the extension of the degeneration concept to all major pathologies that were supposed to develop in an hereditary context, the onset of one of these diseases in any given individual could immediately be translated as the characteristic of a certain type of family medical history.

Under this theory, the disease is not fateful, but the product of a past disorder, the present mark of an older transgression. At a time when official religions had given up disputing medicine's right to interpret diseases, the latter (temporarily) occupied this vacancy through theories such as Morel and Féré's. Similar to the oldest ways of explaining evil, *La Famille névropathique* gave meaning to 'evil' by scholarly and materialistically shaping common sense.

The shift from the sick to his/her family made by generalized hereditary etiology is particularly striking within the late 19th century context where medicine, as practices outside of hospitals, was being shaped around the ideal figure of the family doctor (Pinell and Steffen, 1994). For it was family doctors who could best observe the effects of degenerative heredity since they were in a far better position to do so than hospital doctors. While patients (or their parents) 'often only admit[ted] with great difficulty to a nervous, hereditary defect in their progenies, especially when this defect [was] intellectual' (Déjerine, 1886, p. 30), family doctors were in a position to be aware of hidden defects since they treated each generation. Therefore they had to systematically look for, and record, the manifestations of degenerative heredity, as advised by student handbooks.[2]

MORBID AND PHYSIOLOGICAL HEREDITY: ADJUSTMENT AND PROBLEMS

One would vainly look for any reference to contemporaneous work on non-'pathological' heredity in *La Famille névropathique*. Féré built his patterns without taking into account existing theories in this field, at least explicitly. It is true that, whereas morbid heredity owed much to the alienists' approach, 'normal' heredity was the subject matter of naturalists, botanists and animal biologists who, due to the very nature of their object of study, had little interest in transmission pathology. Hence—and and this is crucial-a deep gap between two types of heredity developing in relative independence from each other, each referring to utterly different worlds of practices. Jules Déjerine tried to build a bridge between these two worlds. *L'Hérédité dans les maladies du système nerveux* analyzed the various theories on 'physiological' heredity. Among them Weismann's theory was the most compatible with medical approaches, in his view. With the most recent development of biological knowledge on insemination (namely Strassburger's and

Van Beneden's work on the fertilization of *ascaris megalocephalus* and Hertwig's and Fol's work on ovum maturation) Weismann's concept of the 'germinative plasma' was no more an abstraction. It could be materialized in the nucleus substance of the cells, the only vector of hereditary tendencies. Fertilization being based on the 'copulation of the male and female pronuclei', the substance of the nucleus should include the germinative plasma from both parents as well as the generative plasma from grandparents, great-grandparents, etc. The persistence of the ancestors' generative plasma in the nucleus thus sheds light on the phenomenon of 'atavism,' *i.e.* the reappearance in an individual of long-lost characteristics; it does not explain it, but it suggests why the phenomenon is possible at all. It also accounts for an apparent paradox: the fact that the continuity of germinative plasma may go together with creating individual hereditary differences. Whenever sexual reproduction occurs, it entails combining hereditary characteristics, thus creating diversity, a diversity gradually amplified in each generation, whence an increase of possible combinations.

There was, however, one element in Weismann's theory which did not convince Déjerine (and most of the physicians): Weismann's refusal to admit that, under specific conditions, acquired characteristics may be hereditary. Without questioning the theory of germinative plasma's continuity, Déjerine amended it to adjust it to a medical argument on morbid heredity. 'One could willingly admit, with Weismann, that reproduction, since it brings two individuals with distinct tendencies together, enhances the odds for hereditary and progressive varieties, but this is no indication as to why these varieties develop in the first place' (Déjerine, 1886, pp. 15–16), and this is where the hereditariness of acquired characteristics may be meaningful.

Unlike the previous issue, here a speculative field dominated by 'theories,' namely the one dealing with the laws of heredity was supposed to bring physicians back onto the solid ground of observable facts. Déjerine borrowed these facts and laws from Darwin and stated them as formulated by Darwin.

Heredity could be direct and immediate (from parents to children) but always with the predominance of either parent in the transmission. Heredity could also respect the law of atavism (descendants inherit the physical and mental dispositions of their ancestors whom they resemble, whereas they do not resemble their own parents). Finally, heredity could arise in a 'matching life period', in which case, the occurrence of the caracters are manifested at the same age they occurred in prior generations (Déjerine, 1886, p. 17). The combination of these laws accounts for observable facts in terms of transmitting morphological characteristics and even the anatomico-physiological structures that, in mankind and higher animal species, have reached a high level of differentiation and sophistication, including the nervous system. But if the hereditary transmission of lower nervous functions is a hardly dubitable fact, what about the higher functions such as intelligence? The thought process being a property of living matter only, it too can be transmitted hereditarily. Déjerine expressed no doubt on this issue, though how the transmitted physiological characteristic manifests itself in the descendants appeared more complex than for physical characteristics. 'It is not enough to be born with a brain gifted with above-average qualities; the individual with this brain must also be in the situation to use it ... At birth, psychological heredity is only a probability and never a certitude; environmental influences are significant' (Déjerine, 1886, pp. 21–22).

The transmission of a predisposition whose expression depends on environmental conditions, psychological heredity is the aspect of 'physiological heredity' which serves as a connection to 'pathological heredity.' Since what is transmitted is only a morbid predisposition as opposed to a specific pathological trait, the disease affecting the descendants will either have a (rarely) similar or (most frequently) dissimilar form, whatever the type of laws transmission abides by. The main contribution of *L'Hérédité dans les maladies du système nerveux* to Féré's argument was Déjerine's successful adjustment (at least from a rhetorical viewpoint) to 'physiological heredity.' From that moment on, pathological nervous heredity was explicitly put in a Darwinian, evolutionistic perspective.

At the end of the 19th century a new paradigm prevailed by which, apart from disorders due to accidental injuries of the nervous system or the consequence of a long immobilization, all diseases affecting the mind, motion or sensitivity referred to degenerative heredity. This paradigm led to reformulating the issues related to those few diseases which were transmitted within a family in a 'similar' form, *i.e.* those that were so far the only ones to be considered as hereditary. I will use the example of 'progressive muscular distrophy' (MD) to clarify my argument.

In 1885 Charcot still deemed MD to be one 'symptom' frequently found when analyzing family history (Charcot, 1885). 'Heredity' was part of the typical clinical picture of MD and was considered a sign leading away from differentially diagnosing spinal amyotrophies, which were said to be 'rarely hereditary.' Yet, 'heredity' was not a characteristic falling under the very definition of the disease and MD could occur erratically within one family; it could be acquired and not hereditary, without being of a different sort. A few years later, isolated cases of MD were no longer approached as acquired forms, but as cases of 'simple' degenerative heredity caused by the action of determining causes (intoxication, trauma, etc.) upon a familial morbid predisposition (Jouffroy, 1902; Raymond, 1889). A connection was sought to other manifestations of nervous heredity in a patient (mental impairment, twitches, etc.) and in his/her relatives (Pillet, 1890). In other words, isolated MD was a form of degeneration falling within the 'dissimilar' hereditary category. We may find it ironical, but the most difficult cases to interpret were those where MD was typically a familial disease since, as it occurred within the 'similar' mode, it seemed to escape the aggravation of the defect that is typical of degeneration dynamics. Certainly there were many families where MD could fit within such a process when, for instance, the clinical form affecting the descendant(s) had more serious clinical features than for the ancestor(s) (Troisier and Guinon, 1889) or when it was included in a family clinical picture that was typical of degeneration (Londe and Meige, 1894). But these situation were not the rule, and what is true for MD was also true for all diseases with similar heredity.

In addition, as long as familial transmission of similar heredity was approached from the general viewpoint of nervous heredity, physicians were not able to further discriminate its features. Hence their difficulty in clarifying the connection between familial diseases and 'infectious and hereditary' diseases such as tuberculosis.

The action of microbes in infection, and Koch's discovery of the tuberculosis bacillus in 1884, did nothing to weaken the dogma of tuberculous heredity. Koch himself contended that the influence of an infectious agent clarified the mechanics of heredity (Koch, 1884, pp. 197–224). If the presence of the microbe was necessary to cause tuberculosis, the disease would only break out on favorable, *i.e.* hereditarily predisposed ground. Tuberculosis may affect several members of a family from various generations. Certainly, contagion phenomena could not be excluded (it was recognized that they even played a significant part) but they did not explain everything, and cases of familial tuberculosis where infection may be due to a contagion were common. Should then tuberculosis be considered as a familial disease in the same way as MD? The debate was unclear and, while a majority of people agreed that tuberculosis should not be considered as a familial disease, some believed that the arguments supporting this opinion were specious. Were there not forms of familial tuberculosis with a marked predisposition toward a specific localization on the body? Were there not families with coxotuberculosis or tuberculous lymphadenopathies (Raymond, 1905, p. 179)? For insurance companies, the case was clear. Prone to minimize the risks, they made the most of a possible 'familial defect.' 'Coverage is refused to any healthy individual who, in addition to either parent, has also lost over one fourth of his siblings to phtisis ... Paternal or maternal heredity adds tens years to the insurance applicant's age when assessing his or her life expectancy, even if he or she is healthy' (Londe, 1895, p. 24).

Paul Londe was the first, in 1895, to significantly clarify the properties of familial diseases. He first pointed out that familial diseases deserved their own place in nosography. Their significant growth in number over the previous ten years (Londe mentioned over twenty cases) meant that it was no longer relevant to just record them without further characterization, especially since they had specific characteristics that were not taken into account. 'Pathology books do not include a specific section dealing with these ailments—they are classified as common diseases. Yet a familial disease is not just another disease with its only distinctive features being that it can be found in several members of one family; *it is a disease which tends to create, next to the normal strain of the species, an abnormal type and almost a degenerate strain of the species* [our italics]. Thus the occurrence of a familial disease can be traced over several and often many generations' (Londe, 1895, p. 19). Adopting a Darwinian viewpoint, Londe explained that familial diseases could potentially create a separate species if reproduction were possible, 'but organic diseases such as those we study extinguish the possibility of reproduction, and these diseases remain familial diseases, the degenerate type dying out due to natural selection' (Londe, 1895, p. 21). The idea that this new strain was a sketch, a degenerate draft of a new species led to the concept that the environment had no influence on pathological familial heredity, that morbid traits were transmitted like any other normal physiological characteristic and expressed themselves independently from any determining external cause.

'Admittedly the environment may have a major influence on a disease that is *just* hereditary. But when dealing with familial diseases strictly speaking (diseases that are common to many members of a generation who are sometimes influenced by very distinct environments, diseases that occur at the same age among siblings, and diseases that have the same symptoms and succession order for each sibling) and when heredity increases over several generations, it is difficult not to see the effects of fate as something unconcerned with environmental differences.' There was a dividing line between 'simple' morbid heredity and familial diseases, the latter occurring as a 'development disorder, actually independent from any acquired infection (which excludes tuberculosis) or intrauterine accident; it is a germ cell disease or a disease resulting from a combination of germ cells' (Londe, 1895, p. 41). This conceptualization of familial diseases became a reference from which other physicians were to continue, and to go deeper into, their specificities. In a paper of 1897, Dr. Pauly and Dr. Bonne from Lyon (Pauly and Bonne, 1897) used Londe's argument to further refine the distinctive characteristics of familial diseases and listed five of them:

1) Familial diseases are diseases affecting several children in one generation without changing forms;
2) They manifest themselves more or less at the same development periods for all children within one family or, more specifically, one generation;
3) Familial diseases are clinically independent from any acquired ailment, infection or intrauterine accident;
4) The familial nature of the disease must be the golden rule and not the exception. It is not enough to observe the same disease affecting several members of one family for it to be a familial disease, especially when the disease does not usually have this trait;
5) Lesions resulting from familial diseases are generally localized. They are more likely to be systematic than erratic, and their symptomatology, usually identical among male and female siblings, varies from one family to another.

Based on this definition, the kinship between familial diseases and teratological ailments was reasserted. 'Familial nervous diseases are caused by a transmitted conformation defect, the arrested development of a nervous terrain; they are *analogous* to general malformations and may be considered as a teratological accident ... The same anatomical characteristic, deviating from the normal type, leading to the same pathological physiology, *i.e.* the same disease, is the characteristic of one single family and triggers similar pathology' (Pauly and Bonne, 1897, pp. 201–202).

ÉMILE APERT AND MENDELIAN HEREDITY

In 1907, Émile Apert, a hospital-appointed physician in Paris, published a book with new insights into familial diseases, an issue dominated so far by specialists in nervous pathology and their specific viewpoint (Apert, 1907). His approach was that of a doctor for children who, because he had worked in Paris' five hospitals and in Paris' Maternity Home, had accumulated clinical data 'on a whole range of congenital diseases and familial diseases.' His overview of these various pathologies as presented in his book was directed by a clear, social medicine perspective. It referred to a logic, expressed by a popular movement, that saw 'defective birth rates' as a national catastrophe. Wanting to add to the work of his master Budin 'who devoted his efforts and talent to decrease infant mortality,' Apert aimed at dealing with another aspect of the decreasing birth rate, 'seed improvement', and co-founded French eugenics. The fight against 'racial degeneracy' supported his analytical approach in its very premises: to clarify the study of degeneration mechanisms which an overhasty generalization in pathology had only confused. A critical vision on his contemporaries' fancy, Apert's book was based on his experience as a hospital pediatrist and his sound knowledge of contemporary works in genetics. His 'seed improvement' concerns went together with a whole set of practices whose central subject matter was the mother/child pair and consequently everything that could have an impact on either or both 'protagonists' and their relationship before, during and after conception, at the time of delivery as well as during infancy. Hence his interest in the hereditary and the accidental but also in what distinguishes one from the other. It was a practical interest guiding doctors in their initial actions: to prevent accidents capable of altering the seed, the fetus or the child. His distinction, announced in the introduction, between congenital anomalies falling within the province of 'fetal diseases' and those resulting in an innate tendency of the germ cell had, from this viewpoint, a major nosographic importance.

After reintroducing the hereditary *vs.* accidental distinction in the origins of malformation, Apert suggested a new approach to hereditary familial diseases and their distinction from 'simple' hereditary diseases by translating into medical pathology the pattern developed by botanist De Vries, whose work he presented in detail (Apert, 1907, pp. 192–197).

Oenotheria lamarckiana is an American plant that, when naturalized in Europe, suddenly started to show morphological changes that were different from those observed and described as variations. These were 'abrupt mutations' creating new species that were different from the old species in several characteristics and unrelated to it by any intermediary form. A change in general living conditions, Apert insisted, brings about a tendency to 'mutate' in some descendants of those 'subjects' submitted to the change. These abrupt variations were not related to adapting to the new environment; they seemed to occur haphazardly, were most frequently adverse, and only exceptionally and accidentally did a variation of this type prove favorable in terms of the new living conditions of the plant. Apert suggested that innate, hereditary and familial diseases 'are of the same order as variations and mutation observed by naturalists ... especially when the living conditions of the species are modified, thus probably causing minute changes in the very makeup of the cells' protoplasma in general and sexual cells in particular; hence a morphological modification of the being coming from the development of this sexual cell' (Apert, 1907, p. 208). Using this reasoning, hereditary diseases as well fell within the domain of variation and mutation phenomena. Hereditary familial diseases had their origin in mutations that were transmitted according to the laws of heredity described in the 19th century by Gregor Mendel, re-discovered and confirmed by De Vries and validated by biologist Cuenot as applicable to the animal kingdom. Apert was thus the first French physician to introduce the concepts of 'dominant characteristics,' 'dominated characteristics,' 'latency of dominated characteristics' and the law of dissociation of characteristics between gametes. He considered these concepts to be the first explanation for 'atavism' and for why inbreeding could be harmful by possibly manifesting atavistic variations that had been latent over several generations (Apert, 1907, p. 338). But what is true for hereditary familial diseases is not true for just any morbid heredity since all morphological characteristics are not 'Mendelian.' Only those morphological characteristics creating a formal difference between two races without a gradual shift abided by these laws, such as albinism for mice or dogs. But size is not a Mendelian characteristics for dogs. This difference matched the 'abrupt mutations' (a Mendelian characteristic) *vs.* 'simple variations' (a non-Mendelian characteristic) distinction. Whereas hereditary familial diseases reflected an hereditary, mutation-type change in a 'Mendelian' characteristic, 'simple' hereditary diseases were due to a variation mechanism on non-Mendelian characteristics.

This interpretation allowed Apert to have 'Mendelian genetics' co-exist with the degeneration theory as amended by Féré. To do this, he only had to call forth the old Hippocratic conception of temperaments as modernized by the knowledge at that time.

'These temperaments, whose hereditary transmission has been acknowledged from time immemorial, are mild, familial variations of the physiological type; they are similar to mild variations of the physiological type which make up physiognomies and family types ... The overemphasis on physiologically-related variations can lead to abnormal physiological types with morbid predispositions and already pathological, chronic conditions ... It may be systematically said that each temperament can be matched to a condition of diathetic predisposition: the sanguine temperament to arthritism; the nervous temperament to a neuropathological predisposition; the bilious temperament to familial choalemia; the lymphatic temperament to familial dysthyroidism ... Familial morbid dispositions are triggers to familial diseases; familial diseases of the nervous system are heralded by a neuropathological predisposition ... ' (Apert, 1907, pp. 353–354).

Articulating the most current (De Vries and Mendel) with the most ancient (Hippocrates), Apert placed the new conceptualization of familial diseases within the continuity of morbid heredity theories developed in the second half of the 19th century. Degeneration was redefined, readjusted in its subject matter, but its principle remained the same; it was made, Apert hoped, just more operative.

NOTES

1. One only has to see Charcot's use of it, such as in two cases of peripheral, facial paralysis with 'nervous heredity' presented at one of his 'Tuesday lessons.' The first case was a young girl whose facial paralysis was clinically objectivized in front of the audience. Her mother 'revealed' their family's history: her own mother, an aphasiac at forty-six, was paralyzed on the right-hand side; one of her uncles had been paralyzed since he was nine; her husband, *i.e.* the child's father, had been institutionalized for two and a half years at the Vaucluse asylum, his own father, an hemiplegic being now deceased after a 'very irregular' life. The second case was a man who, in addition to his 'peripheral,' facial paralysis, had a pronounced stutter. When asked by Charcot, he explained that it was, a familial disease, both his father and grandfather being stutterers. Charcot's comments: 'Here is, Gentlemen, an unexpected and very interesting revelation as the stuttering is in the neuropathological family and has a distinguished rank therein.' Prompting the patient about other nervous diseases in his family, Charcot received another 'revelation': the man had a dead brother, who had been treated at the Bicêtre facility in Bourneville's department, and a retarded child with convulsive attacks. Charcot's conclusion seemed to be stating the obvious: 'Well, Gentlemen, what do you think? Here are two cases of random, peripheral facial paralysis: aren't they significant enough from nervous heredity's point of view?' (Charcot, 1889, pp. 78–79).
2. 'The habits and the diseases of the parents should be carefully determined, for of the ills the child is heir to, not a few are due to alcoholism, to syphilis, to hysteria, to psychic disorders of a parent,'

wrote Prof. Sachs in his handbook on the nervous diseases of children. According to Sachs, clinicians must be aware of the parents' resistance to admitting familial defects and should not hesitate to insist upon these points during his questioning while remaining tactful. They should also conduct a physical examination as thorough as possible, searching for possible degenerative stigmata such as prognathism, malformations of the ears and lips, and abnormal hair (Sachs, 1905, p. 2).

REFERENCES

E. Apert, *Traité des maladies familiales et des maladies congénitales* (Paris: Baillère et Fils, 1907).

F. Bing, 'La théorie de la dégénérescence,' in *Nouvelle Historie de la psychiatrie* (Paris: Dunod, 1994).

R. Castel, *L'Ordre psychiatrique* (Paris: Ed. de Minuit, 1979). Charcot, 'Révision nosographique des atrophies musculaires progressives,' *Progrès médical* (1885), 10, pp. 181–186.

Charcot, *Leçons du mardi* (Paris: Lecrosnier et Babé, 1889).

J. Déjerine, *L'Hérédité dans les maladies du système nerveux* (Paris, 1886).

C. Féré, 'La famille névropathique,' in *Archives neurologiques*, Nrs. 19 and 20 (Paris: Delahaye et Lecrosnier, 1884).

Jouffroy, 'Discussion de la présentation des docteurs Lion et Gasne,' *Bulletin mensuel de la Société médicale des hôpitaux de Paris* (1902), XIX, e S, pp. 12–13.

R. Koch, *Aetiology of Tuberculosis. From Consumption to Tuberculosis* (1884, reprinted New York and London: Garland Publishing, 1994)

P. Londe, *Les Maladies familiales*, Dissertation, Paris, 1895.

P. Londe and H. Meige, 'Myopathie primitive généralisée,' *Nouvelle Iconographie de la Salpêtrière* (1894) VII, pp. 142–159.

Pauly and Bonne, 'Maladie familiale à symptômes cérébello-medullaires,' *Revue de Médecine* (1897) pp. 201–216.

D. Pick, *Faces of Degeneration* (Cambridge: Cambridge University Press, 1989).

A. Pillet, 'Un cas de myopathie pseudo-hypertrophique avec troubles de l'intelligence,' *Revue de médecine* (1890), pp. 399–408.

P. Pinell and M. Steffen, 'Les médecins français. Genèse d'une profession divisée,' *Espace social européen* (1994), 258, pp. 41–55.

F. Raymond, *Maladies du système nerveux, atrophies musculaires et maladies amyotrophiques* (Paris: Doin, 1889).

F. Raymond, *L'Hérédité morbide* (Paris: Vigot Frères, 1905).

Roques, 'Des dégénérescences héréditaries produites par l'intoxication saturnine lente,' *Comptes rendus de la Société de Biologie* (1872), IV, pp. 243–245.

Sachs, *A Treatise on the Nervous Diseases of the Children for Physicians and Students* (New York: Wood, 1905).

P. Troisier and Guinon, 'Deux cas de myopathie progressive chez le père et la fille,' *Revue de médecine* (1889), IX, pp. 48–59.

Chapter 8

HEREDITARY DISEASE AND ENVIRONMENTAL FACTORS IN THE 'MIXED ECONOMY' OF PUBLIC HEALTH: RENÉ SAND AND FRENCH SOCIAL MEDICINE (1920–1934)

Patrick Zylberman[1]

In the early 19th century, heredity and infection provided alteratives explanations for diseases such as tuberculosis and syphilis, and thus spurred conflict in the field of public health and hygiene. Meanwhile, the old notion of 'heredity-degeneration' was evoling into a new concept of 'heredity-evolution'.[2] Slowly too, the old hereditarian ideas were giving way to genetics, or to what we might call medicalized eugenics. But are the notions of heredity and infection actually opposed to each other? The more knowledge we have of genetics, the more we must inquire into the meaning of this opposition inherited from the past. Herein, I would like to preent a special chapter out of the history of French *hygiène,* a concept that evolved into the idea of social medicine. Attention will thus be focused on the state and family as providers of health care within the 'mixed economy' of public health (to borrow Jane Lewis' phrase).[3]

THE SURVIVAL OF THE OLD HEREDITARIANISM

In French medical opinion, the ideas of Pasteurian hyiene were widespread during the inter-war period. By the 1920s, Léon Bernard, Robert Debré and Marcel Lelong proved that tuberculosis was neither hereditary nor congenital, even though its transmission was bound up with hygiene in the family.[4] When new-born children in tubercular families were separated from sick parents, they were sheltered from infection. The mother or wet nurse was the primary vector of transmission for the bacillus. For doctors in soial medicine, the theories of heredity dominant in the late 19th century—*e.g.*, the heredity of predisposition (*hérédité de prédisposition*) centered around the idea of a congenital breeding ground for certain diseases; or seed heredity (*hérédité de graine*), around the idea that the mother transmitted the bacillus through the placenta)—were no longer tenable.[5] Despite ongoing doubts about whether syphilis was hereditary,[6] Léon Bernard declared at the 1927 Hygiene Congress that, for most diseases, 'Heredity carries less weight than contagion' both as a cause of pathology and as a matter of public health. Rightly so, Bernard wanted 'to place limits on morbid heredity'.[7] As a consequence, hygienists saw the family less as a cause of disease than as an environment fostering the conditions where diseases breed.

Of course, this new view of the family could fit in with the old hereditaianism, which doctors had not forsaken. Family doctors still found no other explanation for disease than heredity.[8] They were obsessed by the idea of abnormal offspring and congenital defects. Their ideas resembled folk beliefs. Knowing nothing about the newly born but barely taught science of genetics, and very little about microbiology, they held to disproven ideas that, dating from the 19th century, confused heredity with inheritance. Given that doctors lacked knowledge about the scientific theory of heredity, the ideas so widespread before 1914 did not disappear for a good while afterwards.[9]

Not surprisingly, hereditary disorders became the very secret that both families and doctors had to keep, the doctors bound by law.[10] This unrelenting silence about hereditary defects or disorders undoubtedly accounted for the survival,—in both the general population and among medical practitioners—of the idea that diseases had hereditary causes. For family doctors, hereditarianism was a means of defense against encroachments by (bacteriological) laboratories and social medicine into the field of family health. Reduced to making diagnoses and prognoses about diseases caused by heredity, family doctors were not to blame if they could do little or nothing to cure patients. Little wonder that they supported the Société française d'eugénique, which viewed tuberculosis, syphilis, alcoholism and rickets as hereditary diseases![11] After all, were the advocates of the contagious theory of disease not calling for a public hygiene that would have the right to probe the heart of secrecy, namely family life?

However surprising it might be, the alliance between private practitioners who objected to state interference and eugenicists who called for state interventions was not at all illogical.[12] French eugenics, it should be pointed out, was much closer to pediatrics than to raciology.[13] It sets its aims less on the purity of the population than on the preservation of the family. Except for a few eccentric personalities (among them: two Nobel prize winners and one minister of Health), French eugenicists did not call for sterilizing the 'genetically degenerate'.[14] French opinion never accepted such an action, nor for that matter, 'sanitary marriage certificates', divorce on the grounds of ill heath, or the punishment of persons who spread diseases (*délit de contamination*).[15] The medical profession might have accepted such actions in principle, but it raised objecions about putting them into practice (*e.g.*, that blood tests too often yielded falsely positive results).[16] Hence 'bio-politics' never took shape in France. For both eugenicists and doctors, supporting hereditarianism meant othing more than sustaining the family.[17]

SOCIAL MEDICINE UNDER THE DOMINANCE OF HEREDITY

But the struggle between hereditary and the infectious causaion of disease had not yet ended. In 1925, a 'short bibliography about the causes of decreasing mortality due to tuberculosis' (drawn up for the League of Nations' Health Section) still maintained a baance between the two.[18] And such was the case for not just tuberculosis but also rickets, diabetes and even leprosy. Hereditarians were used to seeing the family as a private domain reserved for them, whereas social medicine tended to see in it the origins of a host of diseases involving both genetic and social factors. Not only did the family spread infection and germs, it also provided the grounds for phenotypes.[19] Whether of hygienist or eugenic suaion, social medicine's 'therapeutic' actions mainly concentrated on family life. Evidence of this were the anti-tuberculosis dispensaries[20] and 'positive' eugenic measures, such as early screening for defects and educating mothers about the dangers of germs. The measures were to protect society for being overwhelmed by a 'hereditary underclass'.[21] Social medicine was tinged with eugenics in French-speaking lands.

René Sand (1877–1953) was one of the more prominent repre-entatives of this social medicine. Trained as a doctor in pathological anatomy in Brussels, Vienna and Berlin, he took up social meicine in 1911. He was soon asked to hold offices in his country's top administration first in the Belgian Ministry of Labor and then the Ministry of Health. In addition, he chaired the Belgian Medical Association; and his international career led him up to the General Secretariat of the League of the Red Cross Societies, the Health Section of the League of Nations (henceforth LN) and, finally, the World Health Organization.[22]

Does poverty cause disease? or disease poverty? Links between the two have never been easy to establish, and the presumed direction of causality has changed from one period to another. These questions interested both hygiene and eugenics. In England, Newsholme and Ryle, on whose work Sand would draw extensively, dwelled on them.[23]

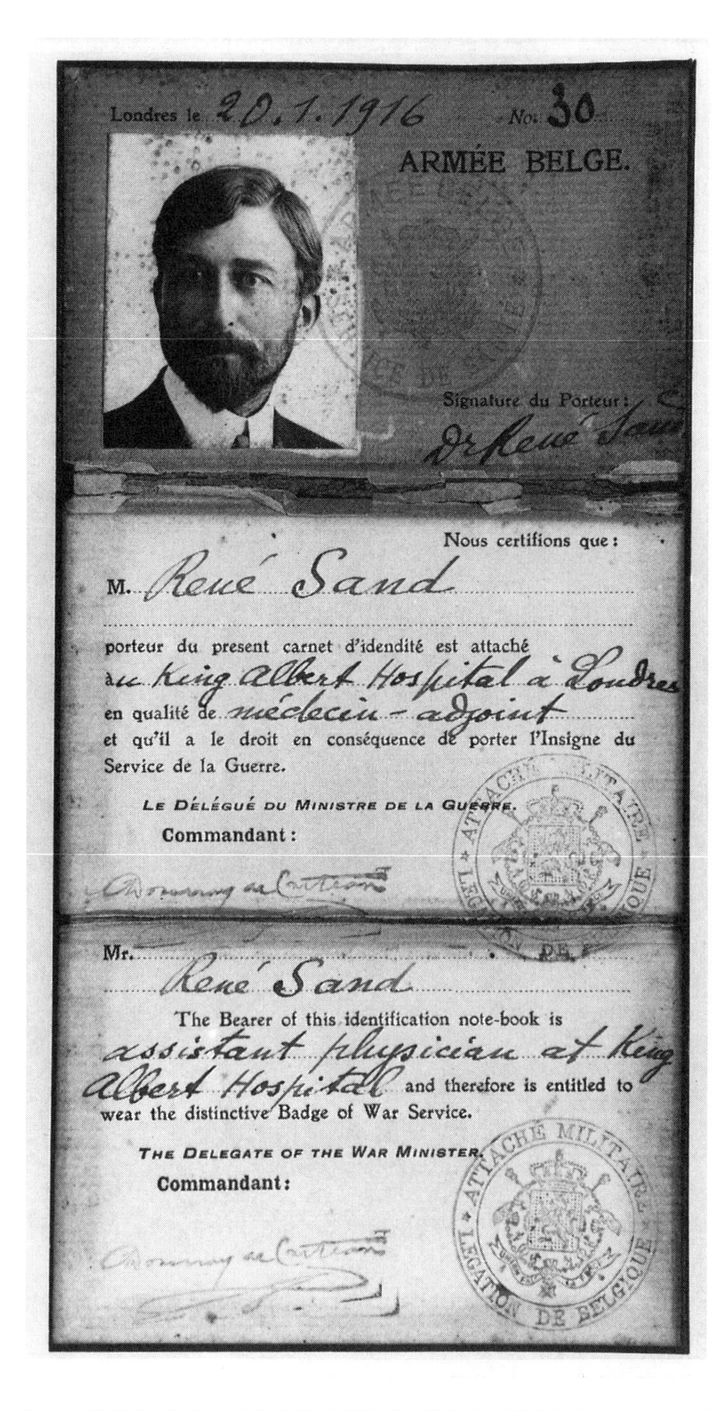

Londres le 20.1.1916 No. 30

ARMÉE BELGE.

Signature du Porteur: Dr René Sand

Nous certifions que:

M. René Sand

porteur du present carnet d'idendité est attaché au King Albert Hospital à Londres en qualité de médecin-adjoint et qu'il a le droit en conséquence de porter l'Insigne du Service de la Guerre.

LE DÉLÉGUÉ DU MINISTRE DE LA GUERRE.

Commandant:

ATTACHÉ MILITAIRE LÉGATION DE BELGIQUE

Mr. René Sand

The Bearer of this identification note-book is assistant physician at King Albert Hospital and therefore is entitled to wear the distinctive Badge of War Service.

THE DELEGATE OF THE WAR MINISTER.

Commandant:

ATTACHÉ MILITAIRE LÉGATION DE BELGIQUE

FIGURE 8.1 RENE SAND IDENTITY CARD WHEN SERVING IN THE BELGIAN ARMY (ARCHIVES DE L'UNIVERSITÉ LIBRE DE BRUXELLES).

FIGURE 8.2 RENE SAND TAKING THE CHAIR, WHO, 1951 (ARCHIVES DE L'UNIVERSITÉ LIBRE DE BRUXELLES).

Hygienists switched back and forth on this issue, from considering socio-economic factors to be determinants of 'health inequalities' (thereby claiming there was no evidence for a class-based biological inheritance) to seeing them as having no effect (thereby playing with the idea that heredity or, at least, intra-family factors were an overriding determinant of these inequalities).[24]

But how to detect and measure health inequalities? Mortality and morbidity rates did not suffice. What was needed were positive indicators of health such as height, weight and body build. Both hygiene and eugenics were interested in statistics on morphological traits, such as Galton's 'family correlaions' or 'social anthropoetric traits' (which Sand, in agreement with Quételet, referred back to Charles Roberts and Alfredo Niceforo).[25] Along with this common empirical data, hygiene and eugenics also shared the same positivist principles. Before 1930 when they diverged because eugenics came to see heredity as the 'cause of causes',[26] both these disciplines had the same subjects, methods and attitudes, in short, the same ways of observing and talking about health. Nonetheless, anthropometry served to emphasize the impact of social factors (child mortality, nutrition, housing) on morphology; it thus refered disease to economic class in addition to biological considerations.[27] But as long as hygiene and eugenics dwelled on phenotypes, they avoided any conflict about choosing between heredity and infection.[28]

Furthermore, hygiene and eugenics resorted to a single type of explanation. Sand more often than not held the family to be responsible for pauperism.[29] He thus echoed Sydney Webb's *Prevention of destitution* (1911), which he had read closely.[30] Were it not for the deleterious impact of morbid heredity, how could science account for the existence of a *résidu social* (Sand's literal rendering of the word 'residuum') unaffected by social improvements, medical progress or a rising standard of living? Sand remained much more moderate than his British predecessors when estimating the proportion of this 'social problem group' in the general population: he mentioned 3% in contrast with 10% according to A. F. Tredgold and E. O. Lewis.[31] Much like the Wood Report (1929) issued by the Mental Deficiency Committee,[32] Sand made clear that he was willing to see epidemiological, economic and social conditions as operating together. But more importantly, he drew, out of these findings, an image of society as biologically stratified in three layers: at the top, a very narrow layer, the physical and intellectual elite; then a huge mass making up nine-tenths of the population; and at the bottom, dysgenic families of the insane, the retarded, epileptics, the deaf and dumb, the blind, criminals, alcoholics and vagabonds.[33] In this so-called 'engulfed group' (or genetic underclass, to use a phrase that had not yet been coined), which 'almost fatally' passed on its defects from generation to generation, heredity overrode social causes of disease.[34]

FROM A SOCIOLOGICAL BIOLOGY TO A SOCIAL EPIDEMIOLOGY

But after 1930, Sand seemed to have given up his belief in eugenics. The Great Depression struck a blow at this discipline: how could anyone imagine that these millions of jobless persons were feeble-minded? Phthisiologists strongly rejected morbid heredity as an explanation. Speaking before an audience of eugenicists in 1926, Maurice Letulle, a well-known phthisiologist, reasoned: 'All of us, or nearly all of us, are carrying the TB bacillus. But does this mean eugenics should forbid nine-tenths of humanity from marrying?'[35] Although Sand was convinced that, higher wages and more social services notwithstanding, 'progress would not be complete as long as we limit our efforts to improving the social environment without tackling hereditary factors', he did come to see 'misery as a vicious circle [...] A lack of means, physical and moral exhaustion, the impossibility of finding gainful employment, all this ties together and affects not just the parents but the children too'.[36] The part due to heredity was being whittled down. This moderation would mark Sand's next book, which would make him known among hygienists.

As proven by his *L'économie humaine par la médecine sociale* (1934), Sand's understanding of social classes shifted radically. In sharp contrast with his past ideas, he now stated that 'pauperism had lost ground in all industrialized countries' thanks to social progress, which had gradually won ground against disease, infirmity and death.[37] Biological determinism was abandoned for a 'social demography', a sort of quantitative sociology based on the Registrar General's statistics.[38] The three strata were not even mentioned. Instead, there were now five groups defined by the head of family's occupation.[39] Sand cast doubt on the part played by the social Darwinian notion of 'selective disease'.[40] The year 1934 marked a turning point for social medicine, as it looked away from biology and toward the social sciences.

And Sand's conception of epidemiology shifted in line with his acceptance of sociology. Morbidity, he admitted, varied according to occupation and social group. He distinguished between 'diseases of the rich' (such as diabetes, prostatitis, appendicitis, arteriosclerosis and suicide) and 'of the poor' (tuberculosis, venereal diseases, anemia, bronchitis, pneumonia and rheumatoid arthritis), and diseases not dependent on social position (neoplasia, peritonitis, atherosclerosis and neurodegenerative affections).[41] 'Inequality between social classes with regard to health is obvious, *except among the poorest classes* [Let us not forget this restriction!], this inequality does not exist at birth."[42] Social and economic factors as well as nutrition and infection now stood at the top of the hierarchy of causes.[43] There was no doubt that the new technical advisor to the Socialist minister of Health in Brussels had slowly shifted to a reformist vision that blended eugenics and socialism. This version of eugenics was also spreading over England.

Such a conception did not fail to affect how Sand understood hereditary disease. While hesitating about the way alcoholism and syphilis were transmitted (and although, with regard to tuberculosis, he had not yet entirely given up on the idea of *in utero* transmission),[44] Sand distinguished between 'hereditary diseases' and 'family diseases'. The approximately two hundred diseases in the first group were not of much interest to social medicine, since they were passed on following Mendel's laws or due to genetic alterations (*e.g.*, blastophthoria) or else from a family member (*e.g.*, TB, syphilis, alcoholism, lead poisoning or nicotine addiction). 'Family diseases' encompassed infections, food-poisoning and 'probably too' nutritional deficiencies, as well as general conditions (such as diabetes, heart problems or arteriosclerosis) transmitted via the mother to offspring through congenital accidents (*e.g.*, syphilis passed on to the baby in the womb).[45] Sand was even prepared to cast doubt on heredity as the cause of half the cases of mental deficiency—an issue over which geneticists had been wrangling since 1930.[46] He held that 'a rational eugenics could be based only on genetics, the science of heredity'[47]—'genetics' was the word he used at that time. All the ideas he had adopted in the 1920s (from Galton, Pearson and raciological eugenics, or about hereditary syphilis and tuberculosis) were now foreign to his thinking.[48]

Nonetheless, Sand's new ideas were not open to the possibility of the balance between genotype and environment shifting over the life span. On the contrary: his conceptions were static, amounting to scarcely more than a typology. Sand was still caught in eugenics.[49]

INTRODUCING MENDELISM INTO FRENCH SOCIAL MEDICINE

Having opened a new course of thought, Sand's *L'économie humaine par la médecine sociale* was warmly received in France, where its author was already well-known. In fact, President Édouard Herriot had written the preface. In the early 1930s, Sand renamed the Belgian Eugenics Society, which he presided, the Belgian Society for Preventive Medicine and Eugenics. What he was proposing fit well with the French hygienists' most cherished theses.

In France, much like elsewhere, differential fertility was a major topic.[50] But the old hereditarian theories were still proudly making a stand. Psychiatry itself was bound to the notion of 'degeneration' through heredity.[51] For French hygienists, eugenics was to serve as a means for coping with poverty, itself associated with immigration. Some of them still believed poverty was hereditary, whereas Sand had laid aside this medical dogma, perhaps after having been convinced by Edgar Sydenstricker's studies on the impact of the economic recession on mortality and morbidity.[52] But among these very hygienists, eugenics had to, we might say, hide under the guise of pediatrics (*puériculture*) or public health in order to gain admission in the circles of social medicine.[53] This had nothing to do with politics or moral conscience, since Pasteurians stood in the way of Galtonian or Pearsonian eugenics.[54] At the 1927 Hygiene Congress, Léon Bernard declared, 'In fact, the immense majority of diseases are acquired: apart from syphilis, nervous degeneration and dystrophic defects (especially those due to drug-addiction) and certain constitutional conditions such as gout, the domain of morbid heredity is extremely limited.'[55] At the time, Bernard chaired the Société de Médecine Publique, and he would soon preside the Conseil Supérieur d'Hygiène Publique de France. Opinion was strongly opposed to eugenics. It was backed up by widespread 'pro-natalism'[56] and, too, by a deep-seated aversion to statistics and Mendelian genetics.[57] Here and there, dissenting voices were raised, like Étienne Burnet's in the *Quarterly Bulletin of the Health Organization of the League of Nations*.[58]

Sand introduced French medical circles to a sort of eugenics that they could tolerate. His 1934 book presented Mendelism, a seldom broached subject among French-speaking hygienists, to French doctors.[59] Perhaps convinced by the Webbs, Sand carefully refused to choose between heredity and environment, since, at the phenotype level, it was quite hard to distinguish them from each other: except for the cases related to dire poverty, differentiating between hereditary and social factors 'depends on the degree of control' exercised over the latter.[60] Sand's explanations provided for feedback between environment and heredity. For instance, infection or poisoning could, by altering gametes, cause ailments like alcoholism to be passed on to offspring. Nonetheless, there was no *a priori* reason for assuming that the genotype bears no relation whatsoever with the environment.[61] Mendelism was not incompatible with the theory that diseases (in particular TB, leprosy and anemia) had social causes.[62] In other words, the genetic predisposition of an individual was to be distinguished from the incidence of a disease, which mainly depended on the social environment.[63]

Nothing in all this contradicted Pasteurian hygiene. Like major French hygienists, Sand was repelled by the idea of sterilizing or segregating the abnormal. Arguing that recessive genes would eventually be expressed, he refused such practices less for moral or economic than for practical reasons.[64]

What specific factors in 1934 accounted for the reception given to Sand's book and to the multifactorial theory of disease? Neither Guyénot (1924) nor Blaringhem (1928) nor Jean Rostand (1928 and 1930) nor even L'Héritier (1934) had been well received in hygienist circles.[65] Alone, Maurice Caullery's *Les conceptions modernes de l'hérédité* (1935) was listed in René Martial's bibliography in 1938.[66] William Schneider suggests Martial, a hygienist turned raciological anthropologist, was the first to present the multifactorial theory. But owing to his many arguments with colleagues, he was soon ousted from his position as secretary of the Syndicat des Médecins Hygiénistes Français in 1925. His professional reputation probably did not ensure wide diffusion to the multifactorial theory...[67] What about Eugène Apert, his predecessor? Given the titles of his articles, we can hardly conclude that he had introduced Mendelian theory into French social medicine.[68] Presumably, Raymond Turpin might have exercise more influence, given his study of the blood group distribution among mongols.[69]

But what does it mean to 'introduce' a new idea or theory? According to Conry, it means more than exercising influence, for it entails making a discourse compatible with an already structured environment.[70] Owing to his authoritative 1934 book, might René Sand not have played a major role in introducing Mendelian genetics and the multifactorial theory in French social medicine? He was scientifically and professionally qualified for this role. His inluence was all the greater in that he increasingly emphasized the advantages of social medicine and used the family as a frame for explanation and interventions.[71] Morally too, he had evolved so as to agree with geneticists and physicians who opposed Hitlerist eugenics.[72]

SPECIAL CIRCUMSTANCES

But beyond this internal evidence, what external evidence supports the claim that René Sand introduced Mendelism in French *hygiène*?

Had Sand not been a Belgian, would he have had as much influence? The first book in French on Mendelian genetics was an American handbook translated in Brussels in 1923.[73] Just as Geneva had,[74] in the 18th century, been the gateway through which ideas about inoculation and vaccination entered France, Brussels was the gate for Mendelian genetics in the 20th century. In fact, French science had access to English sources were through Switzerland and Belgium. Besides, the Belgian government had offered eugenics an excellent opportunity for carrying out its programs. Under the so-called 'Social Defense Act' of 9 April 1930, the Bureau of Penitentiary Anthropology (originally set up by Émile Vandervelde in 1907) was extended to every prison in the country.[75] Thus was officialized the idea that inmates, being 'biologically inferior', were incapable of mending their ways. In Catholic, French-speaking Belgium, eugenics was not just a theory: public authorities had widely endorsed it. All this undoubtedly lent credence to Sand when he addressed French practitiotners. Sand's French colleagues actually listened to him.[76]

Geneva, too, enhanced Sand's reputation.[77] Sand was a member of the LN's Child Welfare Committee from 1932 to 1936, then of the LN's Health Committee from 1934 to 1936. At his government's request, he remained associated with the Health Committee when he took office at the Belgian Ministry of Health.

During the great depression, the science of nutrition, of which Sand had extensive knowledge, was developed.[78] This focus on nutritional deficiencies diverted attention from eugenics.[79] In their 1935 paper on nutrition and public health, a paper that marked a turning point, Burnet and Aykroyd wrote, 'What we call race is not of major importance.'[80] With nutrition in the limelight, medical circles admitted that tuberculosis, rickets or goiter—to cite dieases that had oft been traced back to heredity—had social causes.[81] There was a swing back to Pasteur and Claude Bernard; and the determination to severely limit the causality attributed to morbid heredity.[82]

Although conditions called for 'permanent and systematic' government interventions in public health,[83] interest in the family did not fall off. On the contrary, studies often used nutrition as a positive indicator of health (dubbed *biotypologie* in France),[84] specifically an indicator of the condition of the individual within the family, which statisticians now took to be a consumer unit.[85] Thus placed in an epidemiological perspective, eugenics was absorbed into social medicine and would find a public—in fact, a market—among French public health officials. This mixture of public actions oriented toward the family helped create a variety of eugenics that no longer owed anything to its Galtonian heritage.

Let us relate the foregoing trends to Sand's career. Though trained in Austria and Germany, he came into contact with England and the English-speaking world during WW I. While organizing relief for Belgian refugees at King Albert's Hospital from 1914 to 1916, he learned how British charities worked. For the rest of his life, he belonged to several British and American medical societies, and above all to the British Medical Association (henceforth BMA) from 1917 to 1947. Not surprisingly, his career bore a visibly British (as well as American) imprint by 1934. In his writings, he often referred to papers submitted at the Medical Research Council. There was something in the British approach that helped him evolve toward social medicine.[86]

In the field of eugenics in Great Britain, dramatic changes had occurred in the early 1930s. Lancelot Hogben's *Genetic principles in medicine and social science* (1932) had transformed the methodology used to study heredity, and thus turned eugenics away from its overreliance on anecdotic pedigree studies.[87] In 1933, the report by the BMA's Mental Deficiency Committee came out criticizing the 'science' of heredity and insisting on social factors.[88] Penrose published *Mental Defect*. And Fisher, taking over from Pearson as head of the Galton Laboratory that same year, changed the sub-title of the *Annals of Eugenics* from 'a journal for the scientific study of racial problems' to 'a journal devoted to the genetic study of human populations'.[89] During these years, eugenics was forced to abandon its old fantasies and was medicalized.[90]

Such events could account for Sand's change of mind; but Sand's 1934 book did not mention Hogben, Haldane or Penrose. Leonard Darwin, Lidbetter and Tredgold were quoted at length, all of them leading campaigns for sterilizing the mentally retarded (a practice, as we have seen, opposed by Sand).[91] Sand drew extensively on German eugenicists such as Goldscheid, Grotjahn and Ploetz. What then led Sand to give up the dominant strain of eugenics? Did he do that under the BMA's influence? or because of his close ties with leading French hygienists? Or was he under pressure from the new 'ideology' of the world health system, from which Germany had just been excluded.[92]

This ideology focused attention on the social and infectious causes of disease instead of morbid heredity. Whatever Sand's motives might have been, we are at a loss to explain the variance with his sources. Perhaps, after all, the explanation lies in his personal inclinations. Besides its scientific framework, the new ideas had a moral and political core: British geneticists, French hygienists and members of the Société Belge de Médecine Préventive had a 'distaste for Manchesterism and a market economy' (in line with Goldscheid, who had coined the phrase 'human economy' that Sand fondly adopted).[93] Could ideology have spurred Sand to adopt theories about which he knew very little at the time?

FROM MEDICAL SECRETS TO GENETIC COUNSELLING: THE CONTINUITY IN MEDICAL EUGENICS

So, by 1934, social factors, infection and nutritional deficiencies were outweighing heredity as causal explanations. In a way, it was a victory for Léon Bernard, who had died that year.[94] Hereditary causes were not fully dismissed from consideration but restricted to a 'residual' class of persons biologically associated with utter poverty.[95] According to Hazemann in 1935, 'It is very likely that eugenism and *biotypologie* form, along with political economy, the bases for teaching hygiene in the future.'[96] A year later, as medical advisor to the Minister of Health, he prophesied that health would eventually stand 'on two feet: eugenics, which will teach people how to have healthy children, and vocational counseling.'[97] As what we might call a recessive explanation, heredity remained bound up with ideas about the inheritance of poverty, even in Sand's thought.[98]

Eugenics was like the ague: it had subsided, but that did not mean it would not recrudesce. Out of pre-marital counseling during the interwar period in France and Belgium arose a sort of public service of eugenics. But it would soon vanish? and then come back to life when scientific diagnoses enabled doctors to satisfy families' legitimate demands for knowledge.[99] Suspicions about state interference in family life were exaggerated. Some historians have mistakenly claimed that competition between state and family has constantly characterized Western societies.[100] It is more accurate to affirm that both the state and family were deemed essential, that both were to share the concern for morbid heredity.[101] For this reason alone, the family is still considered to be a major provider of health care and not just a beneficiary of public health measures.

Translated from French by Noal Mellott, CNRS, Paris

NOTES

1. CERMES (INSERM-CNRS-EHSS), 182 Bd de la Villette, F-75019 Paris. This chapter is drawn from a forthcoming study, in collaboration with Lion Murard, on public health in France from 1919 to 1940.
2. Burian, Gayon, Zallen (1988).
3. Lewis (1995), 1.
4. Bernard and Debré (1920). See Piéry and Roshem (1931), 200; and Delarue (1954), 34.
5. Bezançon (1935), 186–87, 204. J. Grancher had tried separating new-borns from tubercular parents as early as 1903 (Œuvre Grancher), just as Bernard and Debré did in 1921 (Œuvre du Placement des Tout-Petits): see Murard and Zylberman (1996), 498–500. Piéry and Roshem (1931), 94, 187, 193; Delarue (1954), 10. The idea of a breeding ground survived into the inter-war period, for instance in A. Lumière's theory of morbid heredity passing on a "tuberculosis ultra-virus': Delarue (1954), 33–4; and Guillaume (1986), 128–29.
6. According to medical opinion, syphilis was transmitted by both parents. *Cf.*: Quétel (1986), 208–13.
7. Léon Bernard (1928), 35.
8. Hildreth (1994), 198–203; and Roger (1989), 120.
9. Conry (1974), 321; Jones (1980), 103–05; Froggatt and Nevin (1971), 2. Léonard (1992/1983), 148–56. Teaching of genetics : Burian, Gayon, Zallen (1988), 366–67; and Limoges (1982), 324-27.
10. Murard and Zylberman (1996), 304–07.
11. Conry (1974), 327–28.
12. Léonard (1992/1983), 153, 175–77; Clark (1984), 155; Roger (1989), 129; Schneider (1990), 82, 93, 264–65.
13. Roger (1989), 128-29; Schneider (1990), 63–83.
14. Léonard (1992/1983), 166; Murard and Zylberman (1990), 83 : the Minister of Health (1936) was Henri Sellier.
15. About marriage certificates and grounds for divorce, see Sutter (1950), 75–83. About punishing persons who spread disease, see Majerczak (1995), 75–82.
16. Sutter (1950), 82–3.
17. See, for instance: Léonard (1992/1983), 163, 175; and Schneider (1990), 37–45.
18. League of Nations (1925), appendix 51 (Dr Yves Biraud).
19. Hazemann (1935*b*), 239; and Sand (1920), 436–37.
20. Hazemann (1935*b*), 257; Murard and Zylberman (1996), 515-25.
21. The phrase *sous-prolétariat héréditaire* comes from Sand (1931), 32.
22. René Sand: house doctor in Brussels, 1899; medical degree, 1900; junior registrar, 1901; assistant professor (pathology, prof. Rommelaere) at the Université Libre in Brussels; medical adviser (industrial injuries) to a huge insurance company; assistant professor (general practice, prof. Stiénon), 1903–1910; lecturer in occupational hygiene, vital statistics and forensic medicine; founder of the Association Belge de Médecine Sociale, 1912; chief inspector in industrial hygiene at the Ministry of the Interior, 1920; general secretary of the League of the Red Cross Societies, 1921 and then technical adviser, 1927; professor in social medicine at the École des Sciences Criminologiques in Brussels, 1936; technical adviser for the newly created Ministry of Health, 1936–1937; founder of the Institut d'Hygiène et de Médecine Sociale in Brussels, 1938; chairman of the Fédération Médicale Belge during the inter-war period. With the support of the Rockefeller Foundation, he was appointed professor of social medicine and of the history of medicine at the Université Libre in Brussels in 1945. The following year, he became chairman of the technical committee in charge of laying down the principles for a World Health Organization. For biographical details, see Anciaux (1988), 1-5, 11-3, 23, 29; Aujaleu (1981), 22; and Schepers (1993), 389.
23. On Newholme, see: Sand (1934), 298, and (1950), 596; and Eyler (1993), 199–202. On Ryle see: Sand (1950), 330, 598–99, 614, 652; and Porter (1993), 252–54.
24. Sand (1934), 140.

25. Illsley (1990), 233–235; Sand (1920), 436; Searle (1981), 234, has shown eugenics was drawing mainly on physical appearence. In Britain, since the mid-18th century, new born babes were systematically measured and weighed: Loudon (1994), 526. About Galton's "family correlations" *cf.*: Froggatt and Nevin (1971), 6; Schwartz Cowan (1977), 150, 180–81; and Roger (1989), 122–23. On social anthropometry, see: Sand (1920), 726–27 and (1950), 328–30, wherein he quotes : Ch. Roberts, *Manual of Anthropometry, or a Guide to the Physical Examination and Measurement of Human Body* (London, 1878), and A. Niceforo, *les Classes pauvres* (Paris, 1905).
26. Mazumdar (1992), 148.
27. Sand (1920), 445, 446, 450.
28. Sand (1920), 434–35. Something very similar *apud* Grotjahn [Mazumdar (1992), 147–48], whose doctrine Sand had a full cognizance.
29. Sand (1931), 28.
30. Sand (1934), 302 and *passim*. *The Prevention of destitution* was translated into French in 1913 with the title *La Lutte préventive contre la misère*.
31. Sand (1931), 31–2 and (1934), 139, 302; and Macnicol (1987), 302 [Lewis]. Sand did not quote Tredgold in 1931. At the time, his source was still Rowntree's survey of New York, where this social problem group was estimated at 18% of the population (65% of the poor): Sand (1931), 32; and Eyler (1993), 201.
32. Mazumdar (1992), 198, 200.
33. Sand (1920), 438. Sand's image of society looks very much like the 'drop' or 'top' of income analysts : see Mendras (1994), 66.
34. On this 'engulfed group', see : Sand (1920), 438; and Paul (1984), 576. The phrase 'genetic underclass' comes from Nelkin (1992), 189–90. See, too: Mendras (1994), 74, Fassin (1996), 47–54, and Lewis (1995).
35. Letulle (1927), 100.
36. Sand (1920), 435 and (1931), 30.
37. Sand (1934), 82–3.
38. Sand (1934), 33. *Registrar General's Decennial Supplement,* England and Wales 1921, Part II: Occupational mortality, fertility and infant mortality (London, 1927), quoted in Sand (1934), 46 and 289.
39. Sand (1934), 94, 103.
40. Nonetheless, a sociobiological pattern would crop up again in Sand (1950), 322–26.
41. Sand (1934), 108–15.
42. Sand (1934), 116, 120, 282–83.
43. Sand (1934), 9–10, 96.
44. Sand (1934), 137 n. 1, 137–138. Strange enough, because Sand might have heard Maurice Letulle's lecture to the joint meeting of the Société française d'eugénique and the Fédération internationale des sociétés d'eugénique in 1926 at the Paris headquarters of the League of the Red Cross Societies. In that lecture, Letulle strongly objected to *in utero* transmission. See: Letulle (1927), 101. La Société française d'eugénique was the French section of the Institut international d'anthropologie chaired by Louis Marin, leader of the very conservative Fédération républicaine, who was to be minister of Public Health in 1934 : *Le mouvement sanitaire,* 6 (1930), 694.
45. Sand (1934), 137–40, 143; Sand (1950), 399.
46. Jones (1986), 93; but still see the 1934 Brock Report (Tredgold seating on the committee with Lewis and Fisher): Trombley (1988), 122–24.
47. Sand (1934), 132; compared with Sand (1920), 435.
48. Sand (1934), 134, 140, (1934), 29–30, 124–25; to be compared with Sand (1920), 435 and Sand (1927), 38-9 (this lecture had previously been published in *Eugénique,* 14 (1926), 261 sq.); and see also Sand (1934), 119–22, and 140 n. 1.
49. Baird (1994), 134, 137-38.
50. Béjin (1988), 482; Dequidt and Forestier (1927), 161; and Sicard de Plauzoles (1932), 12–14.
51. Toulouse (1930), 666–68; and Heuyer (1935), 631-32.
52. Sicard de Plauzoles (1932), 11 n. 1; Sand (1934), 116 and 301–02. Sydenstricker headed both the Statistics Bureau of the United States Public Health Service and the Public Health Department of

the Milbank Memorial Fund. He was also chief of the Statistics Service of the League of Nation's Health Committee.

53. Sicard de Plauzoles (1927), 42, 196.
54. See Ichok (1930*b*), 275–76, who quoted R. Penel, "La stérilisation eugénique en Amérique et l'expérience californienne", *Le concours médical,* 51 (1930), 51–9: "State eugenics will drive us towards a barbarity such as history has never known".
55. L. Bernard (1928), 34–7.
56. Dequidt (1931), 646.
57. French medicine traditionally rejected statistics owing to Claude Bernard. About this and the connection with hereditarianism, see Carol (1995), 131–33.
58. Burnet (1933), 704; and also Vigne's two articles in *La presse médicale,* 5–6/1934, quoted in Hazemann (1935*a*), 161–62.
59. This presentation of Mendelian genetics also contains the error of 48 (instead of 46) chromosomes, an accepted view passed on from Painter (1923), before Tijo and Levan (1955) established the correct number: Sand (1934), 133 n. 1; Kevles (1985), 238–39. Sand (1900:73–83) began his career under the influence of hereditarianism. See Anciaux (1988), 3. In this early work, which borrowed heavily from Le Dantec, Sand turned out to be a Lamarckian.
60. Sand (1934), 132, 117, 125. Webbs: Searle (1981), 241; and Sand (1950), 396–400.
61. Sand (1934), 124, 126.
62. Sand (1934), 113–15.
63. Baird (1994), 143.
64. Sand (1934), 142–43; see also Hazemann (1934*b*), 327.
65. Burian, Gayon, Zallen (1988), 367. About Rostand, see Hazemann's ironic reviews of: *De la mouche à l'homme* (1930), in *Le mouvement sanitaire,* 6 (1930), 34; and of *Du nouveau-né à l'adulte, l'aventure humaine* (1934) in *Le mouvement sanitaire,* 10 (1934), 695.
66. Martial (1938), 202.
67. Schneider (1990), 244. Dequidt (1926), 613; and Syndicat des médecins hygiénistes français, 11/3/1927, *Le mouvement sanitaire* (1927), 458; see also Selskar Gunn's Diary, Rockefeller Foundation Archives, 27/5/1927.
68. *E.g.*, Eugène Apert, "Les lois de Naudin-Mendel dans l'espèce humaine, en particulier dans l'albinisme humain', *Eugénique,* 2 (1914); on Naudin, see: Piquemal (1993), 95; and Schreiber, Eugénique et mariage, *Eugénique,* 9 (1921), 287.
69. See Schneider, Doctors, hereditary disease and eugenics in France (paper for the Conference 'Transmission: human pathologies between heredity and infection', Paris, May 23–25, 1996).
70. Yvette Conry (1974), 20–4.
71. Sand (1934), 147, 83–4, 275–85, 26-7.
72. See for instance Dequidt (1933*a*), 153–55 ("The criminal error of racism"); Dequidt (1933*b*), 352–55; P. Lereboullet, (1933), 272–74; pro-Hitlerist: *e. g.* Schreiber (1935), 668 *sq*.
73. Morgan, Sturtevant, Muller and Bridges, *Le mécanisme de l'hérédité mendélienne,* translated by M. Herland (Brussels: Lamertin, 1923), quoted in Burian, Gayon, Zallen (1988), 367.
74. Dunbar (1941), 638; and Miller (1957), 197–203.
75. Ichok (1933), 161–72.
76. 'Dr Sand, whom we deeply admire", wrote Hazemann (1933*a*), 119–20. See: Sicard de Plauzoles (1927), 87 and 139; Hazemann (1934*a*), 126, and (1933*b*), 217. Sand had been a correspondent of the Société Française de Médecine Légale since 1910. During the interwar period, he was appointed lecturer at both the École Pratique du Service Social and the École d'Application du Service Social in Paris. *Cf.*: Anciaux (1988), 5, 11.
77. Anciaux (1988), 21. League of Nations, Council (1938*a*), 15 (in French). Sand also participated in the 1933 conference, jointly organized by the Health Committee and the International Labor Office, on public health and the recession. See *Bull. Org. Hyg.,* 2 (1933), 310.
78. Sand's list of references in his 1934 book: Paton, Boyd Orr, Bigwood. Weindling (1995*b*), 319–32.
79. Weindling (1995*a*), 146.

80. Burnet et Aykroyd (1935), 334; to be compared with Gigon's declarations at the 29th session: League of Nations (1938*b*), 23 et 26; a physiologist, Gigon was professor at the Basel Medical Faculty.
81. *Bull. Org. Hyg.* (1932), 450; Burnet, League of Nations (1935), 54: Kevles (1985), 203.
82. Burnet and Aykroyd (1935), 331.
83. Burnet and Aykroyd (1935), 332. The first National Institute for Nutrition was set up in Japan in 1921. In 1929, the USSR created its own connected to its Public Health Commissariat. In France, Paul-Boncour, France's LN minister, set up a Comité National de l'Alimentation on 2 April 1936 "at the invitation of the Mixed Committee on Nutrition created by the 25th Assembly of the League of Nations'. Note pour M. Léon Blum, président du Conseil, 3 p., n. d., anon (probably J. Parisot), and Comité National de l'Alimentation (preliminary committee reports), n. d., 16 p., anon, Archives J. Parisot, Office d'Hygiène Sociale, Vanduvre-lès-Nancy.
84. League of Nations (1935), 54 (Burnet) and (1936), 3–4.; as well as Laugier (1936), 554 and (1935), 168 *sq*.
85. Rajchman critically quoting Gigon: League of Nations (1938*b*), 13.
86. See the obituaries: Sand (1953*a*), 571, Sand (1953*b*), 612; and Anciaux (1988), 7.
87. Mazumdar (1992), 162; Kevles (1985), 197–98.
88. Mazumdar (1992), 201–02.
89. Kevles (1985), 211.
90. Searle (1979), 163.
91. Sand (1934), 293, quoted Darwin's *The need for eugenic reform* (1926); *What is eugenics?* (1928) was translated into French in 1931. Sicard de Plauzoles (1932), 5. Sand (1934), 141, 297, quoted Lidbetter's *Heredity and the social problem group* (1933). Lidbetter was not unknown in France: Heuyer (1934), 6–8. About Tredgold see: Sand (1934), 139 and 143, 302; and Mazumdar (1992), 201. Leonard Darwin favored sterilization for the "social problem group' [Trombley (1988), 129; and Mazumdar (1992), 213], a view to which Sand could well have subscribed. See: Sand (1934), 138–39.
92. Wallerstein (1995), 163; Arnould (1934), 406–14.
93. Weindling (1989), 139–41; and Sand (1941), 13. Ichok (1934), 780.
94. Nevertheless, a French Committee for the Scientific Investigation of Population chaired by Léon Bernard had "officially joined" the International Congress for Population Problems held in Rome, in September 1932, under the honorary chairmanship of the Duce himself. This congress featured eugenics: *Revue d'hygiène* (1931), 383. The following congresses took place in Berlin (1935) and Paris (1937). See Kühl (1994), 32–3, 81.
95. Sand (1950), 329.
96. Hazemann (1935*a*), 162.
97. Hazemann (1936), 32.
98. Sand (1934), 97–8, 119, 138–39, 141–42, 155, and (1950), 155, 329; as well as Anciaux (1988), 64.
99. Sand (1933), 94; and (1950), 399. For Belgium, see: Govaerts (1927), 178–79, 182, and March (1925), 241. And for France: Schreiber (1927), 23-4, and Ichok (1929), 697–98. The first pre-marital visits occurred at Baudelocque Maternity in March 1930, but Alexandre Couvelaire put an end to this counseling three years later because demand was short. Edouard Toulouse (Paris, Henri-Rouselle) and Jules Leclercq (Lille) were also unsuccessful. *Cf.* Carol (1995), 326-28.
100. *E. g.* Burguière (1972), 800.
101. Ichok (1930*a*), 227; and Sand (1950), 158, 330, 387. See, too: Baird (1994), 139, 144.

REFERENCES

Anciaux A. (1988), *René Sand et la culture des valeurs humaines* (Bruxelles: Conseil International de l'Action Sociale)

Arnould E. (1934), review of Ickert F., Rassehygiene und Tuberkulosebekämpfung, *Beiträge z. Klin. der Tuberk.* (1933), and of de Scholtz H., Vererbung und Tuberkulose, *Die med. Welt* 48 (1933), *Rev. phtiio. thérap. soc.*, 15

Aujaleu E. (1981), *les Lauréats européens du Prix Léon Bernard* (WHO: Copenhague)

Baird P. A. (1994), The role of genetics in population health, in Evans R. G., Barer M. L., Marmor T. R. (1994), *Why are some people healthy and others not?* (Berlin, New York: De Gruyter)

Béjin A. (1988), Néo-malthusianisme, populationnisme et eugénisme en France de 1870 à 1914, in Dupâquier J. ed., *Histoire de la population française — 3. de 1789 à 1914* (Paris: Presses Universitaires de France)

Bernard L. (1928), Quelques aspects de l'évolution de l'hygiène, *Revue d'hygiène,* 50

Bernard L. and Debré R. (1920), *Les modes d'infection et les modes de préservation de la tuberculose chez les enfants du premier âge*, Académie de Médecine, 5 October

Bezançon F. (1935), La notion de contagion de la tuberculose, *Le mouvement sanitaire,* 11

Burian R. M., Gayon J., Zallen D. (1988), The singular fate of genetics in the history of French biology 1900-1940, *J. Hist. Biol.*, 21

Burguière A. (1972), Famille et société, *Annales ESC*, 27

Burnet E. (1933), L'enseignement de la médecine et la réforme des études médicales, *Bull. Org. Hyg.*, 2

Burnet E. and Aykroyd W. R. (1935), L'alimentation et l'hygiène publique, *Bull. Org. Hyg.*, 4

Carol A. (1995), *Histoire de l'eugénisme en France* (Paris: le Seuil)

Clark L. L. (1984), *Social Darwinism in France* (University of Alabama Press)

Conry Y. (1974), *L'introduction du darwinisme en France au xixe siècle* (Paris: Vrin)

Delarue J. (1954), *La tuberculose* (Paris: Presses Universitaires de France)

Dequidt G. [bdm] (1926), Chronique, *Le mouvement sanitaire,* 2

— (1931), Nataliste et eugénisme, *Le mouvement sanitaire*, 7

— (1933*a*), Persécutions hitlériennes et racisme, *Le mouvement sanitaire*, 9

— (1933*b*), la Loi hitlérienne sur la stérilisation, *Le mouvement sanitaire*, 9

Dequidt G. and Forestier G. (1927), Laboratoire et médecine préventive, *Le mouvement sanitaire*, 3

Dunbar R. G. (1941), The introduction of the practice of vaccination into Napoleonic France, *Bull. Hist. Med.*, 10

Eyler J. M. (1993), The sick poor and the state: Arthur Newsholme on poverty, disease and responsibility, in Porter and Porter

Fassin D. (1996), Exclusion, underclass, marginalidad, *Rev. fr. Socio.*, 37

Froggatt P. and Nevin N. C. (1971), Galton's 'law of ancestral heredity": Its influence on the early development of human genetics, *Hist. Sc.*, 10

Govaerts A. (1927), La pratique de l'examen médical avant le mariage en Belgique, in Sand *et al.* (1927)

Guillaume P. (1986), *Du désespoir au salut: les tuberculeux aux 19e et 20e siècles* (Paris: Payot)

Hazemann R. H. (1933*a*), review of Sand, *La Belgique sociale* (1933), *Le mouvement sanitaire*, 9

— (1933*b*), Existe-t-il des possibilités de développement de l'hygiène publique en France, *Le mouvement sanitaire*, 9

— (1933*c*), review of Lahy J. M., Sur la validité des tests exprimés en 'pour cent' d'échecs, *Le travail humain* 1 (1933), *Revue d'hygiène*, 55

— (1933*d*), Les résultats de la stérilisation eugénique telle qu'elle est pratiquée dans l'état de Californie (The Human Betterment Foundation), *Le mouvement sanitaire*, 9

— (1934*a*), review of Sicard de Plauzoles J., *Le problème sexuel* 1 (1933), in *Le mouvement sanitaire*, 10

— (1934*b*) review of Herd H., *Stérilisation des mentaux* (*Lancet* du 30/9/1933), *Le mouvement sanitaire*, 10

— (1935*a*), review of Vignes H., *Les lois de la stérilisation eugénique*, *Presse médicale* of 19 May and 13 June 1934, *Le mouvement sanitaire*, 11

— (1935*b*), Médecine et service social dans la lutte contre la tuberculose, *Le mouvement sanitaire*, 11

— (1936), review of Husson P., *La sélection psychotechnique des travailleurs et les méthodes statistiques*, *Bull. stat. gén. France* (July 1935), *Le mouvement sanitaire*, 12

Heuyer G. (1934), review of Lidbetter E. J., *L'hérédité et le problème social du groupement* [*sic*], *le Problème sexuel*, November

— (1935), Hygiène mentale de l'enfance, *Le mouvement sanitaire*, 11

Hildreth M. (1994), Doctors and families in France, 1880–1930: The cultural reconstruction of medicine, in La Berge A. and Feingold M., ed., *French medical culture in the nineteenth century* (Amsterdam: Rodopi)

Ichok G. (1929), review of Vervaeck L. and Leclercq J., Le certificat prénuptial, *Ann. méd. lég. pol. sc.*, 9 (1929), *Revue d'hygiène*, 51

— (1930*a*), review of Hazemann R.H., *Le service social municipal* (1928), *Revue d'hygiène*, 52

— (1930*b*), La stérilisation des indésirables aux Etats-Unis d'Amérique, *Revue d'hygiène*, 52

— (1933), Annexes psychiatriques et laboratoires de biocriminologie dans les prisons belges, *Revue d'hygiène*, 55

— (1934), review of Imianitoff F., *La médecine préventive* (Bruxelles, 1933), *Revue d'hygiène*, 56

— (1935), Médecine et service sociale dans la lutte contre la tuberculose, *Le mouvement sanitaire*, 10

Illsley R. (1990), Comparative review of sources, methodology and knowedge, *Soc. Sc. Med.*, 31

Jones G. (1980), *Social Darwinism and English thought: The interaction between biological and social theory* (Brighton: The Harvester Press)

— (1986), *Social hygiene in twentieth-century Britain* (London: Croom Helm)

Kamminga H. and Cunningham A. (1995), *The science and culture of nutrition 1840-1940* (Amsterdam: Rodopi)

Kevles D. J. (1985), *In the name of eugenics: Genetics and the uses of human heredity* (New York: Knopf)

Kevles D. J. and Hood L. (1992), *The code of codes: Scientific and social issues in the human genome project* (Cambridge: Harvard University Press)

Kühl S. (1994), *The Nazi connection* (Oxford: Oxford University Press)

Laugier H. (1936), Programme général de recherches sur les mesures et épreuves biologiques permettant de définir les états de sous-nutrition, *Bull. Org. Hyg.*, 5

League of Nations (1925), Comité d'hygiène, minutes of the fourth sesion, 20-25 April, C.224.M.80.1925.III, annexe 51 (Dr Y. Biraud)

— (1935), CH/22nd session/P.V. (dactyl.), Archives J. Parisot, Office d'Hygiène Sociale, Vandœuvre-lès-Nancy

— (1936), Organisation d'hygiène, Rapport des experts réunis à Genève du 8 au 10 décembre 1936 pour discuter des méthodes d'appréciation de l'état de nutrition de la jeunesse, Genève 10/12/1936, ch/com.exp.Etat Alim/8 (1), 3-4 (dactyl.), Archives J. Parisot, Office d'Hygiène Sociale, Vandœuvre-lès-Nancy

— (1938*a*), C./100th session/P.V.3(1)

— (1938*b*), CH/29th session/P.V. révisé (dactyl.), Archives J. Parisot, Office d'Hygiène Sociale, Vandœuvre-lès-Nancy

Léonard J. (1992 [1983]), Eugénisme et darwinisme. Espoirs et perplexités chez les médecins français du xixè siècle et du début du xxè siècle; suivi de Le Premier Congrès International d'Eugénique (Londres, 1912) et ses conséquences françaises, in *Médecins, malades et société dans la France du xixè siècle* (Paris: Sciences en Situation)

Lereboullet P. (1933), À propos de la loi allemande sur la stérilisation, *Rev. All. nat. accrois. pop. fr.*, September

Letulle M. (1927), Tuberculose et mariage, in Sand (1927)

Lewis J. (1995), Presidential address: Family provision of health and welare in the mixed economy of care in the nineteenth and twentieth centuries, *Soc. Hist. Med.*, 8

Limoges C. (1982), A second glance at evolutionary biology in France, in Mayr E. and Provine W. B. (1982 [1980]), *The evolutionary synthesis: Perspectives on the unification of biology* (Cambridge: Harvard University Press)

Loudon I. (1994), review of Peter Ward W., *Birth weight and economic growth: Women's living standards in industrializing West* (1993), *Soc. Hist. Med.*, 7

Macnicol J. (1987), In pursuit of the underclass, *J. Soc. Pol.*, 16

Majerczak J. (1995), Les hygiénistes et la lutte anti-vénérienne sous la IIIè République, aspects politiques, mémoire of the Institut d'études politiques de Paris

March L. (1925), La septième session annuelle de la Commission Internationale d'Eugénique, *Eugénique*, 3

Martial R. (1938), *Race, hérédité, folie, étude d'anthropo-sociologie appliquée à l'immigration* (Paris: Mercure de France)

Mazumdar P. M. H. (1992), *Eugenics, human genetics and human failings: The Eugenics Society, its sources and its critics in Britain* (London: Routledge)

Mendras H. (1994), *La seconde révolution française 1965-1984* (Paris: Gallimard-Folio)
Miller G. (1957), *The adoption of inoculation for smallpox in England and France* (Philadelphia: University of Pennsylvania Press)
Murard L. and Zylberman P. (1990), De la maladie comme crime, in Job-Spira, Spencer, Moatti, Bouvet (eds.), *Santé publique et maladies à transmision sexuelle* (Paris: John Libbey Eurotext)
— (1996), *L'hygiène dans la République. La santé publique en France, ou l'utopie contrariée 1870-1918* (Paris: Fayard)
Nelkin D. (1992), The social power of genetic information, in Kevles and Hood
Paul D. (1984), Eugenics and the left, *J. Hist. Ideas*, 45
Piéry M. and Roshem (1931), *Histoire de la tuberculose* (Paris: Doin)
Piquemal J. (1993), *Essais et leçons d'histoire de la médecine et de la biologie* (Paris: Presses universitaires de France)
Porter D. (1993), John Ryle: Doctor of revolution? in Porter and Porter
Porter D. and Porter R. (1993), *Doctors, politics and society: Historical essays* (Amsterdam: Rodopi)
Quétel C. (1986), *Le mal de Naples, histoire de la syphilis* (Paris: Seghers)
Roger J. (1989), L'eugénisme 1850-1950, in Bénichou C. ed., *L'ordre des caractères, aspects de l'hérédité dans l'histoire des sciences de l'homme* (Paris: Vrin)
Sand R. (1900), Les causes de l'hérédité, *L'humanité nouvelle,* 1
— (1920), *Organisation industrielle, médecine sociale et éducation civique en Angleterre et aux Etats-Unis* (Paris-Brussels: Baillière et Lamertin)
— *et al.* (1927), *L'examen médical en vue du mariage* (Paris: Flammarion)
— (1931), *Le service social à travers les monde. Assistance, prévoyance, hygiène* (Paris: Armand-Colin)
— (1933), *La Belgique sociale* (Brussels: Office de publicité)
— (1934), *L'économie humaine par la médecine sociale* (Paris: Rieder)
— (1941), *L'économie humaine* (Paris: Presses Universitaires de France)
— (1950), Histoire de l'assistance, suivi de l'Histoire des sciences de l'homme, et de l'Avènement de la médecine sociale, *Arch. belg. méd. soc., hyg., méd. trav. et méd. lég.*, 8
— (1953*a*), Obituary: René Sand, M. D., LL. D., *BMJ*, 21
— (1953*b*), Nécrologie: Le Docteur René Sand, *Rev. méd. belge*, 8
Schepers R. (1993), The Belgian medical profession, the order of physicians and the sickness funds 1900-1940, *Socio. Hlth & Illness*, 15
Schneider W. H. (1990), *Quality and quantity. The quest for biological regeneration in twentieth-century France* (Cambridge: Cambridge University Press)
Schreiber G. (1927), L'examen médical prénuptial dans les différents pays, in Sand *et al.* (1927)
— (1935), la Stérilisation eugénique en Allemagne. Une année d'application de la loi du 14 juillet 1933 pour l'enraiement de l'hérédité moride, *Rev. anthropo.*
Schwartz Cowan R. (1977), Nature and nurture: The interplay of biology and politics in the work of Francis Galton, in Coleman W. and Limoges C., ed. (1977), *Studies in history of biology* (Baltimore: Johns Hopkins University Press)
Searle G. R. (1979), Eugenics and politics in Britain in the 1930s, *Ann. of Sc.*, 36
— (1981), Eugenics and class, in Webster C., ed., *Biology, medicine and society, 1840-1940* (Cambridge: Cambridge University Press)
Sicard de Plauzoles J. (1927), *Principes d'hygiène sociale*, cours libre à la Sorbonne 1922-27 (Paris: Editions Médicales)
— (1932), L'avenir et la préservation de la race: L'eugénique, leçon d'ouverture du cours d'hygiène sociale faite à la Sorbonne le 12 janvier 1932, excerpt from *la Prophyl. antivén.*, 4, April
Sutter J. (1950), *L'eugénique: problèmes, méthodes, résultats* (Paris: Presses Universitaires de France)
Toulouse E. (1930), La tuberculose et la psychopathie, *Revue d'hygiène*, 52
Trombley S. (1988), *The right to reproduce: A history of coercive sterilization* (London: Weindenfeld and Nicolson)
Wallerstein I. (1995), *Impenser la science sociale. Pour sortir du xixe siècle* (Paris: Presses Universitaires de France; Amer. ed. 1991)

Weindling P. (1989), *Health, race and German politics between national unification and Nazism, 1870-1945* (Cambridge: Cambridge University Press)
— (1995*a*), Social medicine at the League of Nations Health Organization and the International Labor Office compared, in Weindling P. (1995), *International health organizations and movements 1918-1939* (Cambridge: Cambridge University Press)
— (1995*b*), The role of international organizations in setting nutritioal standards in the 1920s and 1930s, in Kamminga and Cunningham

Part 4
TRANSMISSION AND MEDICAL PRACTICES

Chapter 9

SPECULATIONS ON CANCER-FREE BABIES: SURGERY AND GENETICS AT ST. MARK'S HOSPITAL, 1924–1995[1]

Paolo Palladino

INTRODUCTION

On 5 November 1995, there appeared in the *Sunday Times* an article announcing dramatically that,

> British doctors will for the first time use a test to select cancer-free babies next month. The procedure raises the prospect of designer babies that will be allowed to develop only if their genetic make-up is approved. In next month's test, embryos of a woman with a high risk of passing on a form of bowel cancer will be screened and only healthy ones will be reimplanted. The same technique is likely to be used within two years to screen test tube embryos for a predisposition to inherited breast cancer.[2]

The woman involved in this revolutionary development was once operated to remove polyps from her intestine because she came from a family susceptible to such polyps. Tragically, these polyps often led to cancer of the colon, and, in the case of this woman's family, the cancer killed her mother and two sisters. The removal of these dangerous polyps had left her unable to conceive, but she desperately wanted a child. The artificial fertilisation of her ova and screening of the resultant embryos, which she was then offered by the increasingly insolvent National Health Service, promises a bright new world where polyposis and its accompanying cancer will no longer take their toll of human lives.[3] However, this and the further developments presaged by the announcement in the *Sunday Times* are evoking public fears that we are entering a world in which babies will be 'designed' like any other material good to be greedily and wastefully consumed. Designing babies free from cancer, though, had already been envisioned and made technically feasible over sixty years ago by Percy Lockhart-Mummery, a leading figure in the development of medical understanding of polyposis as a familial pre-cancerous condition and an ardent eugenist as well. The problem I then wish to address in this essay is why it seems to have taken so long to realise Mummery's dream of human perfection, a dream which some view instead rather contradictorily as a nightmarish return to eugenic planning and occasion for the celebration of human diversity (Shakespeare, 1995).

While an important part of my approach to this question will be to focus on the historical conflict between clinicians and medical scientists over the definition of medicine and its mission, I also want to consider the role of patients and their relatives in shaping the understanding of polyposis as a heritable and possibly preventable disease.[4] Ultimately, I hope that by examining the history of polyposis and its treatment from this particular perspective, I will begin to provide some insight into a further, troubling problem, which seems to be overlooked in the current debates over reproductive technologies.[5] If one of the main criticisms of modern medicine is that it has persistently favoured therapy over prevention, why do we expect with so much trepidation that the newest reproductive technologies will overturn this situation? I want to suggest, very speculatively, that the evolution of these technologies is so dramatic because it is a symptom of a much wider, historical reconfiguration which we are ill-equipped to understand, except nostalgically.

WHAT IS POLYPOSIS?

Until the 1920s, general medical opinion about the causes of cancer was that it was most probably due to some form of irritation.[6] Surgeons specialising in diseases of the colorectal tract, furthermore, had long been aware that the history of some cancers of this tract could be traced back to polyps which eventually became irritated and cancerous. Once the existence of these polyps was revealed by the symptoms of their cancerous transition, such as diarrhoea and anal bleeding, the most common intervention was a difficult and painful removal of the colon. Even after the introduction of radiotherapy as an alternative solution to the problem of cancer, it was generally agreed among surgeons at St. Mark's Hospital for Diseases of the Colon and Anus, the leading British hospital to specialise in the treatment of these organs, that surgery was still the most effective mode of intervention.[7] Unfortunately, between 1918 and 1945, nearly a quarter of those undergoing such surgery died in the surgical theatre or immediately thereafter (Lockhart-Mummery *et al.*, 1956). More tragically, successful operations were followed by life with a colostomy bag and the attendant stigmatisation of those who cannot hide the most hideous impurities of the body. Thus, in one particular case, a patient treated at the Royal Cancer Hospital in 1935 was confined upon his return home to a newly built bedroom in the attic, which his wife would not share and his children were never allowed to enter.[8] Such experiences were so dreaded that some patients preferred to die, even by their own hand.[9] The introduction of ileorectal anastomosis, pioneered by surgeons at St. Mark's Hospital in the 1940s, dramatically reduced operative mortality and removed the need for the colostomy bag, if not the pain from constant diarrhoea. This more radical operation, however, did not displace colostomy until the 1970s, and did nothing to eliminate the need for intrusive, degrading periodic inspections of one of our most private bodily parts to check that the polyps or cancer had been effectively controlled.[10] In sum, polyposis often led to a particularly horrific form of cancer and always left its mark on those affected by it in one way or another (Goffman, 1968).

SOCIALISING CANCER, PROFESSIONAL INTERESTS AND PATIENTS' PERSPECTIVES

The ultimate cause of all this pain and suffering, polyposis, was first claimed as a familial condition in 1882, by Harrison Cripps, a surgeon at St. Bartholomew's Hospital who specialised in colorectal diseases, following the observation of two sisters he had operated for rectal cancer (Cripps, 1882). Perhaps reflecting the endurance of histological explanations of disease in the age of germs and genes, British surgeons did not explore this claim any further for another forty years, when Percy Lockhart-Mummery began to articulate a genetic theory of carcinogenesis grounded in a systematic study of his patients and their families (Lockhart-Mummery, 1925).[11]

Mummery was a leading figure in the organisation of British Empire Cancer Campaign in 1923, as a more clinically orientated centre for investigations of cancer than the Imperial Cancer Research Fund, thus attracting much criticism about his abilities from the champions of medical science Walter Morley Fletcher and Frederick Gowland Hopkins.[12] He had obtained this important, if controversial, position through his work as senior surgeon at St. Mark's Hospital and his private practice on Harley Street: it was said that there was no one among the aristocracy and rich of London who he had not treated, but his indelicate area of specialisation meant that his successes were not widely discussed. These very effective surgical interventions depended on detecting polyps as early as possible, before they had become cancerous and deadly. This was facilitated by Mummery's improvement of the electric sigmoidoscope in 1904, which he celebrated in the *Lancet* by writing that,

> the human mind is so constituted that we always experience a certain pleasure when we have succeeded in obtaining access to some spot hitherto difficult or impossible to explore. And in medicine and surgery the perfection of a method by which we are enabled accurately to explore hitherto inaccessible portions of the human body always arouses our interest and helps to perfect our art (Lockhart-Mummery, 1904).

Another way to view these inaccessible parts of the body, and further perfect the surgical art, was to study the healthy relatives of cancerous patients and assess the risks of the cancer manifesting itself in them. That is, the patients' family histories could perhaps allow Mummery to extend the clinical gaze beyond the reach of the sigmoidoscope.[13] However, not all technologies of visualisation are the same and they can have some unexpected consequences.

At least from 1923 onward, Mummery believed that polyposis was a familial condition, a belief bolstered by Maud Slye's work on the genetic nature of cancer more generally.[14] Despite geneticists' criticism of Slye's work, he greatly praised her, warning however that knowledge of a genetic predisposition was not very useful without understanding what caused the actual manifestation of the cancer.[15] At first he had believed that it arose from some form of irritation, but from the early 1930s he embraced the theory of predisposition more fully and began to build a more complex theory of somatic mutation (Lockhart-Mummery, 1923, 1932). Significantly, he celebrated this theory in *Origin of Cancer* by writing that,

> at last medical science is getting down to the fundamental facts of life and growth, and this will lead inevitably to that knowledge of the primary causes of disease and abnormality which is so necessary if we are able to satisfactorily control disease in human beings (Lockhart-Mummery, 1934, p. 8).

This theory about 'the fundamental facts of life', inspired by Leo Loeb's lecture at the International Conference on Cancer in 1928, which Mummery himself had helped to organise, was based on familial data collected in the polyposis register he had established in St. Mark's Hospital in 1925.[16] From this data he argued that families suffering from polyposis shared a gene that specified an instability of the somatic genetic material, which then led to excessive cell proliferation and increased chance of malignant mutations among these anomalous cells. Importantly, the collection of familial records on which Mummery built his explanation relied not just on formal referrals to St. Mark's

Hospital, but also on a national network of consultants, who had passed through the hospital as registrars and had learned from Mummery a new mode of medical practice. As they dutifully exchanged medical records and access of their patients, the encounter with persons afflicted by polyposis ceased to be that private, personal relationship between individual physicians or surgeons and their patients, upon which members of the medical élite built their prestigious practices on Harley Street. These patients became instead exemplars of a shared, larger diseased group, whose investigation revealed something about the mechanics of carcinogenesis more generally. The social interconnection entailed in Mummery's method and explanation also meant that cancer was comparable to an infectious disease: as genes were passed on from one generation to the next, and thus spread throughout the population, so did cancer. Cancer, like other infectious diseases, thus became a disease of the social body.[17]

This radical transformation of cancer, which marked effectively a point of historical transition from a clinical to a social gaze, was reinforced by Mummery's more explicitly metaphorical discussions of the origin of cancer, in which it was a disease of degenerate, modern civilisation.[18] As he sought to come to terms with the contemporary passage from political individualism to collectivism, cancer became nothing more than the descent of the human body into 'bolshevist' and 'communist' chaos, which could only be remedied by the adoption of 'controlled breeding'.[19] In *After Us, or the World as it Might Be*, a collection of essays on the future of humanity, Mummery then imagined a utopian world where his dream of an orderly and efficient collective would be realised by having all men, except those approximating the ideal citizen, sterilised, and by then allowing women to procreate with the remaining fertile men. He was even prepared to contemplate euthanasia, so that these brave, new parents might eventually engender perfection (Lockhart-Mummery, 1936). In 1935, this commitment to eugenics led Mummery to publicly defend Lord Dawson's claims about the importance of eugenics in the future of medical practice before a very sceptical medical profession, by arguing that 'human genetics must inevitably become the most important social and scientific problem of the next few decades, since it must be solved if the human race is to make any serious progress towards something better' (Lord Dawson of Penn, 1935; Lockhart-Mummery, 1935). However, as he argued in *After Us*, 'sloppy sentiment' stood in the way of such progress, and humanity would then have to await the emergence of 'some kind of autocratic government which has the intelligence, the courage, and the power to enforce the necessary decrees for man's advantage' (Lockhart-Mummery, 1936, p. 220). These views were so radical that even Mummery's close friend and president of the Eugenics Society, Lord Horder, found them too adventurous for his own taste.

Paradoxically, except for these eugenic speculations and a casual comment in a review Mummery co-authored with his close collaborator, Cuthbert Dukes, to the effect that 'one may hope that polyposis families will remain small and finally die out as the result of celibacy or the adoption of eugenic principles', Mummery never sought to translate his ideas about genetic predisposition to cancer into eugenic practice.[20] In fact, he dismissed contemporary research on the 'extrinsic' causes of cancer and the accompanying, largely eugenic proposal for the management of the disease articulated by William Cramer, in that haven of medical science which was the Imperial Cancer Research Fund, by writing that,

> the chances of being able to prevent cancer on the lines suggested by Dr. Cramer is [sic] not a very hopeful one. There is, however, one point worth noting. Where it is known that certain individuals have possibly inherited a susceptibility to develop cancer in a certain organ, then if such individuals are carefully examined, as regards that organ, at regular intervals, there is an excellent chance of the lesion being detected during the early stage, when it is curable Lockhart-Mummery, 1934a, p. 155).[21]

When it came to practical engagement with cancer, eugenics was for Mummery much less important than challenging rival theories of carcinogenesis, such as the viral theory proposed by William Gye.[22] He and Gye twice engaged in acrimonious debate on the pages of *British Medical Journal* and *Lancet*, first in 1932, following Lockhart-Mummery's first presentation of his theory of somatic mutation, and then in 1932, following Gye's review of advances in cancer research and defence of the viral theory against Mummery's alternative (Lockhart-Mummery, 1932, 1938; Gye, 1932, 1938). Mummery was also involved at the same time in a debate over the significance of his theory for the chemotherapeutic approaches to the treatment of cancer proposed by Arthur Todd, which he dismissed as nothing more than palliative measures (Todd, 1932). In sum, Mummery was interested maintaining the primacy of surgical practice, or, to be more explicit, the primacy of his professional interests, over rival approaches to both explaining and treating cancer, even if it contradicted his ideological commitment to eugenics and the possibilities he created to translate such ideology into a radically new way of treating cancer.[23]

Importantly, patients played an equally critical role in shaping the exclusion of cancer from eugenic discourse. Establishing the familiality of polyposis and building practical measures on this hereditary basis called for constant, and often unwanted, surveillance of relatives. For example, Peter Brasher, once a research fellow at St. Mark's Hospital, recalled that in 1935 one of Mummery's colleagues treated a woman for polyposis and then sought to investigate her children, but,

> the co-operation of this part of the family was never freely given, mainly because the father believed that all treatment was meddlesome. They would not communicate with their relatives or give their addresses; the father carried his opposition to the extent of preventing his son having treatment (Brasher, 1954, p. 789).

Such opposition to the entreaties of Mummery and his successors continued long after this time. In 1949, for example, Dukes' journeys to Blackburn to visit one patient's relatives failed when the latter wilfully forgot to keep his appointments.[24] While the archival record provides no explicit explanation for this unwillingness to co-operate, other than dislike for Dukes' and others' condescension, it is clear that these patients and their relatives sometimes had very different ideas from Mummery and Dukes about the reasons for their condition.[25] Firstly, they may not have been very aware of the suspected familial dimension since cancer was something very private about which one should not talk, even within the family, which interestingly and much to the annoyance of Mummery and Dukes included people not related by birth. Secondly, when they were aware of a familial dimension, for example when one patient knew that four of his siblings had died of the same disease, they attributed the coincidence to shared bad diet, specifically eating too many potatoes, rather than to invisible, shared genes. The answer was then to eat more fibre, perhaps in the form of artichokes which this particular patient assiduously cultivated in his allotment.[26] Interestingly, Mummery himself had once believed that the lack of fibre in the modern diet was an important cause of cancer, but once he committed himself to a genetic explanation of cancer he summarily dismissed such explanations and remedies.[27]

In sum, the failure of the genetic explanation of cancer to be translated into eugenic interventions was vastly over-determined by opposing professional interests and patients' unwillingness to collaborate in the fuller elaboration of the explanation.

CONDITIONS OF POSSIBILITY

The modernisation of medicine implied in Percy Lockhart-Mummery genetic explanation of cancer, and the effective transformation of cancer into a social disease, established nonetheless the conditions of possibility for a eugenic approach among those less committed professionally to the primacy of surgery. One such person was, Cuthbert Dukes, who collected much of the data on which Mummery's theory of carcinogenesis was based.

In 1922, after training as a bacteriologist and obtaining a diploma in public health, a form of medical discourse quite different from that celebrated by Mummery, Dukes was appointed by Mummery to the newly created post of clinical pathologist to St. Mark's Hospital. He was actively involved in Mummery's efforts to establish the Pathological and Cancer Research Departments from his very first arrival at the hospital. One of his responsibilities in this context was to follow the fate of patients operated for cancer, to both improve prognoses and settle a long-standing argument between Mummery and Ernest Miles over the merits of their respective operations to remove colorectal cancers.[28] This work was funded in part by the Medical Research Council, which in 1933 brought Dukes' work on familial polyposis to the attention of the Council's newly established Committee on Human Genetics.[29] Around the same time, Dukes also became a member of the Eugenics Society, whose orientation was then being transformed by Carlos Blacker to bring the Society and the suspicious medical profession closer together.[30] One of Blacker's initiatives to achieve this alliance was to arrange the publication of a collection of essays on the inheritance of more familiar clinical pathologies than usually considered by the members of the Eugenics Society (Blacker, 1933).[31] Ironically, in one of these essays, Alfred Piney, a clinical pathologist and practising physician, discussed at length the inheritance of polyposis, writing that 'difficult though it may be to give advice, marriage among sufferers from polyposis coli should certainly be discouraged', but then denied that polyposis was at all significant for any more general understanding of the relationship between cancer and heredity (Piney, 1933, p. 376). Dukes, in other words, was interested in a rare and clinically insignificant disease. Nevertheless, Dukes' connections with human geneticists, and the accompanying recasting of Mummery's theory in mendelian terms, established him as a major figure among medical scientists, a position confirmed in 1948, when he was appointed with E.B. Ford, Alan Greenwood, and Alexander Haddow, to chair the joint international symposium organised by the Genetical Society and the British Empire Cancer Campaign on 'the genetics of cancer'.[32] This importance was also captured more symbolically by the geneticist J.B.S. Haldane's inclusion of Dukes' pedigrees in his critical essay for *New Biology* on 'the prospects of eugenics' (Haldane, 1956). Significantly, Haldane wrote that,

> [i]t is the duty of a physician or surgeon to tell [anyone carrying the gene for polyposis] that about half his or her children will at worst die of cancer, at best be condemned to a life of semi-invalidism ... Certainly, to my mind, such persons should be taught methods of birth control; perhaps they should be given the opportunity of voluntary sterilization (Haldane, 1956, pp. 9–10).

This link between polyposis, genetics and eugenics was facilitated further by the reorganisation of medical practice following the establishment of the Service in 1948. At its outset the Service was committed, at least in principle, to preventive intervention, and that social medicine figured prominently in this commitment.[33] While epidemiology and medical sociology are perhaps the most notable disciplines associated with this form of medicine, genetics was also very important since its approach to understanding disease was fundamentally social (Crew, 1954). Thus, just one year before the establishment of the National Health Service, Dukes had no choice but to argue that polyposis was a condition that did not lend itself to preventive, eugenic intervention and that surgery was the only solution (Dukes, 1947). Once the National Health Service had been established, however, the future of St. Mark's Hospital as an 'independent special hospital' was called into question since 'proctology', the specialism established in 1913 by Mummery and other surgeons at St. Mark's Hospital, was not viewed as legitimate by those who were reorganising dramatically the national system of medical provisions.[34] This threat prompted the creation in 1951 of a financially independent Research Department, with Dukes as its first director, to reinforce the notion that St. Mark's Hospital was an important centre for medical research and not just a haven for clinical consultants.[35] This response was also accompanied by tighter connections with the Imperial Cancer Research Fund, to the detriment of those with the more clinically orientated British Empire

Cancer Campaign.[36] The increasing importance of medical research and its greater institutional independence from the clinical hierarchy at St. Mark's Hospital that was thus achieved meant that Dukes was freer to establish more active links with Lionel Penrose and the new world of human genetics Penrose was trying to forge from the Galton Laboratory at University College.[37] Dukes was now in the position to request far more formally that his colleagues at St. Mark's Hospital should collect blood samples from any patients suffering polyposis and forward them to Penrose to establish linkages between the ABO blood groups and this now more widely recognised genetic disease.[38] He also requested them to allow Ian Aird, at Hammersmith Hospital, access to their medical records to extend his investigations with John Fraser Roberts on blood groups from stomach to colorectal cancers.[39] Furthermore, perhaps instigated by Penrose, Dukes sought to expand the polyposis register to include families from well outside the domain of St. Mark's Hospital, by asking through the *Lancet* that consultants across the country, and the world, report to him any cases of polyposis they might encounter in their practices (Dukes, 1953). This stronger institutional connection with the world of medical research, and human genetics in particular, was translated into greater openness toward preventive, eugenic interventions. Thus, when one patient, apparently better educated than most, expressed some concern about the implications of his condition for reproduct-ion, asking Dukes and Mummery's son, Lynn Lockhart-Mummery, whether he should be sterilised to avoid bringing into the world similarly affected children, they, like Haldane, recommended the more moderate course of visiting a 'good family planning clinic' for advice on contraceptives.[40] Others, such Tom Rowentree, a young surgical registrars at St. Mark's Hospital, were not at all averse to thinking about the more drastic course of action envisioned by this patient and many eugenists (Rowntree, 1950). Importantly, while Penrose supposedly renewed the legitimacy of eugenic language by removing the social biases which had undermined its viability before the second world war, the language used by Dukes to describe the collection of the family histories necessary for genetic explanation and eugenic intervention suggests a persistent, great social division between the investigators and their subjects, often working-class families referred to St. Mark's Hospital from humble cottage hospitals.[41] He wrote that,

> In each family one individual is selected who is called the collaborator. This person is chosen with care. The collaborator may be of either sex, an affected individual or only a relative, old or young; none of these distinctions matters. The essential quality of the collaborator must be that one or she is 'tribal' in outlook, is the sort of person who knows nephews and nieces or aunts and uncles. Having chosen the collaborator, I make note each year in my diary of his or her birthday and write annually so that the birthday letter arrives on the right day. Before writing the letter I consult the family chart, making note of the members about whom information is most needed. The after expressing the usual birthday greetings I inquire after little Alice or Sister Susie or Uncle Tom or whomever it may be, enclosing, of course, a stamped addressed envelope for reply (Dukes, 1952, pp. 2–3).

Apart from the unfortunate connotations that the word 'collaborator' may have had in post-war Britain, this language is very similar to that articulated in the anthropological surveys of Mass Observation and the closeness is nowhere more clearly revealed than in the remarks of Richard Bussey, Dukes' long serving assistant, when he wrote to one consultant that, 'we have sent our beaters out after some polyposis children who have not been seen for a while or not at all. One of these patients has apparently been caught in your net'.[42] The objectification of the human subject in the new language of human genetics, to which Dukes contributed so importantly, continued to be inflected by the discourses of class and racial difference which had shaped rather more visibly and controversially the language of eugenics.[43]

THE ENDURING POWER OF THE SURGEON

While the establishment of the National Health Service, and the emphasis placed in this context on medical science, including the new discipline of human genetics, made possible greater openness to a eugenic approach to polyposis, this did not in fact translate into a systematic eugenic approach to its treatment. Therapy rather than prevention continued to dominate medicine even with the advent of the National Health Service, which was only made possible by compromises with the clinical establishment and its professional ambitions.[44]

Cuthbert Dukes' efforts to collect more data on familial polyposis were often frustrated by the surgeons at St. Mark's Hospital, who, notwithstanding all his requests for collaboration to advance the work of the increasingly important Research Department, admitted, treated, and discharged cases of polyposis before 'reliable' family histories had been drawn. Dukes volunteered to draw these histories himself, assuring these surgeons that communication with the members of the affected families, who might some day also become patients, would not result any uneven 'distribution of these cases throughout members of the surgical staff'.[45] Even those surgeons who were interested in these histories, such as Lynn Lockhart-Mummery, had some difficulty in collecting the required data. For example, in 1951, he contacted a probable member of one of the families on the register at St. Mark's Hospital, apparently disquieting the latter's doctor. Three years later, when he sought to visit the person in question, he had to reassure the doctor that he had 'written an entirely noncommittal letter ... saying that I am getting in touch with you'.[46] In other words, Dukes, the director of the Research Department, remained a medical scientist on the periphery of the clinical hierarchy, with insufficient authority to overcome the possible threat to established institutional relationships between clinicians and their patients entailed in his genetic methods of investigation. Drawing up reliable familial records required Dukes to very carefully negotiate his way around these institutional relationships, negotiations often reflected in the obsequious language he used to acknowledge the collaboration of surgeons such as Mummery and Henry Thompson. Significantly, Dukes' awareness of his proper position coincided with his resistance to calls for an independent college for clinical and academic pathologists, calls which eventually resulted in the formation of the Royal College of Pathologists (Foster, 1982). He was duly rewarded for his loyalty to the surgical profession with a fellowship in the Royal College of Surgeons, and then celebrated his membership by writing the following poem, modelled after an exchange in *Alice in Wonderland*, a poem which stressed the foolishness of those who would challenge the authority of surgeons even in matters of heredity:

'You are old, Father William' the young surgeon said,
'and your colon from polyps is free
Yet most of your siblings are known to be dead
A really BAD family tree.'

'In my youth,' Father William replied with a grin,
'I was told that a gene had mutated.
That all who carried this dominant gene
To polyps and cancer were FATED.

'I sought for advice from a surgical friend,
Who sighed and said 'Without doubt

Your only escape from an untimely end
is to have your intestine RIGHT OUT'

'It seemed rather bad luck, I was then but nineteen,
So I went and consulted a quack,
Who took a firm grip on my dominant gene
And promptly MUTATED IT BACK.'

'This' said the surgeon, 'is something quite new
And before we ascribe any merit
We must see if the claims of this fellow are true,
And observe what your CHILDREN inherit!'[47]

This careful subordination of genetics and eugenics to the imperatives of the surgical profession persisted even after Dukes' retirement. When a 'small genetic clinic' for the investigation of cancer of the colon was established at St. Mark's Hospital in 1960, with funding from the Medical Research Council, it was placed under the joint supervision of the younger Mummery, by this time a senior surgeon, and Basil Morson, a clinical pathologist and Dukes' successor as director of the Research Department.[48] The function of research in this clinic was primarily to improve the use of familial pedigrees to identify susceptible individuals so that they would most benefit from surgical intervention. Thus when one patient asked whether he and his wife should have children because they too might carry the gene for polyposis, the answer was that they should not worry because 'if they've got it ... we'll cure it', presumably by surgical removal.[49] This function remained unchanged as late as in 1980, when Morson discussed the role of the clinic in the prevention of cancer by writing that,

> [t]here are five principles which govern our attempts to prevent cancer in clinical practice. There are the identification of persons who are at increased risk of developing cancer, the registration of such persons, the estimation of the magnitude of cancer risk, the appropriate investigation and last but not least, the meticulous follow-up of patients at increased risk ... We aim by th[ese] method[s] to select out from the population at increased risk only those individuals who show evidence of being at exceptionally high risk. We offer them surgical removal of the colon and rectum at the price of a permanent ileostomy. The cancer prevention rate in these patients is about 50% but among the other half cancer is detected at such an early stage of development that the chance of cure by surgery is high (Morson, 1980, pp. 13–14).

Morson's notion of 'prevent[ing] cancer' was rather different from that imagined by Dukes, or by Percy-Lockhart-Mummery in his eugenic speculations.

FROM REGISTRIES TO MOLECULAR MARKERS

The relationship between genetics and surgery which blocked the possibility of translating the genetic explanation of polyposis into preventive, eugenic measures, however, began to crumble as genetic discourse was molecularised.

One of the respondents to Cuthbert Dukes' call in the *Lancet* for records of cases of polyposis was Arthur Veale in New Zealand. After training to become a surgeon, and studying mathematics to advance his clinical investigations of blood pressure, be became very interested in a large family affected by polyposis (Veale, 1958). He then began to correspond with Dukes, who kept him informed on Lionel Penrose's mathematically demanding work on the linkage between polyposis and genetically determined haematological markers.[50] With the help of these two, Veale was appointed in 1960 as research fellow in genetics at St. Mark's Hospital and in the Galton Laboratory.[51] His work during the tenure of this fellowship resulted in a doctoral dissertation which established polyposis as an autosomal, dominant genetic disorder (Veale, 1965). Significantly, Veale stressed in one of his first reports on the progress of his doctoral work that 'the construction of linkage maps of man is not the ultimate objective but merely an incidental by-product of the investigation' (Veale, 1961, p. 173). Unlike Penrose, who was interested in the demographic and evolutionary significance of polyposis, or, in other words, in understanding whether the condition was dysgenic, Veale was interested in throwing some light on the process of carcinogenesis, again in competition with the newly resurgent viral theory. He concluded very problematically that,

> if the onset of malignancy at a particular site is determined by the completion of a 'partial' mutation, the existence of such a mutation could be proved by demonstrating that it was linked with some other genetically determined factor. This would contribute more to a theory of carcinogenesis than any number of associations or family studies merely showing an increased risk of carcinoma among relatives (Veale, 1961, p. 178).

Veale, in sum, viewed family histories of polyposis as not very useful to understand the genetics of carcinogenesis once the condition was linked to a genetically determined haematological marker. There was no need for further, controversial collections of family records.

Veale's conclusion begged questions about the function of the register at St. Mark's Hospital as the tool of medical research championed by Dukes and Percy Lockhart-Mummery. Not surprisingly, the British

Empire Cancer Campaign, which had funded the collection of data for the register since 1924, began to ask whether the collection of family histories amounted to clinical follow-up rather than clinical research, and whether it should therefore discontinue its funding.[52] Efforts to expand the polyposis register beyond the domain of St. Mark's Hospital were then renewed by appointing Richard Bussey to systematise the collection of records and transform the register into a reference collection for researchers beyond St. Mark's Hospital. In 1968, it became an official reference centre for the World Health Organization. However, just as with any museum of natural history, the museological work of the registry was fundamentally out of step with the now dominant molecular approach to understanding the causes of disease. It thus returned to its more limited, original role as a diagnostic tool, and also became a ready resource for clinical trials, such as those to test whether ascorbic acid retarded or arrested the growth of polyps. Even then, bringing different families together for examination involved interference with established relationships between clinicians and their patients, which greatly offended older figures such as Charles Mann. It required persistent reassurances that any new cases of polyposis detected in the course of the trial would be referred back to the family's official consultant.[53] It was much easier to work with the new diagnostic procedures promised by Veale's successors, molecular geneticists. Thus, when John Northover, a senior surgeon at St. Mark's Hospital educated to have a very different attitude toward the relationship between the clinic and medical science, introduced a new unit of the Imperial Cancer Research Fund to study the genetics of colorectal cancers, he stressed that,

> St. Mark's has played an important part in the evolution of the surgery of colorectal cancer, but surgery alone has reached its limits as a curative measure, and other methods of treatment must be explored ... New pathological techniques are being developed which reveal clinically important information on the biology of the disease, and these need to take their place in the assessment of patients at St. Mark's (Northover, 1984, p. 53).[54]

Of course, Northover overlooked how the search for molecular genetic markers, to develop these 'new pathological techniques' would have been impossible without family records such as those collected by the registry.[55] In fact, the recent controversies over DeCode Genetics and Hoffman-LaRoche's attempt to purchase of the medical records of the entire population of Iceland should remind us that the genealogical approach and all its problems are intimately, if invisibly involved in the achievements of molecular genetics celebrated in the article in the *Sunday Times*.[56] Far more importantly, the discovery of molecular markers was particularly significant for a hospital under considerable financial straits and facing closure during the contemporary effort to radically reorganise the national provision of medical care. This financial significance of pathological markers was already evident in 1980, in Basil Morson's discussion of cancer and its prevention, when he also wrote that,

> We require clinical, genetic and laboratory markers of increased risk in a patient ... Measurement of magnitude of cancer risk has an immense political, economic and professional importance because health resources, in terms of money and expertise are finite in every society, including the richest (Morson, 1980, p. 14).

However, the pressure on financial resources was now much more intense. The new molecular markers promised more a definite identification of patients at risk than could possibly be afforded by statistical calculus of mendelian genetics. This meant greater certainty that any particular member of an affected family did not carry the gene, which would then result in very important savings by relieving the staff at St. Mark's Hospital from the need to call them in for periodic examinations.[57] This strengthened immeasurably the position of geneticists at St. Mark's Hospital, such as Penrose's daughter, Shirley Hodgson, who were now able to establish a 'genetic counselling' clinic for the families registered in the now formally independent Polyposis Registry (Hodgson, 1991). Hodgson thus wrote to prospective mothers and fathers that,

> there have been some exciting new advances in our understanding of the genetic basis of polyposis, and this can provide us with new methods of ... testing an unborn baby for the disease.[58]

While Hodgson and many others involved in genetic counselling, aware of the ethical problems involved in their enterprise, denied that the most appropriate response to positive results should be to terminate the pregnancy, some families assumed that such a result meant inevitably termination and refused therefore the offer of a test.[59] Thus, after meeting with a prospective father, Hodgson wrote to John Nicholls, a surgeon at St. Mark's Hospital, that she had explained to his patient that acceding to a blood test 'could enable us also to do a prenatal test, but he is not thinking of having any further children and did not think that a termination of pregnancy would be justified'.[60] The overwhelming majority of patients, however, were favourably inclined to testing. As Northover pointed out in his discussion of the limitations of surgery, patients at St. Mark's Hospital were 'far more in favour of early diagnosis and application of linkage data to family affairs than their medical attendants might have thought'.[61] Interestingly, though, these patients were, like those more dubious about the merits of testing, unwilling to terminate a pregnancy in the case of a positive result. For example, after a meeting with one such patient, Hodgson reported that this patient had no reservations about the test 'because she does not fees [sic] that the psychological burden to her of having an affected child would be quite worrying'.[62] Of course, as the article in the *Sunday Times* suggests, some patients were favourable to both screening and eliminating defective embryos. Hodgson and Northover, as well as Theresa Marteau, a clinical psychologist at Guy's Hospital, were very puzzled by the diversity of these responses and so launched a series of studies of how patients respond to the results of prenatal genetic screening (Whitelaw, Northover and Hodgson, 1996; Vines 1996). Marteau concluded rather contradictorily that these patients simply misunderstood the nature of their condition, telling a reporter that,

> even though they have been attending clinics and know quite a bit about [the inherited nature of polyposis], they still conceptualise it as being a multi-factorial condition ... it is not that they are ignorant, it's just that people have a sense that a gene may be necessary but not sufficient—there are environmental triggers. Scientifically I don't think it's a bad way of thinking about it (Vines, 1996).

In other words, Marteau thought that, though scientifically astute about the genetic determination of disease, these patients hoped against hope that a positive result did not mean that the child would in fact be affected by polyposis. Scientific knowledge, however, is not the only factor involved in decisions about genetic screening, and both Hodgson and Marteau are well aware of this. The subjects of their studies also understood that, even if affected, their children could eventually be operated and live a relatively normal life, just like its father or mother, and thus they exercised the freedom granted to them by the emphasis on consumer choice which accompanied the demand by the reformers of the National Health Service for greater financial stringency (Jacobs,

1992).[63] That is to say, the genetic information the patients received from Hodgson and her colleagues was being used to manage the risks and inevitable complications of everyday life, and sometimes, but only sometimes, this included decisions to eliminate defective embryos.

Significantly, for Martin Richards, from the Centre for Family Research at the University of Cambridge, studies such as those conducted by Northover, Hodgson and Marteau belie the ingenuity of genetic counsellors' claim to impartiality with regard to patients' decisions about the best response to the results of genetic screening: these patients' rationality is interesting because it runs counter to the counsellors' own tacit presupposition that a defective embryo or foetus should be aborted.[64] From this perspective, the encounter between genetic counsellors and patients was an encounter between the normative system of humanist, preventive medicine and a more contradictory and individualised consumer culture which defies modern notions of the norm (Rapp, 1998). One must wonder, then, what will happen as the emphasis on choice, consumption and economic efficiency further erodes the power of the medical profession and diagnostic test for genetic disorders, such as those developed by LabCorp to test for polyposis, are made more freely available on the open market. As might be expected, the medical profession is very worried about this situation, but the British government is opposed to professional calls for the regulation of this new market.[65] Without the implicit norms of this increasingly weaker profession, it may become impossible to articulate an oppositional language, a language which very significantly is of little importance to those patients who are trying to establish more congenial forms of support than are offered by St. Mark's Hospital: knowing that one carries a gene for polyposis can be an effective resource for the creation of communities and identities never imagined by the discourse of social medicine.[66] To put it another way, the norms of the profession guarantee the contemporary, vigorous debate over the nature of human life as it is profoundly transformed by the realisation of Mummery's dream, but one must begin to ask for how much longer? and what will take its place? See Rabinow (1996, 1999).

CONCLUSION

St. Mark's Hospital, one of the leading hospitals in Britain to specialise in the treatment of the colon and rectum, has figured prominently in the history of intestinal polyposis, a condition which usually leads to cancer of the colon. In particular, from 1923 onward, clinicians and medical scientists in this hospital have articulated, with the aid of an ever more extensive genealogical register of patients and their relatives, a genetic explanation of the condition which transformed cancer into a fundamentally social disease. Significantly, for much of this history, work at St. Mark's Hospital was noticeably devoid of any meaningful eugenic dimension, even though remedying this social disease by preventing the birth of affected children was eminently feasible. This was true even when interest in polyposis was expressed by an ardent eugenist such as Percy Lockhart-Mummery, senior surgeon at St. Mark's Hospital and first proponent of the genetic explanation. Interest in the familial nature of polyposis at St. Mark's Hospital was instead driven until the last thirty years by questions about the process of carcinogenesis and its implications for the position of the surgeon in the political economy of cancer and efforts to control its harvest of death. To be more precise, knowledge of hereditary patterns gleaned from the genealogical register effectively extended clinical insight beyond that offered by other technologies, such as Mummery's electric sigmoidoscope. It thus allowed earlier and more effective surgical interventions against cancer of the colon, and in so doing bolstered the surgeon's pre-eminent position in the institutional organisation of St. Mark's Hospital. In sum, the discursive displacement of the clinical gaze by a social one that took shape during the first decades of this century did not lead inevitably to changes in medical practice, changes toward a preventive, if not eugenic approach. It was instead resisted, even by those who did much to articulate the new language, and sometimes it was also resisted by those it objectified, patients and their relatives.[67]

The new language of polyposis and cancer of the colon established nonetheless conditions of possibility which could be developed further by newer, differently educated generations of clinicians such as John Northover and medical scientists such as Shirley Hodgson, all less attached to the older institutional formations in which it contradictorily took shape. Their success in aligning the new language and medical practice, however, was underdetermined by changes in the education and institutional organisation of the medical profession that took place following the establishment of the National Health Service. It was instead contingent on unexpected contemporary developments outside St. Mark's Hospital. As the article in *The Sunday Times* pointed out, the discovery of molecular markers for polyposis opened the way to medical practices in which planned reproduction figured very prominently. While the article suggested a causal relationship between technological innovation and changes in medical practice, it was not the availability of these new technologies by themselves which made the transition possible, but rather it was made possible by contemporary reforms of the National Health Service which demanded greater financial efficiency within St. Mark's Hospital: knowing with greater certainty who was not a carrier of the gene for polyposis meant greater savings by eliminating the need for periodic screening and thus opened the path for a central role of genetic screening in St. Mark's Hospital. The historical relationship between surgeons and geneticists was thus fundamentally reconfigured. The surgeon and the geneticist alike now see the future of medicine in genetic screening, even if a world of 'cancer-free babies' threatens to remove the need for the old art of surgery sometimes celebrated by Mummery. Importantly, some experts in the field of genetic counselling have suggested that these surgeons' and geneticists' interest in explaining the variety of responses articulated by their patients to prenatal genetic screening and diagnosis reveals the normative nature of this vision of the future of medicine. The norm is the perfect human imagined by Mummery in his more modern moments. However, this neat conclusion seems to me rather unsatisfactory since the convergence on genetics in the medical practices at St. Mark's Hospital was realised by policies predicated on the idea that the modern discourse of society is no longer.

Critics of genetic screening have assailed the norms of the new medicine for being profoundly dehumanising and fear the institutional power which supports it. I think, however, that the criticism is misplaced. As the current debate over genetically modified organisms suggests, professional power is no longer what it used to be, and the greater threat to these critics' concern with the integrity of human life may lie in the further articulation of those policies which first began to undermine this power and coincidentally opened the way for a central role of genetics in medical practice.[68] The view that genetic counselling is normative may be quite perceptive. The link between genetic explanations of cancer and normative eugenic interventions, however, is not a necessary one, but contingent on a particular organisation of health care which is now changing very profoundly. Patients are no longer enthralled by paternal, professional authority, for as Hodgson, the 'vendor' of genetic information, has put it, 'they are adults, you know', and the information they obtain from St. Mark's Hospital and elsewhere is useful simply because it provides them with the wherewithal to manage the complications of life more efficiently by minimising the accidents of chance.[69] This information may even help them to construct new forms of community and identity. Importantly, there is no reason not to believe, and in fact it seems very reasonable, that at least some of those providing genetic information also participate in the more heterogeneous contemporary consumer culture in which their patients articulated their very diverse responses.[70] Some of these surgeons and geneticists may in fact not be at all disingenuous when they argue that the new emphasis on choice vitiates the historical link between genetics and eugenics: they, like any retailer of consumer goods, may have ideas about how one should respond to the information that they carry the gene for polyposis, but they also know that, just as consumers often do very unexpected things with the goods they purchase, those informed that they carry the gene for polyposis often do very unexpected, but nonetheless legitimate things with their information. In sum, as the discourse of social medicine, its subjects and its truths, loses its hold, just as the cognate discourse of society is losing its hold, the production of 'cancer-free' or 'designer' babies may not mark the final realisation of the humanist eugenic dream or nightmare, but a break with the past. It may mark instead the birth of a world in which we will each strive in our different ways to be more efficient self-acting and regulating machines as an end in itself. We will sometimes even buy genetic information, perhaps from the successors of LabCorp, to so refashion ourselves, and the modern discourse of the human will be no longer. See Pearson (1997). Since I do not want to be nostalgic, I prefer to leave the judgement of whether this will be for good or bad to a future, no longer human historian, perhaps the robot historian who will look upon us as 'little more than industrious insects pollinating an independent species of machine-flowers that simply did not possess its own reproductive organs during a segment of its evolution'. (De Landa, 1991, p.3).

NOTES

1. Acknowledgments: I wish to thank Kay Neale for granting me access to the records of the Polyposis Registry and enabling me to interview members of the families on the register, to whom I also very thankful.
2. 'Doctors select cancer-free babies' Sunday Times (5/11/1995).
3. For a systematic discussion of assisted conception, which also informs this essay, see Franklin (1997).
4. For a review of the conflict between clinicians and medical scientists, and of the attendant historiographical problems, see Warner (1985, 1995).
5. For an informative account of many issues at stake in these debates and an introduction to the extensive literature on these issues, see Kerr (1999).

6. See 'Cancer and the public' *Lancet* (1921), II:347–348; and Austoker (1988).
7. For a history of St. Mark's Hospital, see Grenshaw (1985).
8. Interview with members of a family listed in the Polyposis Registry (St. Mark's Hospital), 28/6/1996.
9. Polyposis Registry: Family 4: Letter to member of parliament, 7/6/1991.
10. Interview with members of family 4, 28/6/1996.
11. For contemporary views on the genetic explanation, see 'Genetics and cancer' *Lancet* (1927), II:925–926; and (1928), II: 1137–1138.
12. Public Record Office (London): FD1: 2039: Hopkins to Fletcher, 11/4/1923, and Fletcher to Hopkins, 12/4/1923; see also 2040: Fletcher to Sir Edward Marshall Hall, 12/4/1923. Morson (1962).
13. The only explicit discussion of the study of hereditary patterns as revelation of what is hidden to eye that I have been able to find is in Jacob (1988). More generally, see Aepers (1991) and Cartwright (1995).
14. See Lockhart-Mummery (1925) and Lockhart-Mummery (1923). On Slye, see Gaudillière (1999) and Rader (1995).
15. For a contemporary review and criticism of Slye's, see Cockayne (1927).
16. Lockhart-Mummery (1934), pp. 88–89.
17. Significantly, while cancer never became 'notifiable' like conventional infectious diseases, there was much discussion in the early part of this century on what should be the proper attitude of governmental health authorities toward the disease. The discussion eventually resulted in the Cancer Act which first established a national policy to combat the disease and thus established cancer as a fundamentally social disease.
18. Palladino, (1999).
19. Lockhart-Mummery (1934), pp. 1, 135.
20. Lockhart-Mummery and Dukes (1939), p. 589.
21. See also Cramer (1934).
22. On Gye's viral theory, see Austoker (1988).
23. For a more general discussion of the ambiguous attitude of the medical profession toward eugenics, see Mazumdar (1992).
24. Polyposis Registry: Family 3: C.E. Dukes note, 29/12/1960; and Family 16: Patient to Tom Rowentree, 14/12/1949.
25. Interview with members of family 4, 28/6/1996.
26. Interview with members of family 4, 28/6/1996. On kinship and related question which these patients' understanding of family seems to raise, see Bouquet (1993).
27. On diet and cancer, see Lockhart-Mummery (1936a); cf. Lockhart-Mummery (1923).
38. See *St. Mark's Hospital for Diseases of the Rectum and Colon, Annual Report* (1929), p. 31; see also St. Bartholomew's Hospital, Archives Department: St. Mark's Hospital Papers: uncatalogued accessions: 'The origin of the A. B. C. classification and histological grading of intestinal carcinoma', manuscript, no date.
29. St. Mark's Hospital Papers: KMC 1/1: Minute book of the medical committee, 21/3/1933.
30. Contemporary Medical Archive Centre (London): *Eugenics Society: Annual Report* (1936–7).
31. See Mazumdar (1992).
32. See 'The genetics of cancer' *British Medical Journal* (1948), II:86–88.
33. See Oswald (1991). On the influence of eugenics and genetics on the evolution of social medicine and institutions of the welfare state, see Oakley (1991).
34. St. Mark's Hospital Papers: KMC 1/2: Minute book of the medical committee, 11/11/1947.
35. St. Mark's Hospital Papers: KMC 1/2: Minute book of the medical committee, 18.12.1951.
36. Archives of the British Empire Cancer Campaign (Oxford): Box XC1018: C.E. Dukes folder: C.E. Dukes to F.B. Tours, 19/5/1950.
37. Basil Morson, personal communication, 16/11/1995; and Polyposis Registry: Family 14: A.M.O. Veale to Cuthbert Dukes, 27/12/1953. On Penrose and human genetics, see Paul (1995).
38. St. Mark's Hospital Papers: KMC 1/2: Minute book of the medical committee, 19/2/1952.
39. St. Mark's Hospital Papers: KMC 1/2: Minute book of the medical committee, 16/3/1953. See also Aird, Bentall and Roberts (1953).

40. Polyposis Registry: Family 17: Patient to H.E. Lockhart-Mummery, 6/11/1952; and Family 22: C.E. Dukes to Patient, 7/8/1951.
41. Interview with members of family 4, 28/6/1996.
42. Polyposis Registry: Family 30: H.J.R. Bussey to M. Orr, 23/2/1977. On the anthropological outlook of Mass Observation, see Baxendale and Pawling (1996).
43. This contradicts Pauline Mazumdar's claim that the birth of the National Health Service, and the accompanying extension of medical provisions to the problematic social groups targeted by eugenists, marked the end of the eugenics movement; see Mazumdar (1992).
44. See Oswald (1991).
45. St. Mark's Hospital Papers: KMC 1/2: Minute book of the medical committee, 3/6/1950.
46. Polyposis Registry: Family 38: H.E. Lockhart-Mummery to I.M. Macleod, 16/4/1954.
47. St. Mark's Hospital Papers: Un-catalogued Items.
48. St. Mark's Hospital Papers: KMC 1/3: Minute book of the medical committee, 15.3. 1960. See also *St. Mark's Hospital for Diseases of the Rectum and Colon, Annual Report* (1966), p. 15; and Bussey and Morson (1967).
49. Interview with members of family 14, 10/7/1997.
50. Polyposis Registry: Family 14: A.M.O. Veale to C.E. Dukes, 27/12/1953; and Archives of University College (London): Penrose Papers: 176/1: Correspondence with A.M.O. Veale.
51. Penrose Papers: 176/1: A.M.O. Veale to L.S. Penrose, 13/10/1958.
52. Archives of the British Empire Cancer Campaign (Oxford): Box 2015: Basil C. Morson to F.B. Tours, 11/8/1966. On the displacement of the museological approach, see Allen (1985).
53. Kay Neale, personal communication, 17/4/1996.
54. On the transformation of medical education which took place in parallel with the establishment of the National Health Service, see Rivett (1997).
55. On the methodological reductionism of molecular studies of human heredity, see Marks (1994). On the accretion on, rather than displacement of, the museological approach, see Pickstone (1993).
56. See Berger (1999) and Haraldsdóttir (1999).
57. Kay Neale, personal communication, 4/7/1996. See also Cromwell et al. (1998). On the relationship between technological innovation and administrative rationality in the medical sphere, see Sturdy and Cooter (1998).
58. Polyposis Registry: Family 4: S.V. Hodgson to patients, 27/1/1992.
59. S. Hodgson, personal communication, 11/4/1996, See also Kerr, Cunningham-Burley and Amos (1998).
60. Polyposis Registry: Family 4: S.V. Hodgson to J. Nicholls, 25/8/1992.
61. Northover (1984), p. 39; see also Hodgson (1991), p. 26.
62. Polyposis Registry: Family 33: S.V. Hodgson, note, 22/11/1993.
63. On consumer choice and the reform of the National Health Service., see Klein (1995).
64. Martin Richards, personal communication, 11/11/1995. See also Richards (1993).
65. 'Voluntary code for "mail order" gene tests' *Guardian* (24/9/1997); see also Giardiello (1997).
66. Interview with members of family 14, 10/7/1997.
67. On public opposition to the intrusive nature of preventive medicine, see Turner (1995).
68. On the demise of professional authority, see Freedland (1999).
69. Shirley Hodgson, personal communication, 10/6/1999. See also Beck (1992).
70. On consumer choice and the inadequacy of those sociological models which reduce the consumer to a passive respondent to commercial or other institutional agents for understanding contemporary developments in the management of genetic resources, see Lury (1996).

REFERENCES

I. Aird, H.H. Bentall, and J.A. Fraser Roberts, 'A relationship between cancer of the stomach and the ABO blood groups' *British Medical Journal* (1953), I: 799–801.

G. Allen, *Life Science in the Twentieth Century* (Cambridge: Cambridge University Press, 1985).

S. Alpers, 'The museum as a way of seeing' in I. Karp and S.D. Lavine (eds.), *Exhibiting Cultures: The Poetics and Politics of Museum Display* (Washington: Smithsonian Institution Press, 1991), pp. 25–32.

J. Austoker, *A History of the Imperial Cancer Research Fund*, 1902–1986, (Oxford: Oxford University Press, 1988).

J. Baxendale, and C. Pawling, *Narrating the Thirties: A Decade in the Making, 1930 to the Present* (London: Macmillan, 1996).

U. Beck, *Risk Society: Towards a New Modernity* (London: Sage, 1992).

A. Berger, 'Private company wins rights to Icelandic gene database' *British Medical Journal* (1999), I: 11.

C.P. Blacker (ed.), *The Chances of Morbid Inheritance* (London: Lewis, 1933).

M. Bouquet, *Reclaiming English Kinship: Portuguese Refractions of British Kinship Theory* (Manchester: Manchester University Press, 1993).

P.H. Brasher, 'Clinical and social problems associated with familial intestinal polyposis' *Archives of Surgery* (1954), 69: 785–796.

H.J.R. Bussey and B.C. Morson, 'Familial polyposis coli' in R.W. Raven and F.J.C. Roe (eds.), *The Prevention of Cancer* (London: Butterworth, 1967), pp. 141–145.

L. Cartwright, *Screening the Body: Tracing Medicine's Visual Culture* (Minneapolis: University of Minnesota Press, 1995).

E.A. Cockayne, 'Heredity in relation to cancer' *Cancer Review* (1927), 2: 337–347.

W. Cramer, 'The prevention of cancer' *Lancet* (1934), I: 1–5.

F.A.E. Crew, 'The welfare state: a eugenic appraisal' *Eugenics Review* (1954), pp. 81–90.

H. Cripps, 'Two cases of disseminated polypus of the rectum' *Transactions of the Pathological Society* (1882), 33: 165.

D.M. Cromwell *et al.*, 'Cost analysis of alternative approaches to colorectal screening in familal adenomatous polyposis' *Gastroenterology* (1998), 114: 893–901.

M. De Landa, *War in the Age of Intelligent Machines* (New York: Zone Books, 1991), p. 3.

C.E. Dukes, 'Familial intestinal polyposis' *Journal of Clinical Pathology* (1947), 1: 34–37.

C.E. Dukes, 'Familial intestinal polyposis' *Annals of Eugenics* (1952), 17: 1–29.

C.E. Dukes, 'Research into intestinal polyposis' *Lancet* (1953), 1: 44.

W.D. Foster, *Pathology as a Profession in Great Britain and the Early History of the Royal College of Pathologists* (London: Royal College of Pathologists, 1982).

S. Franklin, *Embodied Progress: A Cultural Account of Assisted Conception* (London: Routledge, 1997).

J. Freedland, 'Goodbye to the oracle' *Guardian* (9/6/1999).

J-P. Gaudillière, 'Circulating mice and viruses: the Jackson Memorial Laboratory, the National Cancer Institute, and the genetics of breast cancer, 1930–1965' in M. Fortun and E. Mendelsohn (eds.), *The Practices of Human Genetics* (Dordrecht: Kluwer, 1999), pp. 89–124.

F.M. Giardiello *et al.*, 'The use and interpretation of commercial APC gene testing for familial adenomatous polyposis' *New England Journal of Medicine* (1997), 336: 823–827.

E. Goffman, *Stigma: Notes on the Management of Spoiled Identity* (London: Penguin, 1968).

L. Grenshaw, *St. Mark's Hospital, London: A Social History of a Specialist Hospital* (London: King Edward's Hospital Fund, 1985).

W.E. Gye, 'The origin of tumours' *British Medical Journal* (1932), II: 420–421.

W.E. Gye, 'Some recent work in experimental cancer research' *British Medical Journal* (1938), I: 551–554.

J.B.S. Haldane, 'The prospects of eugenics' in M.L. Johnson, M. Abercrombie, and G.E. Fogg (eds.), *New Biology* (London: Penguin, 1956), 22: 7–23.

R. Haraldsdóttir, 'Icelandic gene database will uphold patients' rights' *British Medical Journal* (1999), 1: 806.

S. Hodgson, 'Clinical genetics' *St. Mark's Hospital for Diseases of the Rectum and Colon, Annual Report* (1991), p. 26.

F. Jacob, *The Statue Within: An Autobiography* (New York: Basic Books, 1988).

M. Jacobs, 'A family affair' *St. Mark's Hospital for Diseases of the Rectum and Colon, Annual Report* (1992), pp. 14–15.

A. Kerr, 'Double trouble: social analyses of the new human genetics' *Science as Culture* (1999), 8: 97–103.

R. Klein, *The New Politics of the National Health Service* (London: Longman, 1995).

J.P. Lockhart-Mummery, 'The diagnosis of tumours in the upper rectum and sigmoid flexure by means of the electric sigmoidoscope' *Lancet* (1904), I: 1781–1782.

J.P. Lockhart-Mummery, *Diseases of the Rectum and Colon* (London: Ballière, 1923).

J.P. Lockhart-Mummery, 'Cancer and heredity' *Lancet* (1925), I: 427–429.

J.P. Lockhart-Mummery 'The origin of tumours' *British Medical Journal* (1932), I: 618–620.

J.P. Lockhart-Mummery, *The Origin of Cancer* (London: Churchill, 1934).

J.P. Lockhart-Mummery, 'Prevention of cancer' *Lancet* (1934a), I.

J.P. Lockhart-Mummery 'Medical science and social progress' *British Medical Journal* (1935), II: 1022.

J.P. Lockhart-Mummery, *After Us, or the World as it Might Be* (London: Stanley Paul, 1936).

J.P. Lockhart-Mummery, 'Recent research on cancer' *Nature* (1936a), 137: 321.

J.P. Lockhart-Mummery, 'Causation of cancer' *British Medical Journal* (1938), I: 810.

J.P. Lockhart-Mummery and C.E. Dukes, 'Familial adenomatosis of colon and rectum' *Lancet* (1939), II: 586–589.

H.E. Lockhart-Mummery *et al.*, 'The surgical treatment of familial polyposis of the colon' *British Journal of Surgery* (1956), 43: 476–481.

Lord Dawson of Penn, 'Medical science and social progress' *British Medical Journal* (1935), II: 829–833.

C. Lury, *Consumer Culture* (Cambridge: Polity Press, 1996).

J. Marks, 'Blood will tell (won't it?): A century of molecular discourse in anthropological systematics' *American Journal of Physical Anthropology* (1994) 94: 59–79.

P.M.H. Mazumdar, *Eugenics, Human Genetics, and Human Failings: The Eugenics Society, its Sources and its Critics in Britain* (London: Routledge, 1992).

B.C. Morson, 'Some prominent personalities in the history of St. Mark's Hospital' *Diseases of the Colon & Rectum* (1962), 4: 173–183.

B.C. Morson, 'Prevention of cancer' *St. Mark's Hospital for Diseases of the Rectum and Colon, Annual Report* (1980).

J.M.A. Northover, 'Imperial Cancer Research Fund colorectal cancer unit' *St. Mark's Hospital for Diseases of the Rectum and Colon, Annual Report* (1984).

A. Oakley, 'Eugenics, social medicine and the career of Richard Titmuss, 1935–50' *British Journal of Sociology* (1991), 42: 165–194.

N.T. Oswald, 'A social health service without social doctors' *Social History of Medicine* (1991), 4: 295–315.

D.B. Paul, *Controlling Human Heredity, 1865 to the Present* (Atlantic Highlands: Humanities Press, 1995).

P. Palladino, 'The president has cancer: Cancer and the modern body politic' *Textual Practice* (1999), 13: 533–549.

K.A. Pearson, 'Life becoming body: on the 'meaning' of post-human evolution' *Cultural Values* (1997), 1: 219–240.

J.V. Pickstone, 'Ways of knowing: towards a historical sociology of science, technology and medicine' *British Journal for the History of Science* (1993), 26: 433–458.

A. Piney, 'Hereditary neoplastic diseases' in C.P. Blacker (1993), pp. 373–377, on p. 376.

P. Rabinow, *Essays in the Antropology of Reason* (Princeton: Princeton University Press, 1997).

P. Rabinow, *French DNA: Trouble in Purgatory* (Chicago: University of Chicago Press, 1999).

K. Rader, 'Making mice: C.C. Little, the Jackson Laboratory and the standardization of *Mus musculus* for research' Ph.D. thesis, University of Indiana (1995).

R. Rapp, 'Refusing prenatal diagnosis: the meanings of bioscience in a multicultural world' *Science, Technology, & Human Values* (1998), 23: 45–70.

M.P.M. Richards, 'The new genetics: some issues for social scientists' *Sociology of Health and Illness* (1993), 15: 567–587.

G. Rivett, *From Cradle to Grave: Fifty Years of the NHS* (London: King's Fund, 1997).

T. Rowentree, 'Three new families of intestinal polyposis' *Proceedings of the Royal Society of Medicine* (1950), 43: 686–688.

T. Shakespeare, 'Back to the future? New genetics and disabled people' *Critical Social Policy* (1995), 46: 22–35.

S. Sturdy and R. Cooter, 'Science, scientific management and the transformation of medicine in Britain, c. 1870–1950' *History of Science* (1998), 36: 421–466.

A.T. Todd, 'The origin of tumours' *British Medical Journal* (1932), I: 910–911.

A.M.O. Veale, 'Possible autosomal linkage in man' *Nature* (1958), 182: 409–410.

A.M.O. Veale, 'Genetics, carcinogenesis, and family studies' *British Surgical Practice and Surgical Progress* (1961), pp. 169–185.

A.M.O. Veale, *Intestinal Polyposis* (Cambridge: Cambridge University Press, 1965).

G. Vines, 'Star of the big screen' *Times Higher Education Supplement* (21/5/1996), p. 17.

J.H. Warner, 'Science in medicine' *Osiris* (1985), 1: 37–58.

J.H. Warner 'The history of science and the sciences of medicine' *Osiris* (1995), 10: 164–193.

S. Whitelaw, J.M. Northover, and S.V. Hodgson, 'Attitudes to predictive DNA testing in familial adenomatous polyposis' *Journal of Medical Genetics* (1996), 33: 540–543.

Chapter 10

VERTICAL ANCESTRIES AND HORIZONTAL RISK: HEPATITIS B AND AIDS

Jennifer Stanton and Virginia Berridge

INTRODUCTION

In this paper, we discuss two diseases (or, in the case of AIDS, clusters of disease) which appear at first glance to fall into the clear cut category of viral etiology, with transmission by intimate contact between persons, or via blood. This is contagion facilitated by what has been termed 'risk groups' and subsequently 'risk behaviour'. For this form of transmission, we use the term 'horizontal risk' because such contagion passes sideways, as it were, mainly between consenting, or ignorant, adults. There is no doubt about the nature of this horizontal contagion for both hepatitis B and for AIDS. But where does the issue of heredity emerge? Initially we could think of no hereditary element in the framing of the two diseases syndromes. But familiarity with their histories enabled us to recognise ways in which other framings co-existed with the horizontal viral-contagion model. These are what we have termed the 'vertical ancestries' in this paper.

Such vertical ancestries did not operate identically for both diseases. For hepatitis B, in the period of the 'discovery' of the virus (or rather its antigen which stood as a marker for a long while) there were genetic connections, though of a somewhat spurious nature. The virus was not identified by a deliberate search, but as an offshoot of a project to find and catalogue genetic variations in disease susceptibility among different racial groups. Such a presumed genetic component did not mark the initial scientific history of AIDS. But, as the paper argues, the early epidemiological detective work on risk groups for AIDS carried with it the symbolic baggage of eighteenth and nineteenth century predisposition to disease with its mingling of genetic, environmental and moral. And it is the symbolic and policy importance of the continuing emphasis on the vertical mode of transmission which we emphasise for both conditions/diseases. Here our focus is on the 'vertical' nature of mother to child transmission. **Viral** mother to child transmission of hepatitis B or AIDS carries with it echoes of **vertical** mother to child transmission, of that late nineteenth century/early twentieth century eugenic cluster of concerns—syphilis, alcoholism, poverty—which became national concerns over the 'future of the race'. We look at how that symbolic framing translated into policy terms. For hepatitis B, this framing of disease has continued in the way in which the disease is dealt with in developing countries. Although explanation is now more likely to include political and economic dynamics, it has still been possible for strong advocacy to flourish for intervention in the mother to child link, which can be termed 'vertical' intervention in 'vertical' transmission. For AIDS, the vertical mother to child link has had considerable policy implications, not least in stimulating a response based, as at the turn of the century, on fears for the future of the nation.

The conceptualisations of 'horizontal risk' and 'vertical ancestries' which we identify have not remained static and discrete, but rather have shifted and coalesced over time. Notions of contagion have co-existed with those of heredity; the terrain between the two has been blurred. Hepatitis B and AIDS have experienced different policy histories, certainly until now in terms of types and levels of response. But they seem to share a common 'staging' in terms of conceptualisations of horizontal and vertical. Both have gone through a circular process, from horizontal to vertical and back to horizontal. That for hepatitis B can perhaps be seen as spiral, with vertical rememerging in new form. AIDS, too, with its more recent history, may be reconceptualised in this way.

HEPATITIS B

> We suggested that hepatitis virus may have several modes of transmission. It can be transmitted horizontally from person to person similar to 'conventional' infectious agents. This is seen in the transmission of hepatitis B virus (HBV) by transfusion. Other forms of direct and indirect horizontal transmission exist ... It has even been reported to have been spread by computer cards, an extraordinary example of adaptation by this ingenious agent! HBV may also be transmitted vertically. If the genetic hypothesis were sustained, then it would imply that the capacity to become persistently infected is controlled (at least in part) as a Mendelian trait. The data also suggest a maternal effect. A re-analysis of our family data showed that in many populations more of the offspring were carriers when the mother was a carrier than when the father was a carrier (Blumberg, 1977, p. 147).

Thus spoke Baruch Blumberg, the 'father' of hepatitis B in 1976, on receiving his Nobel prize for the discovery of the antigen. Many of the notions encapsulated in this quotation appear more or less strongly at different points in time.

This section of the paper (on hepatitis B) aims to unpick these strands, to describe a circular process, or spiral perhaps, from horizontal to vertical to horizontal and back again to vertical in a new form. These phases can be arranged approximately, and of course artificially (for there was much overlap), into the following periodisation:
Horizontal (1) = notions of infectious and serum hepatitis or jaundice, pre-Second World War.
Vertical (1) = Blumberg's search for genetic predispositions, 1950s/1960s.
Horizontal (2) = reconceptualization and new epidemiology of hepatitis B, 1970s/1980s.
Vertical (2) = mother-to-child transmission targetted for vaccination, 1980s/1990s.

Horizontal (1): Pre-Second World War

Hepatitis, or jaundice, was regarded as a mainly infectious disease whose mode of transmission was mysterious before the Second World War, but one form was associated with injections, vaccinations especially for yellow fever, with blood transfusions, and also with therapy for venereal diseases. The latter was referred to in Britain as 'arsenotherapy jaundice' from the arsenic compound used in treatment of VD: in fact this name indicates that this form of jaundice was regarded as poisoning of the liver by toxic chemicals, rather than an infectious form. It was analogous with jaundice in factory workers handling the explosive TNT.[1]

By the 1950s and 1960s hepatitis was also seen as prevalent in institutions for the mentally handicapped.[2] Perhaps unofficially, it was known to occur in prisons. Altogether it bore a stigma by association with illicit drug injecting, mental defect, crime and promiscuity: and before long another association, with homosexuality. But the strong link with medical procedures had another effect—it was a hazard for doctors, nurses, dentists, laboratory technicians, an occasional hospital infection that threatened patients and staff alike. Let us bear in mind this double image, of the 'degenerate' or dirty disease, with another, iatrogenic side arising from benevolent attempts to help the suffering.

Vertical (1): 1950s/1960s

The next step came from an apparently unconnected direction: that is, the search to explain the observed phenomenon that people from different parts of the world showed varying susceptibility to a range of diseases. In the account that follows, it is important for the reader to bear in mind that the aim of the research was to uncover genetic variations to account for these different responses: this is one element of the 'vertical ancestry' claimed in the title—biochemical genetics. Then, in the course of trying to identify the nature of an antigen found by accident (the research also involved antigens native to the blood), the researchers tried to make a link with congenital conditions, another sort of 'vertical ancestry'.

Baruch Blumberg, an American medical student on elective in Surinam in 1950, remarked on the wide variations in response to tropical disease between the many different ethnic groups represented at the bush hospital where he worked.[3] During a period of postgraduate work at the biochemistry department at the University of Oxford (1955–57), Blumberg encountered Ford's work on polymorphic variations (Ford, 1956). Such variations in some part of the immunological chain might explain different disease responses: Blumberg decided to begin by looking at serum proteins.[4] Together with Anthony Allison, a colleague in the department, he organised summer field trips to collect samples of blood from isolated groups, producing papers which reflected the global range of their search.[5]

Blumberg continued this line of research while based at the US National Institutes of Health, where he was joined by Allison in 1960. They developed a hypothesis that multi-transfused patients might display antibodies to the serum proteins they did not themselves possess, a hypothesis they were pleased to confirm on the thirteenth attempt.[6] But Blumberg's greatest find was a reaction to an antigen in the blood of an Australian Aborigine (in fact this sample was sent to him in the States, by an Australian colleague): due to its origin this became known as the 'Australia antigen'. Not knowing what it signified, Blumberg's team tried to match it against other blood samples in the vast samples bank at the NIH.[7] They found a similar antigen in blood from Down's syndrome patients. So already 'Australia antigen' had a double vertical ancestry: Australian aborigines and Down's syndrome people had higher than average prevalence (Blumberg *et al.*, 1967).[8] However, an infection was suggested by the high rate of this antigen in multi-transfused patients. To cut a long story short, American and British studies established a link between Australia antigen and hepatitis B, which was firmly accepted around 1968, though for many years, clinicians and virologists continued to refer to 'Australia-antigen-associated hepatitis'. Transmission was mainly horizontal, but with a kind of vertical element.

Horizontal (2): 1970s/1980s

The practical upshot of the antigen 'discovery' was a test for hepatitis B, which enabled epidemiogical studies to progress. This disease, which had only appeared in its rather rare acute form, seen mainly in Western countries in the 'risk group' pattern described above, turned out to be enormously prevalent in poorer countries, but in an invisible carrier state, now made visible by the test. Thus, two major patterns of transmission of the disease were distinguished. In high-prevalence (mainly poor Third World) countries, nearly everyone caught the disease as babies, and showed no symptoms, while in low-prevalence countries, the notion of 'risk groups' or risky behaviours was developing, to be enhanced in the era of AIDS.[9] By the late 1970s those seen as most at risk for hepatitis B were drug users, gay men, haemophiliacs, and health workers—a distillation of earlier categories—plus infants of mothers from Asia or Africa. This last category was an extrapolation from the worldwide epidemiological findings.

When a vaccine became available in the UK in 1982, the guidelines for prioritization focussed on health workers, and relatives of infected patients. An alliance emerged between clinicians counted as 'experts' in this field, and vaccine manufacturers, to expand the use of the vaccine by including more groups in the guidelines.[10] Supporters argued along lines of special vulnerability, one of which was the 'vertical' mother-to-child link. Even before the vaccine was licensed for use with babies, testing was advocated: '... antenatal serological screening tests are needed to identify those patients of non-Caucasian ethnic origin (and the Caucasian patients whose histories suggest the possibility of hepatitis B virus carriage), whose newborn infants will be at risk and require protection' (Polakoff, 1982, p. 1295).[11] This coincided with the approach from the WHO expert group on hepatitis. How it translated into implementation is difficult to trace; probably at first immigrant mothers were targeted, and testing was later extended to include all mothers.[12]

In countries like Britain with low prevalence, hepatitis B was also conceived as a horizontally transmitted disease affecting certain 'risk groups'. Those most affected (or infected) were drug users and gay men, well into the 1980s, but the vaccine that was available was scarcely targetted on these groups. However, there is evidence that the old

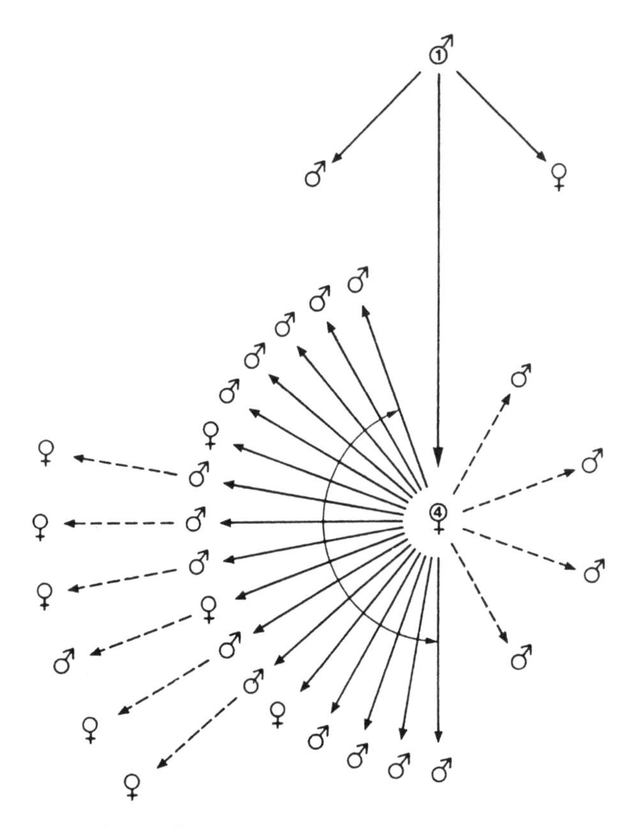

FIG 10.1 'HORIZONTAL' SPREAD OF HEPATITIS B IN THE MEDICAL PRESS

associations with drugs and promiscuous sex still formed an uneasy mélange of epidemiological 'fact' and medical prejudice. A clear instance is provided in a 1987 report on a cluster of cases in Wales (Clee and Hunter, 1987) (see Fig 10.1).

The authors commented that: 'All cases were traced to one man' (case 1) who had caught hepatitis B from a heroin dealer via a shared needle. In the same way this index case passed the disease on to three others. One of these, case 4, was a woman who was '*responsible* for most of the remaining cases', by needle sharing, sexual contact or both. They say: 'The woman in case 4 *admitted to* sexual contact with nine men during one week in early June. This was immediately after a suction termination of pregnancy and while she was still bleeding' (Clee and Hunter, 1987, p. 531).[13] The normally dispassionate language of the report slips into a judgmental tone when discussing this woman ('responsible ... admitted to'). This is reminiscent of a fictional account of an outbreak of hepatitis B in Edinburgh, in the course of which three doctors become infected and two died: at the end of the book the culprit is revealed as a nurse who had sex with these doctors while menstruating (Douglas, 1975).[14]

Vertical (2): 1980s/1990s

As pointed out previously, epidemiological studies established that poorer countries had a far higer prevalence of hepatitis B than developed countries. Transmission was believed to be mainly mother-to-child in Asia, whereas in Africa, babies tended to catch the disease later, as they crawled on the ground. The majority cleared the infection, but perhaps ten or twenty percent became carriers and of these, quite a large proportion might suffer long term liver damage. Estimates of deaths from primary liver cancer due to hepatitis B, along with estimates of numbers of carriers worldwide, climbed all through the 1970s and 1980s. There are said to be 280 million carriers with over 700,000 deaths a year.[15]

In third world countries as in wealthy countries, cost was a barrier to universal vaccination,[16] but there was another problem. To explain this, it would be necessary to rehearse the historical debate concerning the overall WHO/UNICEF immunization programmes. In brief, the major clash in theories of health promotion for developing countries concerns the value of 'vertical' programmes, which are chosen and directed from the centre in a top-down model, versus 'horizontal' programmes which aim to integrate a variety of health interventions and to involve the community.[17] Vertical programmes often focus on one disease. Horizontal ones do not have a disease focus, and tend therefore to produce less easily measured results: however, advocates would argue that the effects are more sustainable.

Thus the programes of childhood immunization for Third World countries have been contentious among health experts. But national immunization days tend to be very popular, and on the whole these programmes have reduced deaths from targetted diseases—but these are childhood diseases familiar to parents. Hepatitis B is very different since although it infects children in poor countries, it does so invisibly, and only kills adults much later. One historian enthused that with hepatitis B we have the first vaccine that may prevent a form of cancer—liver cancer—and on those grounds the integrity of a campaign for vaccination seems incontestable (Muraskin, 1995). There is, however, a problem if hepatitis B immunization is not popularly perceived as necessary: then a campaign can be seen as an archetypal vertical programme imposed from above.

The point for my argument in relation to the cycle of framing of hepatitis B is the way that pilot immunization projects in poor countries in Asia sought out mothers and their newborn babies, often in remote villages that were seldom reached by health facilities. Mode of *transmission* (mother-to-child) and mode of *intervention* (top-down programme) here are both vertical. These may not be the same areas, but they are similarly inaccessible, to those where Blumberg and colleagues travelled in search of blood forty years earlier. Although hepatitis B is seen in Europe as an infectious disease, affecting certain groups, it is countered as though it were an hereditary taint in the poor countries most affected. The fact that a vaccine is being used to break the chain of hepatitis B handed down from mother to child places the disease exactly between heredity and infection.

AIDS

For AIDS, a similar sequence of events can be identified in terms of shifting formulations, which moved from horizontal to vertical and back to horizontal again. AIDS, unlike hepatitis B, did not have a presumed genetic component in its initial scientific characterisation, even though scientific approaches to the new syndrome derived from cancer research, where genetic approaches were becoming increasingly common. Nevertheless, like hepatitis B, it has occupied the blurred late twentieth century terrain between heredity and infection. This section of the paper will examine how scientific formulations drew on concepts which brought together both vertical ancestries and horizontal risk, and how concepts of infection married with those of vertical transmission informed both public and policy reactions. The mingling of horizontal and vertical conceptualisations was the key factor in stimulating a national policy response. So, as with hepatitis B, these stages can be categorised as follows: Horizontal; the limited 'risk group' early to mid 1980s. Horizontal and vertical; the 'risk to all' and hereditary transmission; the policy crisis of 1986–87. Limited horizontal and vertical; the reconceptualisation of AIDS after 1988. Vertical revival?; the scenario shifts in the 1990s.

Epidemiology and the Limited Horizontal 'Risk Group'

The initial characterisation of the syndrome, in the absence of other paradigms, was an epidemiological one. Epidemiologists, as 'disease detectives', in the British context, as in North America, had come to provide a powerful tool in the formation of health policy in the post war period. In both countries, they identified horizontal groups where the syndrome clustered. Gay men were firmly characterised as a 'risk group' and AIDS was a 'gay plague' in the media view. So, too, were Africans and Haitians as early reports began to trickle through in 1983 and 1984 associating AIDS with an origin in 'black' areas.[18] That characterisation was never as strong in Britain as in the U.S. in the early years. But we need to unpick this early conceptualisation of horizontal spread in particular groups; for this notion of confined contagion also comprised within it ideas of hereditary and racial/genetic susceptibility. The strength of the epidemiological model lay in its ability to move away from unifactorial determinism towards consideration of a variety of social and environmental factors. But its weakness lay also in drawing on particular variables which reflected social values and thereby gained the status of 'scientific fact' (Oppenheimer, 1988). The identification of particular groups 'at risk' encouraged ideas of racial susceptibility. The concept of the 'risk group' also built more broadly on the idea of differential susceptibility, the belief that different groups in society were more susceptible to disease, either through inherited, environmental, or behavioural factors. Charles Rosenberg has analysed how historically those threatened with epidemics have attempted to make sense of the random nature of infection by stressing differential susceptibility (Rosenberg, 1992). Some groups in populations were, so it was argued, **predisposed** to disease. For AIDS and gay men, it was argued that frequent sex with multiple partners, late hours, pubs, clubs, the use of 'poppers', were all implicated. Hepatitis B was suggested as a predisposing factor. The notion of predisposition was undoubtedly there, but the emphasis was on lifestyle and behavioural rather than on vertical and hereditarian issues. AIDS was conceived of horizontally, but confined within its predisposed risk group; this, in policy terms, was not perceived as particularly risky.

Horizontal and Vertical; The Future of the Race

It was when this limited and confined horizontal risk became, through potential vertical transmission, a horizontal risk to all, that policy moved to a new and crisis stage. Ironically, it was through the progress of scientific investigation of the syndrome, that the hereditarian and vertical implications became more obvious and the potential crisis clearer. Science solidified throughout 1984–5; and the discovery of the virus in some senses, as other writers have also argued, recategorised

AIDS as a biomedical problem open to chemical resolution. AIDS was a set of opportunistic infections, variously defined. A vaccine might prevent, or treatments, cure. But the discovery of the virus and the technical ability to test brought in its train the discovery of the 'clinical iceberg' of those who were infected but not yet ill. Suddenly, it was no longer gays who were the only 'risk group'. It was infected blood donors, drug users, haemophiliacs, bisexuals. The horizontal gay risk group transmuted through these additional groups into the idea of national risk; the threat was not just to gays, but to everyone. Contagion was something no longer confined but which could affect us all. But it was the hereditary element within transmission which most concerned policy makers. Through the sexual contacts of drug users, AIDS could spread into the general population; the children of bisexuals could carry the syndrome into the next generation. The wider concept of horizontal risk combined with the idea of vertical transmission to produce a sense of national emergency. Here is an illustration from the time. In the *Sunday Times*, two journalists (one later to be a leading AIDS dissident), laid out the nightmare scenario.

> A British family out for a walk: father (39), mother (35), two teenage children and a baby. They look happy and healthy. But from what we now know about the way the AIDS epidemic is developing, all are potential victims. The father could be infected by a prostitute or by a homosexual relationship; his wife through an affair with a casual stranger—or from her husband. And the boy and girl, if they have a drugs problem, could pick it up from an infected hypodermic needle. Even the baby could be infected, it either parent was a carrier. Many like these may soon be under threat.[19]

In 1986–7, it was sentiments such as these, urged on government by a coalition of gays, scientists and medical men, and underpinned by knowledge of actual widespread vertical and horizontal transmission in Africa, which brought a national emergency response.[20] The epidemiological risk group transmuting into national risk, with its vertical and hereditarian implications carried with it clear resonances from that earlier twentieth century concern for deterioration of the race. The role of the infected family in 'national deterioration' carried with it a symbolic legacy.

Within the infected family, the role of women was of particular importance. Early on in the UK, prostitutes had come into the picture. In the autumn of 1985, for example, a Thames TV programme, *AIDS and You* had stressed prostitutes as a 'bridging group' for heterosexual spread, an association with obvious symbolism given the nineteenth century 'double standard' and the history of the stigmatisation of prostitutes as vectors of infection. There was in fact, little evidence that prostitutes were fulfilling that role in the mid-80s, although those injecting drugs were obviously a greater risk.[21]

But the greatest emphasis was placed on women as mothers. These always attracted, as did drug users (and especially drug using mothers), a harsher response than did other 'risk groups'. In the early days, no pregnancy at all or abortion were considered the best response for HIV infected women. In early 1986, the Royal College of Obstetricians and Gynaecologists produced guidelines on pregnancy where the advice, based on existing knowledge, was harsh. Women should not become pregnant, and, if they did, abortion was the best alternative.[22] In Edinburgh, the infectious disease specialist Ray Brettle, who saw drug using women, was 'brutal with pregnant women—they may be dead in three years' time and the baby may be dead in a year ...'[23] Drug using mothers were likely to lose their children to fostering and adoption. This harsher response to those who carried greater hereditary responsibility continued. In subsequent years, screening and testing were seen as appropriate for pregnant women though they might be 'human rights' issues for other groups. The Social Services Committee, which enquired into AIDS and reported in 1987, rejected anonymous screening, but supported the idea that antenatal testing, with fully informed consent, be available for all pregnant mothers.[24] The subsequent history of anonymous screening in the UK context was a tortuous one, rather outside the scope of this paper. A period of 'turmoil on the inner track', little of which filtered through into public discussion, ended in November 1988 when Kenneth Clarke as Department of Health minister, finally gave political authorization to a programme to bring it into effect. But the position of women within that programme remained distinctively different from that of other groups in society. As confidential documents put to the Expert Advisory Group on AIDS (EAGA) admitted, testing of low risk women was being undertaken in some areas for surveillance purposes. The boundaries between screening for epidemiological research purposes and testing for individual identification were blurred so far as women were concerned; so, too, was the idea of 'informed consent' which appears to have been variably interpreted.[25] It may be, that, as some commentators argued, pregnant women were tested on the basis of the 'sitting duck' theory. But women and children also carried with them their own symbolic history, carrying responsibility for the 'future of the race'. As such, their treatment was harsher.

Vertical Becomes Horizontal; The Re-Emergence of the 'Risk Group'

To this notion of vertical transmission was added, in the early 1990s, the issue of race. The results of anonymous screening revealed high rates of sero-positivity in African women attending London antenatal clinics.[26] But this discovery marked a new stage in the interrelationship between horizontal and vertical. For this form of vertical transmission once again confined the notion of horizontal risk behind defined boundaries. Transmission was again relegated to the 'other', to a group with differential susceptibility unlikely to infect the general population. This fitted with a policy perception where the risk of AIDS was redefined; no longer was it a national crisis. Notions of national degeneration and the future of the race no longer informed national policy making. There was a renewed 'limited horizontal' focus, a revival of the 'risk group' approach from another direction as well. This was through the strategy of the 're-gaying' of AIDS which developed in the early 1990s. Gay men, who had actively promoted the national risk/ vertical transmission concept in the mid 1980s, moved to a more focused stance. Some argued that the national crisis response had neglected those suffering most, and that money and service development had gone elsewhere. They argued for a more 'targeted' response which recognised that the majority of actual and potential people with the syndrome were likely to be gay men.[27] This argument had its policy impact, in particular at the local level, where strategies and campaigns were developed. But in general it marked also the policy downgrading of AIDS. As one AIDS Unit civil servant remarked, the policy response would not have occurred at such a high level in 1986–87 if only gay men had been seen to be involved.[28] Limited horizontal risk carried more minor policy implications.

Revival of the Vertical?

In some senses, as with hepatitis B, a change of location and a revived form of the vertical can also be identified in the mid-1990's. For the focus of policy attention had shifted to the developing world, in particular to Africa and Asia, where the potential for heterosexual spread remained strong. The salience of AIDS as a high profile issue was defined on the international stage rather than within the UK and the role of AIDS in developing and part developed countries provided the justification. Here the role of women, and especially of women as mothers remained centre stage. Prevention of infection and especially of sexually transmitted disease 'co-factors' was a major focus of intervention, and the search for a vaccine seemed unlikely to impact on developing countries, except through their controversial use as a natural testing ground for science in clinical trials. In this sense, then, AIDS differed from hepatitis B; there was no vaccine implicated in the interrelationship between the horizontal and the vertical. But in other respects in the Third World it, too, continued to occupy the blurred terrain between heredity and infection.

CONCLUSION

Our argument in essence is this. Both hepatitis B and AIDS were seen at different times as having important **vertical** transmission patterns as well as changing **horizontal** transmission. There was a remodelling over time from predominantly horizontal, to vertical, to horizontal conceptualisations. Hepatitis B 'arrived' earlier, and its story contained a genetic episode not present for AIDS. But in terms of framing, the similarities between these histories are striking.

Above all, the common element in these stories is the shift from a horizontal risk to a vertical emphasis on women as mothers transmitting disease to their infants. With hepatitis B, this emphasis was especially directed towards vaccination of infants in the Third World, though testing of mothers became policy in developed countries. With AIDS, there was differential screening and testing of mothers in the UK, while Third World women were increasingly a focus of concern. With both, echoes of eugenic anxieties about the future of the race are discernable.

The history of the framing of these two diseases demonstrates how symbolic responses derived from much earlier reactions to epidemic threat remained important even in the response to disease in the immediate past. The major difference between the two diseases/ syndromes has lain in their differing policy impact. Here we argue that when, as was the case with AIDS, potential generalised **horizontal** spread was combined with potential **vertical** transmission, the policy response rose to a crisis level. With hepatitis B, this conjuncture has not yet occurred. Both vertical and horizontal patterns of transmission have been perceived as containable; but the future inter-relationship of the two remains to be seen.

NOTES

1. Medical Research Council file 3144/21, Cases of jaundice through being in contact with T.N.T. (1942–1943). For discussion of pre-war and wartime concepts of hepatitis see Stanton (1995).
2. Made notorious by the experiments Saul Krugman conducted on child inmates of a mental health institution in New York during the 1950s and 1960s: Stanton (1995). For contemporary critique, see Beecher (1966). This institutional prevalence was earlier explained as Down's syndrome children being especially 'susceptible' to hepatitis B; later interpretations suggest such children's social compliance renders them vulnerable to sexual exploitation by inmates or staff: author's interview with epidemiologist, 21 November 1990.
3. Baruch Blumberg, interview with author, 22 November 1992.
4. For fuller explanation see Blumberg (1977), pp. 137–58.
5. For example: Allison, Blumberg and ap Rees (1958); Blumberg, Allison and Garry (1959).
6. See for example Allison and Blumberg (1961).
7. According to Blumberg, this began with his NIH team's collection of serum and plasma samples, and moved with him to the Institute of Cancer Research where, as the Blood Collection of the Division of Clinical Research, it numbered 200,000 specimens by 1976: Blumberg (1977), pp. 139–40.
8. This paper points to a link with hepatitis as well as leukaemia and Down's.
9. Of course, it took many studies over many years to establish these patterns; for an early paper by another leading American researcher see Prince (1970). Note Prince's avoidance of the term 'Australia antigen' at a period of rivalry between his and Blumberg's teams. For a later survey see London and Blumberg (1985).
10. See Stanton (1994).
11. At this stage prophylaxis would be by means of immunoglobulin, with vaccine added later.
12. When asked whether mothers were selected for testing on the basis of country of origin, or were all mothers screened for hepatitis B, a leading UK expert (who acted as government advisor and WHO consultant) said he did not know the details of local practice: Virologist, interview with author, 8 June 1992.
13. My emphases.
14. Discussed in Stanton (1995), pp. 138–40.
15. British Medical Association, *Immunisation against hepatitis B* (Report of the Board of Science and Education, 1987), gave an estimate of 200 million carriers worldwide; but Maynard (1990) suggests there are about 300 million carriers: 'Of the [approximately] 122 million infants born in 1985, 1.3 million will be expected to die an HBV-induced liver death'.
16. For rich countries, the disincentive was low prevalence of hepatitis B, plus high cost of vaccine. In poor countries with high prevalence, the essential step was to lower the cost of the vaccine. For fullest account see Muraskin (1995).
17. In fact this language is not limited to the Third World context; see for example Ashton and Seymour (1988) for 'vertical' and 'horizontal' programmes on health of older people, and teenage pregnancy, in a British urban area.
18. For example, de Kock (1984).
19. 'At Risk' *Sunday Times*, 2 November 1986.
20. For a fuller discussion of this period see Berridge (1996).
21. See evidence given to the House of Commons Social Services Committee, *Problems Associated with AIDS* vol. ii, q. 818. (London: HMSO, 1987).
22. Reported in *The Times*, 10 February 1986.
23. House of Commons. Social Services Committee, *Problems Associated with AIDS*, vol. ii q. s 756–62. (London: HMSO, 1987).
24. House of Commons, Social Services Committee, *Problems Associated with AIDS*. Report of the Committee. (London: HMSO, 1987).
25. For more details of this episode see Berridge (1996). This section is also based on the confidential minutes of the Expert Advisory Group on AIDS, which have been made available to Virginia Berridge on an unattributable basis.

26. Summarised in the *AIDS Newsletter*, 6, 12/13 (1991), 792; see also J. Banatvala and others, letter to *The Times*, 17 February 1992.
27. See, for example, letter from Edward King of Gay Men Fighting AIDS in *The Guardian*, 25 June 1992.
28. Former AIDS Unit civil servant, interview with Virginia Berridge, November, 1994.

REFERENCES

A.C. Allison, B.S. Blumberg, and W. ap Rees, 'Haptoglobin types in British, Spanish Basque and Nigerian African populations' *Nature* (1958) 181: 824–25.

A.C. Allison, and B.S. Blumberg, 'An isoprecipitation reaction distinguishing human serum protein types' *Lancet* (1961) i: 634–37.

J. Ashton, and H. Seymour, *The new public health: the Liverpool experience* (Milton Keynes: Open University Press, 1988) pp. 86–87 and 118–20.

H.K. Beecher, 'Ethics and clinical research' *New England Journal of Medicine* (1966) 274: 1354–60.

V. Berridge, '*AIDS in the U.K. The Making of Policy, 1981–1994*' (Oxford: Oxford University Press, 1996).

B.S. Blumberg, 'Australia antigen and the biology of hepatitis B', Nobel Lecture, December 13 1976, in *Les Prix Nobel en 1976* (Stockholm: Imprimerie Royale P.A. Norstedt & Söner, 1977).

B.S. Blumberg, A.C. Allison, and W. Garry, 'The haptoglobins and hemoglobins of Alaskan Eskimos and Indians' *Annals of Human Genetics* (1959) 23: 349–56.

B.S. Blumberg *et al.*, 'A serum antigen (Australia antigen) in Down's syndrome leukemia and hepatitis' *Annals of Internal Medicine* (1967) 66: 924–31.

W.B. Clee, and P.R. Hunter, 'Hepatitis B in general practice: epidemiology, clinical and serological features, and control' *British Medical Journal* (1987) 295: 530–32; diagram on p. 531.

K.M. de Cock, 'AIDS: an old disease from Africa?' *British Medical Journal* (1984) 289: 306–08, 1454–55.

C. Douglas, *The houseman's tale* (London: Hutchinson, 1975).

E.B. Ford, *Genetics for medical students* (London: Metheun, 1956).

W.T. London, and B.S. Blumberg, 'Comments on the role of epidemiology in the investigation of hepatitis B virus' *Epidemiologic Reviews* (1985) 7: 59–79.

J.E. Maynard, 'Hepatitis B: global importance and need for control' *Vaccine* (1990) 8, Supplement, S18.

W. Muraskin, *The war against hepatitis B: a history of the International Task Force on Hepatitis B Immunization* (Philadelphia: University of Pennsylvania Press, 1995).

G.M. Oppenheimer, 'In the eye of the storm; the epidemiological construction of AIDS' in E. Fee and D. Fox (eds.), *AIDS: The Burdens of History* (Berkeley: University of California Press, 1988) pp. 267–300.

S. Polakoff, 'Immunisation of infants at high risk of hepatitis B' *British Medical Journal* (1982) 285.

A.M. Prince, 'Prevalence of serum-hepatitis-related antigen (SH) in different geographic regions' *American Journal of Tropical Medicine and Hygiene* (1970) 19: 872–79.

C. Rosenberg, 'Disease and social order in America: Perceptions and expectations' in C. Rosenberg (ed.), *Explaining Epidemics and Other studies in the History of Medicine* (Cambridge: Cambridge University Press, 1992) pp. 258–77.

J. Stanton, 'What makes vaccine policy? The case of hepatitis B' *Social History of Medicine* (1994) 7: 427–46.

J. Stanton, 'Health policy and medical research: hepatitis B in the UK since the 1940s', PhD thesis, University of London, 1995, chapter 2.

Chapter 11

'PREDISPOSITIONS', 'COFACTORS' AND 'IMAGES OF AIDS'

Ilana Löwy

SCIENTIFIC PRACTICES AND THE MEANING OF A DISEASE

'AIDS,' the historian of medicine Charles Rosenberg proposes, 'remains for us one of the ways in which society has always framed illness, finding reasons to exempt and reassure in its agreed upon etiologies. But it remains for us as well that biological mechanisms define and constrain social responses. Ironically this new disease reflects both elements—the biological and the cultural—in a particularly stark form. (...) When certain immunologists suggest that the predispositions for AIDS may grow out from successive onslaughts on the immune system—this may or may not prove to be an accurate description on the natural world. But to many ordinary Americans (and perhaps to a good many medical scientists as well), the meaning lies in another frame of reference. As in cholera a century and half before, the emphasis on repeated infections explains how an individual has predisposed him or herself. The meaning lies in behavior uncontrolled' (Rosenberg, 1986, p. 51–52). This is undoubtedly true. Scientific knowledge is expressed in an ordinary language, and is laden with cultural and moral undertones. But is the meaning of the affirmation that individuals may have a self inflicted or inborn 'predisposition' to AIDS limited to the moral connotations of 'lack of control' implied in this term? What about its scientific connotations? Rosenberg suggests that diseases can be seen as 'occupying points along a spectrum, ranging from those most firmly based in a verifiable pathological mechanism to those, like hysteria and alcoholism with no well understood mechanism, but with a highly charged social profile', and that AIDS occupies a place on both ends of the spectrum (Rosenberg, 1986, p. 53). This paper attempts to bring together the 'two ends of the spectrum', and to examine the ways 'verifiable pathological mechanism', verifiable, that is, through specific techniques used by scientists, are carrying societal and cultural values. The symbolic dimensions of AIDS are present not only in the interpretations of this disease by lay persons, but they are embedded within scientific discourses and practices. The 'meaning of scientific knowledge' cannot be dissociated from the history of the production of this knowledge.

AIDS was first defined as a poorly understood 'immunodeficiency syndrome' the development of which, scientists proposed, may depend on unspecified 'life-style' elements. From the mid-1980's on, AIDS became a 'virus-induced immunodeficiency'. It was not clear, however, how this immunodeficiency was induced by the etiologic agent of AIDS, the HIV. Two competing 'images of AIDS'—that is, two conceptual frameworks which reflected and shaped the ideas and practices (or the 'thought styles') of professional communities—coexisted in the late 1980s. One, the 'war image' proposed by virologists, was focused on direct effects of the multiplication of HIV in the body. The other, the 'network image' proposed by immunologists and clinicians, stressed complex immune regulations, networks of cells and molecules, and 'predisposing factors'. In the 1980's these two models were seen as useful, but there was a general agreement that neither was able to account for all the pathological and epidemiological manifestations of HIV-infection. In the 1990s, however, a technical innovation, the development of quantitative PCR (polymerase chain reaction, a method developed by molecular biologists), has strengthened the virologists position, and has reinforced the perception of AIDS as a 'classical' rather than 'unusual' viral infection. The 'network image' of AIDS has not been discarded, but it has been increasingly presented as a subordinate explanation which accounts for the secondary manifestations of the 'war' between the virus and the organism. AIDS seems to follow, like tuberculosis or syphilis did, a pathway which leads from a complexity-centered perception of a given pathology to its definition as a 'straightforward infection', that is, as a disease in which the immense complexity of pathological manifestations and the great diversity of individual reactions to an infection should not obscure the basic importance of a single event at the origin of all these phenomena: the multiplication of a pathogen in the body.

AIDS BEFORE HIV: LIFE STYLES AS PATHOLOGY

The acronym 'AIDS' became rapidly a familiar name, so familiar that many among those who use it are unable to say what the letters A.I.D.S stand for. The process of 'naturalization' of this acronym was even faster in France, where SIDA became rapidly 'Sida' (capital S only) and then 'sida' (no caps), a name of disease as self evident as, say, influenza.

The name has been thus dissociated from the original definition of this disease. The first epidemiological observations of this new pathology were reported by an unexplained increase in the frequency of two rare diseases, Kaposi's sarcoma and Penumocysis carinii pneumonia (Fee and Fox, 1988; Grmek, 1990; Epstein, 1997). These observations stressed that the new pathology appeared in populations (homosexuals who participate intensively in the homosexual subculture and intravenous drugs users) known for their high frequency of infections and a generalized 'predisposition' to diseases. In a 'pre-AIDS' article on 'Medical aspects of homosexuality' Selma Dritz, (who later played an important role in the description and the management of the AIDS epidemic—in San Francisco) explained that 'just as pathogenic organisms undergo mutations and environmental influences, so do populations' characteristics and modes of interaction change. (...) With the relaxation of traditional moral restraints and the emergence of more permissive modes of social and sexual interplay in the last ten years, some major cities acquired large, highly visible homosexual communities. This concentration has produced more opportunities for frequent sexual contacts between homosexual men. The frequent practice of oral and anal sex had been accompanied by rapidly increasing incidence of 'classical' venereal diseases, venereal herpes and enteric diseases' (Dritz, 1980).[1] Recent changes in pathological patterns, Dritz proposed, did not reflect changes in pathogens, but a 'mutation' of their hosts' behavior and its consequence, the weakening of their immune system. This view prevailed in the first scientific papers dedicated to AIDS.

Some of the first studies dedicated to severe immunodeficiency in previously healthy individuals, proposed that the new disorder could be induced by the combination of several events which, together, induced a collapse of the immune system. Other studies attributed a predominant role to a single event, such as an inhalation of the drug amyl nitrate or an infection by cytomegalovirus. But even the papers which favored 'monocausal' origins of the new pathology incorporated this hypothesis in a 'multicausal', framework, usually one which combined 'life style' elements and 'predispositions'. Thus a paper dedicated to the development of Kaposi's sarcoma in young homosexual men proposed as an explanation of this phenomenon, 'the combined effects of persistent viral infection plus an adjuvant drug cause which causes immunosupression in some genetically predisposed men' (Durack, 1981). Another study of Kaposi's sarcoma in homosexuals, revealed 'significant association between the incidence of Kaposi's sarcoma and the use of number of recreational drugs (amyl nitrite, ethyl chloride, cocaine, phencyclidine, methqualone and amphetamine), as well as a history of mononucleosis, and sexual activity in the year before the onset of disease. Patients with Kaposi's sarcoma also reported substantially higher rates of sexually transmitted infections than did controls' (Marmor, Laubenstein, Williams *et al.*, 1982). A study which proposes that cytomegalovirus induces immunodeficiency in homosexual men adds: 'we acknowledge the possibility that cytomegalovirus infection was the result rather that a cause of the T-cell defect, and that some other exposure to undetected microorganism, drug or toxin made the patients susceptible to infection with opportunistic organisms, including cytomegalovirus' (Gottlieb, Schroff, Schanker *et al.*, 1981). A study of chronic Herpes simplex infection in homosexuals explains that 'the cause of the immunodeficiency disorder that we observed is undoubtedly complex. (. . .) Since these cases are certainly rare, even among homosexuals, additional factors must be involved in susceptibility. A group may be specifically hypersensitive to the herpes virus, perhaps because of their genetic background, e.g., HLA-D linked immune response genes' (Segal, Lopez, Hammer *et al.*, 1981). Other researches postulated a relationship between Kaposi's sarcoma and HLA-DR5 antigens. This antigen, they claimed, was found in 60% of homosexual men with Kaposi's sarcoma (Mack, 1982).

The interest in multicausal hypotheses of the origins of acquired immunodeficiency declined following the discovery of such immunodeficinecy in groups which—unlike homosexuals and intravenous drug users—were not suspected of developing unusual and bizarre life styles: hemophiliacs, recipients of transfusions and, to a lesser extent, Haitian emigrants. Haitians could be still be represented as the 'other' (Farmer, 1992).[2] Thus an article which described high frequency of AIDS among Haitian emigrants stated that 'to substantiate any hypothesis about the pathogenesis of AIDS with respect to Haitians, we need to learn more about the Haitian life style in both the United States and Haiti. The assumption that heterosexual Haitians and homosexual Americans have little in common may prove erroneous when epidemiological and anthropologic surveys are completed' (Viera, Frank, Spira and Landesman, 1983). Hemophiliacs and recipients of blood transfusion could not, however, be easily subsumed under the heading of 'unusual life styles'. The description, in the summer of 1982, of three AIDS cases among hemophiliacs (one was a child) increased the probability that an unknown virus was the initial cause of AIDS. A review article published in *Science* in

August 1982, strongly supported the 'viral hypothesis' as the origin of AIDS: 'most interest right now is focused on the possibility that an infectious agent causes AIDS. The resemblance of the population at risk for hepatitis B suggests a viral pathogen, as does the discovery of the disease in hemophiliacs' (Marx, 1982). Additional cases of immunodeficiency in hemophiliacs described in the spring of 1983 reinforced the experts' conviction that the new disease should be attributed to an infectious agent, and in all probability a new virus (Lederman, Ratnoff, Scillian *et al.*, 1983; Menitove, Aster, Casper, *et al.*, 1983; Deforges, 1983).

In September 1983 an article 'AIDS—two years later', published in *The New England Journal of Medicine* summed up the causes for the rapid adoption of the viral hypothesis by the experts: 'the lack of consensus regarding the cause continued until July 1982, when three cases were reported among patients with hemophilia who had been treated with clotting factor concentrates. By then, carefully documented cases of AIDS had occurred in heterosexual men and women who were intravenous drugs abusers, suggesting a pattern of transmission reminiscent of that for infection with hepatitis B virus. The hemophilia cases greatly strengthened this interpretation. The sudden occurrence of a new syndrome that affected primarily these three distinct populations who shared only their susceptibility to hepatitis B, convinced many investigators that a transmissible agent was the primary factor responsible for the immunologic defects characteristic of AIDS' (Curran, 1983). The hepatitis B model was attractive because it put together several of the characteristics of the (putative) transmission of the new disease: transmission through sexual relations, blood and blood-products, high prevalence in homosexuals, intravenous drug users and patients who receive blood derivatives (Curran, 1995).[3] The assumption that AIDS was induced by a virus did not eliminate, however, the 'cofactor' hypothesis. Scientists attempted to combine the two: a single transmissible agent, such as a virus, may induce the basic imunologic defect, the ultimate expression of which depends on other variables, such as nutritional status, other infections of chronic illness, environmental agents or genetic predisposition (Viera, Spira and Landesman, 1983).

AIDS AFTER HIV: COMPLICATED EXPLANATIONS FOR A COMPLICATED PATHOLOGY

The conviction that AIDS is a viral disease, opened the search for its etiologic agent. This search was rapidly successful. The controversy which opposed two candidates for the role of the 'AIDS-inducing virus', the French LAV (described in 1983) and the American HTLV-III (described in 1984) ended in a compromise: it was agreed that LAV and HTLV-III were identical, and the virus acquired a new name—Human Immunodeficiency Virus or HIV («Chronology», 1987; Seytre, 1993). An agreement among the experts was slow to be translated into health policy terms. For example, the French blood supply and French-produced clotting factors were still contaminated by the HIV in 1985 (Setbon, 1993; Moulin, 1995). One should not confuse, however, delays induced by shortcomings of health policies with a delay in the recognition of the viral etiology of AIDS. A small minority of experts led by the molecular biologists Peter Duesberg excepted, the relevant professional milieu accepted from 1984 on that AIDS was a retrovirus-induced disease (Epstein, 1997, pp. 170–184). 'Predisposing factors' and 'facilitating elements' continued, however, to be evoked, in order to explain poorly understood aspects of the natural history and epidemiology of HIV-infection (Matur-Waugh, Mildvan and Seene, 1985; Volberding, 1987; Levy, 1988).

Some researchers attempted to uncover a 'genetic predisposition' for AIDS. An author of a study published in 1987 and which proposed that a protein called Gc (group specific complement) or a vitamin D binding protein may play a role in the susceptibility to HIV infection, argued that practically every infectious disease had a hidden genetic element. 'Recall', he explained, 'that genetic diseases and infectious diseases used to be the province of separated specialists, departments and textbooks. But this distinction has been crumbling with each new discovery of genetic susceptibility underlying infectious diseases. For instance, tuberculosis, the paradigmatic infectious disease of the nineteenth century, although undoubtedly caused by an infection with the tubercle bacillus, is also co-determined by genetic factors that modulate the immune response and gastric acidity' (Diamond, 1987).

One of the unexplained aspect of AIDS was the difference between the epidemiological pattern of transmission of this disease in Africa and in Western countries. The pathological manifestations of AIDS and its etiologic agent were similar in Western and in African countries. The pattern of AIDS epidemiology in Africa was, however, different. African AIDS was a sexually transmitted infection which was not limited to specific 'risk groups'. It has a sex-ratio close to 1:1 (in the West, AIDS was in the 1980s mainly a male disease), and an alarmingly high rate of infection (Newmark, 1986; Biggar, 1986; Piot, Plummer, Mhalu, *et al.*,

1987).[4] The differences between the epidemiological patterns of HIV infection in the West and in Africa were attributed to the existence of 'predisposing factors' in African populations: high frequency of infections, in particular sexually-transmitted infections, poor general health status, poor nutrition and unidentified genetic factors (Piot, Kreiss, Ndynia-Achola *et al.*, 1987; Quinn, Piot, McComick *et al.*, 1987 Johnston and Laga, 1988). None of these putative 'cofactors'—with the sole possible exception of genital ulcers in women—was, however, susceptible to an efficient external intervention (Pepin, Plummer, Burnham *et al.*, 1989). The experts' recommendation for health-policy makers oscillated between wishful thinking (it would have been desirable to improve health care, to promote education, to increase food supply, to limit the economic dependence of women) and calls for 'specific prevention' of AIDS through the diffusion of information about the way this disease is transmitted and the distribution of condoms (Pape, 1989).

Another 'mystery of AIDS' was the difficulty to correlate clinical manifestations of this disease with the presence of viral particles in the blood. The number of viral particles in T-cells in the blood of HIV infected individuals, measured with an indirect method, the end-point dilution, was found to be very low. According to some evaluations, only one in 100.000 CD4 lymphocytes was HIV infected. The direct effects of HIV multiplication were thus unable to account for the widespread destruction of CD4 cells induced by this virus (Levy, 1988).[5] Two hypotheses were proposed to explain the damage for the immune system of AIDS patients. One postulated that 'proximal pathogens', probably incomplete variants of the HIV virus which cannot be detected in standard tests, participate in the destruction of cells of the immune system. This hypothesis, based on the feline leukemia model, was not pursued (Haas, 1989). The second and more popular hypothesis postulated that HIV infection triggered a chain of pathological reactions which culminated in the destruction of the immune system. This hypothesis was first made in 1984, when researchers had shown that AIDS patients presented a vast array of immunological abnormalities (Seligman, Chess, Fahey *et al.*, 1984). In 1986, immunologists proposed that AIDS is an autoimmune disease, that is the immune system is stimulated by HIV to destroy itself (Zeigler and Sittes, 1986).[6] This attractive and off-quoted hypothesis obtained experimental support in 1991. Scientists had shown that mice, who were manipulated to develop an auto immune disorder through the injection of lymphocytes from another mouse strain, developed at the same time specific antibodies to one of HIV antigens, the gp 120 protein. This result indicated, the authors of this study proposed, that the virus mimics the cell's structure and in this way starts a chain of self-destruction of the immune system (Hoffman, Kion and Grant, 1989). Their study was hailed by a *Nature* editorial as an important breakthrough in the understanding of AIDS, and a finding which may lead to the development of efficient therapies for this disease (Maddox, 1991). A related hypothesis, published in 1991, attributed the collapse of the immune system in AIDS patients to an 'superantigen' encoded by the HIV. This 'superantigen', it was proposed, is particularly efficient in activating the cellular immune mechanism, and starts a uncontrollable 'chain reaction' which leads to T-cell depletion (Marx, 1991). In the early 1990's, immunologists tried to cure AIDS through adoptive immunotherapy, that is, the transfer of activated immunocompetent cells (here, the CD8 lymphocytes) to patients. The rationale behind these studies was, they explained, that 'thus far most of the treatment for patients with AIDS has been focused on an antiviral approach. It is clear, however, that AIDS and HIV infection is primarily a disease affecting the immune system, and potentially one could use immunological approaches to complement the anti-viral approach in this category of patients' (Heberman, 1992, p. 35). However, attempts at adoptive immunotherapy of AIDS were not very successful, while this treatment was expensive and labor-intensive. Autoimmune hypotheses and immunological investigations were not totally abandoned, but in the 1990's the center of interest in AIDS studies gradually shifted to the dynamics of viral multiplication.

FROM A 'HIDDEN VIRUS' TO A 'MIGHTY WARRIOR': CHANGING PERCEPTIONS OF THE NATURAL HISTORY OF HIV INFECTION

A review article published in *The New England Journal of Medicine* in March 1986 (by June Osborn) warned against the premature abandonment of the hypothesis of 'predisposition' and 'cofactors' in AIDS. 'From the beginning of the epidemic', it explained, 'there have been hints that other infections, drugs, and life style characteristics have contributed to the occurrence of AIDS in the populations at highest risk, and it will be crucial to understand the importance of such factors in order to counsel the hundreds of thousands of people we believe to be asymptomatically infected at present' (Osborn, 1986). A review article published in *The New England Journal of Medicine* nine years later (by David Ho), offered unambiguous advice to asymptomatic carriers of

HIV. They should get as early as possible an aggressive anti-retroviral therapy. The only way to overcome HIV infection is 'to hit the HIV early and hard' (Ho, 1995). The contrast between these two articles illustrates the replacement of 'multifactorial' images of AIDS which stressed complexities by simplified 'war' images. This dramatic shift in the representation of AIDS was directly related to changes in the perception of the natural history of HIV infection.

An 1984 article which presented the growing evidence that a lymphotrophic retrovirus was responsible for immune abnormalities in AIDS, added that the rapid discovery of the 'cause of AIDS' is a triumph of today's biological science. This discovery, 'is also an evidence that society now has the knowledge and the scientists needed to turn medical questions into a predictable series of efforts in problem solving, in the way that we have come to expect of engineers in space exploration and computer scientists in automation. Perhaps such accomplishments partly justify the present excitement about biotechnology. Unpredictability in the timing of advances in the biological and medical investigations was characteristic of the past. Now, however, we are establishing a firmer footing, and the control of virus infection is beginning to emerge as a fairly rational science! (Marigan, 1984).

This optimistic mood did not prevail, however. In 1988, the special issue of *Science* dedicated to AIDS highlighted the difficulty to understand this pathology, and the even greater difficulty to control AIDS pandemics (Koshland, 1988). Efficient therapies for AIDS and an efficient vaccine against the disease were slow to come, while the understanding of the natural history of HIV infection remained incomplete. It was not clear what determined HIV pathogenesis, and what influences the progression from asymptomatic infection to a full-blown disease (Levy, 1988). Studies of the molecular structure of HIV progressed rapidly, but, as the molecular biologist David Baltimore put it: 'we are rapidly learning about the role of each of HIV's approximately 100,000 nucleotides, but remain largely ignorant of rudimentary aspects of the process underlying the development of AIDS in humans' (Baltimore and M.B. Feinberg, 1989).

The understanding of the natural history of HIV infection was seen as central to the development of efficient therapeutic approaches. Circa 1989, the dominant feature of this infection was a long latent period which followed primoinfection and seroconversion. The experts assumed that during this latent period the virus was hidden (in the nucleus, in a small number of CD4 lymphocytes) while the pathological phenomena induced by HIV infection were attributed to its indirect effects on the body (Fauci, 1988; Moss and Bacheti, 1989). The perception of HIV as a 'hidden virus' changed, however, in 1989, following the discovery of the virus in lymph nodes of asymptomatic HIV infected individuals and following development (by David Ho and his collaborators) of more sensitive methods of quantification of HIV in the blood (Ho, Mougdil and Alam, 1989).[7] New studies made with these methods increased the previous estimates of the percentage of HIV infected lymphocytes in the blood of infected individuals. In asymptomatic HIV-infected patients, Ho and his collaborations affirmed, there were 20 HIV containing cells per million peripheral blood white cells (one in 50,000 cells), and in patients with full blown AIDS, between 2000 and 3000 virus containing cells per million white blood cells (that is, about one in 500 white blood cells). These results, Baltimore and Feinberg explained, 'should dispel any lingering doubts about whether HIV is the true culprit in AIDS' (Baltimore and Feinberg, 1989, p. 1675). They did not put end, however, to interrogations on mechanisms of HIV-induced pathology. The HIV was shown to be present in a greater number of CD4 cells than was postulated earlier, but its presence in one in 500 cells (in advanced stages of HIV infection), even less in one in 50,000 cells (in asymptomatic infection) could not account, by itself, for the massive destruction of the immune system in AIDS patients. A 1993 survey on the status of AIDS research organized by *Science*, and published under the heading 'AIDS research: The mood is uncertain', explained that 'what causes the immune system to collapse in AIDS' is the most important question still to be solved (Cohen, 1993).

One of the main problems in AIDS studies was the correlation of 'biological markers' of the disease with its clinical progress. Several 'biological markers' for AIDS were singled out. The first to be described was the decrease in CD4 -lymphocytes (the observation that patients with opportunistic infections or Kaposi's sarcoma had very low levels of CD4 lymphocytes in the blood led to the definition of the new disorder as 'acquired immunodeficiency'). Additional 'biological markers' for AIDS were described later: antibodies to HIV p24 core antigen, β-microglobulin and neopterin (all three are proteins found in the serum). All these 'biological markers' showed partial correlation with the clinical evolution of HIV infection, but none was sufficiently precise to predict outcomes (Moss, 1988). Their lack of accuracy in predicting the progression to AIDS became especially important when these markers were used to evaluate the efficacy of anti HIV drugs. Experts attempted to improve the predictive accuracy of 'biological tests' through the improvement of laboratory techniques, better standardization of tests, and combination of several indicators. All their efforts failed, however, to uncover a more reliable method to evaluate the effects of therapies. A 1991 article which summed up these efforts concluded that the most consistent sign of progression to AIDS is (still) the decrease in the number of CD4 lymphocytes, but recognized that this 'biological marker' was not always trustworthy, and in particular, that it was not well adapted to the evaluation of the efficacy of anti-HIV drugs: its predictive accuracy decreased in treated patients (Cotton, 1991).

While the general mood about AIDS research in 1993 was described as 'uncertain', some scientists proposed that important advances were made in the understanding of the natural history of HIV infection. The most important new finding, a 1993 review stressed, was the confirmation by molecular techniques of the earlier observation that even in the asymptomatic phase of the infection the virus was found in massive amounts in the lymph nodes, and was actively replicating in lymph node cells. The latter finding questioned the existence of a latent phase of HIV infection (Weiss, 1993). In the 1980's scientists assumed that HIV was a truely latent virus, that is, viral DNA proviruses that do not express viral RNA were hidden in CD4 cells. This perception of 'genuine latency', was however replaced later with the conviction that during the so-called 'silent phase' of the infection HIV may indeed be latent in some cells, but it remains active in the lymph nodes. In 1993, the persistent replication of HIV in the lymph nodes was presented as the most important feature of HIV infection. Viral replication 'generates conditions that promote its continued growth. In other words, HIV eventually becomes its own opportunistic infection (...) The emerging view is (...) that a progressive HIV burden involving first activation and eventual destruction of the immune system is what lies behind AIDS (Weiss, 1993, pp. 1277–1278).'

The shift in the representation of HIV infection from a 'latent' one, with a 'hidden' provirus, to an active and continuous attack of a virus on the immune system had direct implications for proposed therapeutic strategies. When the scientists believed that clinical latency was synonymous with low viral burden and microbiological latency, they hesitated about starting an aggressive treatment with antiretroviral agents at this stage. Anti-HIV drugs such as AZT (zidovudine) are toxic, and often cannot be tolerated for a long time. It was better, many doctors felt, to reserve these drugs for an acute stage of the disease. Moreover, clinicians feared that an early treatment would lead to the development of drug-resistant mutants of HIV and compromise the chances of later therapeutic intervention (Cotton, 1990).[8] However, when more sensitive molecular techniques indicated that a relatively high load of HIV is present in the lymphoid tissue during all the stages of the infection, the rationale for an early and vigorous treatment increased.[9] The advantages of such early treatment should be weighted, however, Anthony Fauci, (director of the NIAID, and one of the leading US AIDS experts) warned in 1993, against the possible disadvantages of disrupting delicate physiological equilibria (Fauci, 1993). 'Any comprehensive therapeutic strategy' he explained, 'must consider the complexities of the pathogenic mechanisms of HIV disease (...) Certain stages of the disease may benefit more than others from an intensive regimen of antiretroviral drugs. (...) Certain types of intervention may be appropriate at one stage of the disease and contradicted at another. (...) It is only through the process of carefully conducted clinical trials based on the expanding knowledge of HIV pathogenesis that answers to these important questions will be forthcoming' (Fauci, 1993, p. 1016).

Two years later, a careful investigation of the subtle dynamics of HIV infection was not presented any more as a precondition for the development of efficient anti-HIV therapies. Hence David Ho's statement from 1995 that the only efficient way to deal with an HIV infection is to hit HIV early and hard (Ho, 1995). His statement was based on new findings on the dynamics of HIV infection. In 1995 scientists described the extreme rapidity of viral turnover in the supposedly 'silent' stage of infection. This new understanding of the dynamics of HIV infection was made possible thanks to the development of a new (or rather perfectionned) technique: quantitative PCR. Improvements in the PCR technique (branched DNA signal-amplification technique; in situ ultrasensitive quantitative PCR) increased the ability to produce (trustworthy) measures of viral replication. The application of this technique to studies of the dynamics of HIV infection reversed HIV's original image of a 'lentvirus' which induces a 'latent infection'. HIV, the new data indicated, was anything but slow (Ho, Neumann, Perelson, *et al.*, 1995; Wei, Ghosh, Taylor, *et al.*, 1995). As Ho and his collaborators stressed, 'AIDS is primarily a consequence of the continuous, high level replication of HIV-1, leading to virus- and immune-mediated killing of CD4 lymphocytes' (Ho, Neumann, Perelson, *et al.*, 1995, p. 126). The rapid turnover of HIV could also explain why indirect methods such as end point dilution detected low amounts of this virus in the blood. At any given moment the quantity of the virus in the blood is indeed low, because viral particles relased to the bloodstream are immediately eliminated by immune mechanisms—especially in asymptomatic individuals without major impairment of these mechanisms. This finding should not be interpreted, however, as pointing to a low level of viral activity.

Immunologists rapidly integrated the new vision into data on dynamics of HIV-induced transformations in lymph nodes. The interest in complicated cellular interactions was replaced by investigations—made by PCR in situ—of the rate of infection of specific cellular populations in the blood and lymph nodes: T-cells, macrophages and dendritic cells. In parellel, researchers compared data on viral load (seen as indicating the true status of infection), with measurements of the number of CD4 positive cells, and with lymph node biopsies, and concluded that the level of CD4 is a less precise indicator of the evolution of HIV infection than morphological changes in lymph-nodes, found to be relatively well-correlated with PCR data (Pantaleo, Cohen, Schwartzentruber, *et al.*, 1995). The shift from an interest in cellular regulations of anti-HIV response to the emphasis on the dynamic of cell destruction by retrovirus illustrates the changes brought about by new findings on rapid multiplication of HIV (Novak, 1995). In 1995, this infection was likened to a total war: 'that billions of virions and infected cells can be destroyed every day vividly illustrates the very hostile environment created by the immune system—the meanest of streets are nothing in comparison'. To win this war, a concentrated attack on the rapidly proliferating and mutating virus was therefore needed: 'monotherapies could not succeed. Only a combination of drugs has the potential to outgun the virus' (Wain-Hobson, 1995). Finally, the efficacy of such combination of drugs in the organism could be estimated rapidly. New techniques were able to demonstrate an impressive drop in the level of the virus after the initiation of a therapy: 'when new cycles of infection are interrupted by potent antiretroviral therapy, plasma virus levels fall abruptly by an average of 99% and in some case by as much as 99.99% (10,000 fold) (Wei, Ghosh, Taylor, *et al.*, 1995, p. 122). Such a rapid drop was important because, the experts argued, the success of a given antiretroviral therapy depended to a large extent on its capacity to act faster then the virus: 'current protocols for monitoring the acute antiviral activity of novel compounds should be modified to focus on the first days following drug initiation' (Ho, Neumann, Perelson, *et al.*, 1995, p. 126).

The latter proposal was particularly attractive for pharmaceutical companies which produce anti-HIV drugs (and also to AIDS patients who wish to shorten the time of testing promising therapies). The end-point of clinical trials of anti-HIV therapies was first defined as the improvement of patients' survival. Later clinical trials of antiretroviral compounds were shortened through the introduction of 'biological markers' (or 'surrogate markers') as the main indicator of the efficacy of a given therapy. Several months were necessary, however, to document changes in 'biological markers' such as the number of CD4 lymphocytes. By contrast, changes in virus load measured with PCR could be observed after only a few days, dramatically reducing the length (and the cost) of clinical trials. For some experts, the hierarchy of values was overturned, and the final goal of AIDS therapy was redefined as the elimination of HIV. Thus Joep Lange from the World Health Organization explained that, 'the virus is the real thing. Clinical end points are the surrogates' (Cotton, 1994). Some AIDS activists agreed: thus Xavier Rey-Coquais, from the French group 'Actions—Traitement', explained that 'the aim of a clinical trial should not be to verify if a given combination of drugs increases the level of CD4 lymphocytes or diminishes the frequency of opportunistic infections, but to check if it decreases the viral load and maintains it at the lowest possible level (Rey-Coquis, 1996, p. 41).'[10] The announcement, in January 1996, that a combination of three anti-HIV drugs (two nucleotide analogues and a protease inhibitor an anti-protease), tested for several weeks in HIV-infected individuals, eliminated the virus from their blood, led indeed to a pressure of patients associations to receive quickly this therapy, to a rapid attribution of a temporary marketing permit to its producer, and to a generalization of the new therapy to important segments of AIDS patients in industrialized countries (Fontanay and Chambon, 1996; Cohen, 1996).

In parallel (1995–1996), 'resistance to HIV infection', dissociated from epidemiological considerations and translated into molecular terms, became a major research subject for virologists and molecular biologists. The focus in studies of 'predisposition to AIDS' had shifted from the study of elements ('cofactors') which increase or decrease the susceptibility of populations to this disease, to the investigation of molecules ('coreceptors') which control the entry of HIV into a cell, thus from interest in human populations to investigations conducted in the test-tube. Investigations which led to the description of these 'coreceptors' started as studies on the role of a sub-population of lymphocytes and of molecules secreted by lymphocytes to control infections. Scientists observed that CD8 T-lymphocytes contained HIV infection. They had isolated HIV suppressive factors secreted by these lymphocytes—chemokines (regulatory molecules) RANTES, MIP 1-alpha et MIP 1-beta, and then found that these molecules blocked the entry of macrophage-tropic (virus which selectively infects macrophages) but not T-cell tropic (virus which selectively infects T-lymphocytes) HIV strains into cells (Cocchi, DeVico, Garziano-Demo *et al.*, 1995).[11] At about the same time, researchers who studied the mechanism of entry of HIV virus into cells proposed that a co-receptor, named 'fusin' is necessary (in addition to the previously known HIV receptor the CD4 molecule) for the infection of T-cells with T-cell tropic strains (Feng, Border, Kennedy, *et al.*, 1995). Experts then concluded, by analogy, that similar 'co-receptors' should be found in macrophages too, and proposed that the inhibitory molecules, (RANTES, MIP 1-alpha and MIP 1-beta), may react with this coreceptor. Both suppositions were confirmed. No less than five large groups (a telling testimony to the density of interactions in this area and to perception of this subject as an especially 'hot' topic) published quasi-simultaneously articles which identified a macrophage-specific 'co-receptor' of the HIV virus, named CC CKR-5 (later abbreviated to CCR-5, or CKR-5) and showed that this 'co-receptor' reacted with inhibitory chemokines RANTES, MIP 1-alpha and MIP 1-beta (Alkhatib, Combadiere, Border, *et al.*, 1996; Deng, Liu, Ellmeier, *et al.*, 1996; Dragic, Litwin, Allaway, *et al.*, 1996; Choe, Farzan, Sun *et al.*, 1996; Doranz, Rucker, Yi, *et al.*, 1996).[12] In parallel, investigators studied individuals who, despite multiple exposures to HIV virus through sexual contacts, remained infection-free. They found that in a few cases, T-cells of 'resistant' individuals could be infected by HIV in the test tube while their macrophages (which are seen as the primary 'entrance-gate' of the virus in sexually-contaminated individuals) resisted infection.[13] Two such 'resistant' individuals (out of a cohort of 25) were found to be homozygous (they carried two identical copies of a gene) to a specific mutation in the gene coding for the CCR-5 receptor (Liu, Paxton, Choe, *et al.*, 1996). A correlation between mutant CCR-5 and 'resistance to HIV infection' was also found in a larger epidemiological study (Samson, Libert, Doranz, *et al.*, 1996).

Research on the role of the mutant 'co-receptor' CCR-5 was presented in the media as an important breakthrough in AIDS research: uncovering the biological background of resistance to AIDS and opening new pathways to a prophylaxis of this disease and to a therapeutic intervention in HIV infection. The claim was, however somewhat excessive. The mutation in the CCR-5 gene carried by two 'HIV resistant' individuals, described in the first study may explain (entirely or partly) why they 'escaped' HIV infection, but the basis of resistance of other persons in the 'resistant' cohort remains unknown, while their macrophages were found to be fully susceptible to infection in the test tube (Liu, Paxton, Choe *et al.*, 1996, p. 370). The second study shows that among a population of 'slow progressors' to AIDS, one finds a higher frequency of individuals who are heterozygous for a mutant CCR-5 gene (that is, who carry a single copy of this gene) than in the general population. It does not demonstrate, however, that persons homozygous for the mutation are indeed protected from HIV infection (Samson, Libert, Doranz, *et al.*, 1996, p. 723).[14] A defect in the CCR-5 gene is probably but one among many biological mechanisms which may account for the resistance to HIV infection. This does not diminish, however, the intrinsic interest of these elegant studies which revealed this defect. These studies reveal other things as well. Their spirit was aptly captured by the title of Antoni Fauci's review article 'Resistance to HIV-1 infection: its in the genes' (Fauci, 1996).[15] In this article, centered on the protective role of the mutant CCR-5 gene, Fauci explains that 'it is of particular interest that the defective allele is either absent or extremely rare among black African and Japanese. The gene distribution and hence the lack of protection may contribute in part to the fulminant nature of the epidemics in sub-Saharan Africa (Fauci, 1996, p. 967).'[16]

New developments in studies of 'resistance to HIV infection' (translated into molecular terms), together with the new perception of the natural history of this infection (expressed as a change in viral load measured by PCR), may point to an intensification of a 'technological drive' in AIDS studies. In the mid-and late 1990s, these studies were centered on interactions between viruses and cells, on molecules involved in these interactions, and on a search for ways to affect this process through the use of (other) molecules, engineered and mass-produced by the pharmaceutical industry.

AN EPIDEMICS OF SIGNIFICATION?

Paula Treichler coined the expression 'epidemics of signification' to describe the AIDS pandemic (Treichler, 1987). The success of this expression reflects the recognition of the societal impact of AIDS. AIDS may be interpreted as a 'modern' (or 'post-modern') disease which links sex and blood, drugs, homosexuality, migrant populations, prostitution and third world poverty, and which displays in a dramatic way the strengths, but also the limitations of modern biomedicine, and the advantages, but also the shortcomings, of Western societies. But isn't the expression 'epidemics of signification' a tautology? Could there be an epidemic that is not 'an epidemic of signification'? Epidemics were always the carriers of multiple social meanings, from God's wrath to a large array of human shortcomings (Fleck, 1935; Rosenberg, 1989). This paper does not deal with the general cultural and social meaning of AIDS, a topic extensively discussed by philosophers, anthropologists, historians, literary critics and social scientists, but with meanings embodied in scientific perceptions of AIDS and in practices of researchers who study this disease. It follows two scientific perceptions of AIDS: as a multilayered physiological phenomenon which reflects the deregulation of complicated immune mechanisms rooted in the previous history of the infected individual, and as a 'simple infection', a direct consequence of the struggle between an invading pathogen and the body's defenses. These two perceptions reflect two different approaches to immune mechanisms. The organizers of a meeting on the 'Twentieth century body', proposed that modern biomedicine lead to the development of a series of distinct and non-overlapping perceptions of the human body, one of which is the 'defended body'.[17] This general category may, I suggest, be further sub-divided into three distinct entities: the 'attacked body' (or the 'body at war'), a view which stresses the role of the immune system as fighting infections, the 'immunological body' (or the 'self-aware body'), a view focused on the physiological tasks of immune mechanisms and their role in defining 'biological individuality', and the 'defended social body', a view which follows the collective effects of immunity. The latter is out of the scope of this paper, mainly because (for the moment) there is no vaccine for AIDS. The two other 'bodies'—the 'attacked body' and the 'immunological body'—are, by contrast, central to my argument on the co-development of contrasting 'images of AIDS' and the recent predominance of 'war images' of this pathology.

The 'attacked body' and the 'immunological body' developed together. Immunology have oscillated from its very beginning between two competing sets of images: 'war images' which represented immune mechanisms as an army which resist an attack of invading microorganisms or police forces which eliminate undesirable aliens, and 'physiological images' which stressed the role of immune mechanisms in basic physiological functions such as digestion, elimination, contact with the exterior or aging (Löwy, 1991). The co-existence of these two sets of images, I propose, facilitated the integration of contrasting cultural meanings into scientific practices. The two immunological theories developed in the late 19th century, the cellular theory (Metchnikoff) and the humoral theory (Ehrlich) could be (and were) interpreted in physiological terms. The destruction of pathogenic microorganisms by phagocytes was presented as part of a general mechanism to eliminate old and defective cells, while humoral antibodies were viewed as 'lateral chains' of the protoplasm which usually serve to adsorb food. The humoral and cellular views of immunity could be also (and were) integrated into 'war images', in which an army of phagocytes fought invading troops of pathogens, and antibodies were represented as sophisticated weapons ('magic bullets') that neutralize and eliminate the enemy.

The predominance of 'war images' or of 'physiological images' of immune reactions was related to the role played by immunologists in a given context. Briefly, 'war images' prevailed when immunologists were perceived mainly as suppliers of specialized services to bacteriologists and epidemiologists, while 'physiological images' were often related to the development of immunology as an autonomous branch of biological knowledge and as a discipline which provides a specific language for the understanding of pathogenic phenomena (Moulin, 1992). Physicians traditionally employed terms such as 'terrain' and 'constitution' to explain the great variety of individual reactions to an encounter with a pathogenic microorganism. The phrase, attributed to Claude Bernard, 'le microbe n'est rien, le terrain est tout', aptly sums up the resistance of clinically-oriented doctors to over-reductionistic interpretations developed by new bacteriological theories. In the early 20th century, some scientist proposed to translate the imprecise term 'terrain' into the notion of (chemically—determined) 'biological individuality'. Alexis Carrel, a surgeon and biologist who developed, circa 1910, a surgical method to graft kidneys in dogs, had found that isografts (when a kidney was taken out the body, then regrafted into the same individual) were always successful, while heterografts (when a kidney from one dog was grafted into another dog) invariably failed. He concluded that a unknown biological mechanism, able to recognize small chemical differences between individuals, was responsible for a graft rejection, and proposed to set up a research program to study this mechanism (Carrel, 1910). At the same time, the physiologist Charles Richet, who studied anaphylaxis (violent, occasionally deadly reaction to a second injection of a sensitizing antigen), concluded that the specific reaction of an individual to an external antigenic stimulus depends on inherited 'predispositions', but also on the individual's history: 'Anaphylaxis,' he explained, 'like immunization, creates humoral differences between different individuals. (. . .) Each of us by our chemical make-up, above all by our blood and probably also by the protoplasm of each cell, is himself and no one else. In other words, he has a humoral personality. Each living being, though presenting the strongest resemblance to others of his species, has his own characteristics (Richet, 1911). Richet's and Carrel's propositions to investigate the chemical and physiological mechanisms of 'biological individuality' were not fulfilled, however. The biochemical and the biological techniques available circa 1910 were not adapted to the study of minute structural differences. In addition, from the 1910s on, immunologists often abandoned the investigation of complicated physiological reactions, such as graft rejection or anaphylaxis, in favor of the more 'duable', but also more practice-oriented study of the structure and function of specific antibodies in the serum.

In the interwar period, clinical immunology (serodiagnostic, serotherapy), was perceived mainly as a sub-division of medical bacteriology, while basic immunological (or rather immunochemical) investigations were focused on structure and specificity of antibodies, and were viewed as a part of the larger domain of protein chemistry. Both serology and immunochemistry favored a static approach to the study of antibodies, presented mainly as efficient anti-microbiological weapons. After World War Two, the development of antibiotics diminished the practical importance of serodiagnosis and serotherapy. Immunologists became interested again in immunopathological phenomena (allergies, anaphylaxis, delayed hypersensibility, autoimune diseases, graft rejection) and in physiological approaches to immune mechanisms. Peter Medawar's studies on skin grafts firmly placed graft rejection within immune phenomena, and Macfarlane Burnet's investigation of the dynamics of the formation of antibodies presented the problem of immune response as a biological memory embodied in cells and proteins. Their research led to the redefinition of immunity in terms of self/non self discrimination. This redefinition, and the elaboration, in the mid-1950s of the clonal selection theory, a theoretical framework which reconciliated immunological views with the newest developments in molecular biology, favored the development of a physiological theory of immunity (Medawar, 1946: Burnet and Fenner, 1949). Immunity was viewed as a fundamental physiological mechanism responsible for the 'self/nonself discrimination', that is, for the constitution of 'biological individuality'. This task was conducted by a complex network of cells and molecules—the 'immune system' (Burnet, 1959, Klein, 1982).

The rapid growth of immunology from the 1960's on may be attributed to the capacity of the 'new immunology' to translate the medical question of individual reactions to pathologies into a new vocabulary of biochemistry, genetics and molecular biology. Individual differences between patients, once attributed to 'predispositions' and 'terrain', could now be expressed in terms of the presence of specific HLA antigens, of levels of cellular reactivity, and of quantities of regulatory molecules or their receptors in the serum. In the 1970s and 80s, immunological theories stressed complexity with cascades of reactions and networks of cells and molecules. This trend towards growing complexification was amplified by specific practices of immunologists, such as studies of a large class of immunoregulatory molecules—the cytokins—or the extensive use of a new instrument—the Flourescein Activated Cell Sorter (FACS)—to distinguish between sub-populations of lymphocytes (Cambrosio and Keating, 1992).

The definition of acquired immune deficiency syndrome as an 'acquired deficiency' was directly related to the new developments in immunology. Opportunistic infections were rapidly linked to 'acquired immunodeficiency' and to the disappearance of CD4 lymphocytes because doctors knew that these pathologies were, as a rule, observed in individuals who suffered from severe impairment of immune mechanisms, (patients with agranulopenia; recipients of organs grafts who received immusupressive therapy), and because they were able to verify the immunodeficiency hypothesis by quantifying sub-populations of lymphocytes in their patients' blood. During the 'pre-virus' period of AIDS studies (1981–1982), clinicians usually assumed that the observed immunodeficiency was induced by a complicated interplay between genetic background, lifestyle elements, and, possibly one or several infectious agents. The growing conviction (from 1982 on) that a new virus, the HIV is the single 'cause of AIDS' did not challenge this conviction. The presence of HIV in the body could not account, numerous experts believed in the 1980s, for the pathological manifestations of AIDS. 'Images of AIDS' postulated therefore physiological cascades, complex cellular interactions, and the participation of 'cofactors'. In the 1990s, however, the pathological manifestation of AIDS has been increasingly perceived as a direct result of the multiplication of HIV and the 'exhaustion' of immune mechanisms by successive waves of viral attacks. Multilevel changes in AIDS patients' immune systems were reinterpreted as secondary results of the viral attack. AIDS continues to be perceived as a complicated disease, but the 'interesting' complexity—that is, the one which needs to be understood in order to elaborate efficient therapies—became the complexities of viral multiplication and viral mutagenesis. The focus is now on molecular mechanisms which govern HIV entrance into cells, those responsible for the killing of the cell by the virus, and those who control the elimination of viruses by the body. Studies of well-defined molecules replaced interrogations on the complexities of physiological reactions of the body to HIV infection, or epidemiologically-oriented studies on the complicated role of 'predisposition' and 'cofactors' in this infection (Lever 1996, p. 27; Paxton, 1996).[18]

Why has a 'network'-based physiological model of AIDS, in principle at least better adapted to the description of an exceptionally complicated pathology, been loosing ground in the 1990s to a virological/molecular model, based on 'total war' images and 'single-hit' interactions? Why have immunologists failed to maintain their 'ownership' of a disease first described in immunological terms and still defined, for practical reasons at least, as a decrease in the number of a sub-population of immunocompetent cells, the CD4 lymphocytes (Keating and Cambrosio, 1998). My narrative has stressed the role of a technical innovation—the quantitative PCR—in the successful alignment of practices of laboratory scientists, clinicians, and industrialists, and in parallel, capturing the interest of the various AIDS associations. Other narratives are, however, possible. Thus a 'realistic narrative' may accentuate the role of more sensitive and precise methods for the quantification of viral load and for the study of viral turnover in the development of a better understanding of the mechanism of HIV infection. A 'sociological narrative' may focus on the difference between the professional power and the prestige of retrovirologists and molecular biologists on the one hand, and immunologists on the other, and will attribute the predominance of 'images' proposed by the two first groups to their stronger professional status, and their close relationships with the biotechnology industry. An 'actor network narrative' may accentuate the superiority of the virologists' favored technology—the quantitative PCR—over the one promoted by immunologists—the CD4 lymphocyte. In a 'strength trial' between the two, the PCR was more efficient in mobilizing human and non-human actors, in constructing networks which linked scientists, viruses, polymerase chain reactions, industrialists, anti-HIV drugs, governmental regulatory agencies, AIDS associations and the pharmaceutical industry, and in redefining boundaries between normal and pathological. Finally a 'view from the margin narrative' may link the abandonment of the physiological perception of AIDS, and the lack of interest in the broader context of the development of this pathology to the increasing despair about the possibility to help the large majority of HIV-infected individuals—those who live in developing countries. In these countries, poverty precludes the elimination of the 'facilitating factors' of HIV infection, such as poor nutrition or frequent parasitic and bacterial infections. It is useless to understand better, some expert might have assumed, the complicated interactions between HIV infection and its 'cofactors', if the relevant 'cofactors' cannot be eliminated through appropriate health policies. In contrast, the simplified 'war image' of AIDS, or the interest in 'coreceptors' of HIV, at least holds a promise of efficient action where possible, that is, in the Western countries.

A final remark. This paper argues that the success of the AIDS 'war' model in the mid-1990 has been directly related to the success of therapies based on the elimination of HIV through a coordinated action of molecules which block viral multiplication. The stabilization of the 'war' model may thus depend on the future fate of anti-HIV therapies. If these therapies and especially tri- and quadri-therapies which associate an anti-proteases and replication inhibitors—will continue to be perceived as successful, it is highly probable that AIDS will follow the path traced by infectious diseases such as tuberculosis or malaria.[20] It will become an 'curable' pathology, which nevertheless continues to be an extremely serious health problem for underprivileged populations: the poor of the industrialized countries, but, above all, all the inhabitants of developing countries. The development of an efficient (individual) therapy for AIDS will probably also led to a (relative) neglect of studies of the physiopathology of this disease, akin to the (relative) neglect of studies of the physiopathology of syphilis and tuberculosis which followed the development of antibiotics. By contrast, if therapies based on direct inhibition of HIV multiplication will, in the long run, fail to efficiently control the AIDS epidemics in Western countries, it is possible that experts will return to more physiological perceptions of this disease. They may revert to multifactorial hypotheses, to 'network models', and to an interest in 'predispositions' and 'cofactors',—that is, to a search for an 'image of AIDS' which incorporates the history of individuals and of populations into scientific practices.

ACKNOWLEDGEMENTS

This study was funded by ANRS research grant n° 711–104.

NOTES

1. Dritz was sympatethic to the homosexual sub-culture, and her article did not advocate a moral reform, but the development of new health care measures, adapted to new social phenomena.
2. Farmer studied the racist undertones in the representation of Haitians as carriers of AIDS.
3. Some researchers disagreed, however, with the infectious hypothesis, and proposed that AIDS is above all an immune disorder, not an infectious disease immunologists (Levy and Zingler, 1983).
4. AIDS is associated with two sub-types of the HIV: HIV-I and HIV-II, and HIV-II is mainly isolated from African cases of AIDS. The differences between the epidemiological pattern of AIDS infection in Western countries (pattern A) and in Africa (pattern B) could not be correlated, however, with differences in the sub-type of HIV.
5. The scarcity of viral particles was one of the main arguments advanced by Duesberg when he claimed that HIV was not the cause of AIDS. (Duesberg, 1988).
6. Zeigler and Sittes interpreted this result as pointing to the possibility of activation of idiotypic networks by HIV.
7. Adenopathy—or the swelling of lymph nodes—was, from the very beginning one of the important signs of HIV infection and the HIV (then LAV) virus was first isolated from lymph nodes of a pre-AIDS patient.
8. The 'latency' of HIV observed in the 1980s might have been an artifact of observation—or if one prefers, of the tendency of physicians and biomedical scientists to avoid the use of invasive techniques such as lymph-node biopsies, more burdensome for patients. At first researchers looked for viral replication in an easily accessible site, the blood, and only later did they turned to the investigation of the less accessible lymph nodes.
9. The Concorde trial—a multicenter British/French study which tested the effects of the treatment of asymptomatic HIV infected individuals with AZT—did not show clinical benefit of such therapy. (Concorde, 1994). Other (and more restricted) studies (Kinloch de Löes *et al.*, 1995; Perrin and S. Kinloch de Löes, 1995), indicated that an aggressive therapy with AZT may be efficient, if initiated in the very first stages of HIV infection.
10. Actions-Traitement is an association specialized in the follow up of clinical trials of AIDS therapies. It is a member of the collective TRT5 (Traitement Recherche Thérapeutique), a coalition of five AIDS associations (Actions-Traitment, Act-Up, Aides, Arcat-Sida and Vaincre le Sida) which represents AIDS patients in official debates on clinical trials of AIDS therapies in France.
11. Macrophage-tropic HIV strains can multiply in macrophages, but not in T-lymphocytes. The reverse is true for T-cell tropic strains. There are also intermediary HIV strains able to infect both macrophages and T-cells. (Doranz, Rucker, Yi, *et al.*, 1996).

12. The CC-CKR-5 belongs, the experts explain, to a family of molecules, which may also include molecules which play a role in the entrance of HIV into the cell.
13. Scientists believe that when HIV contamination starts through a sexual contact, the infection is initiatied by macrophage-tropic strains. Later some of the HIV strains adapt to the infection of both macrophages and T-cells. These 'dual strains' may use both fusin and CKR-5 as entry co-receptors. As the infection progresses, there is selection for T-cell tropic cells, correlated with the aggravation of disease. The shorter the time period between the contamination date and the appearance of full blown AIDS in persons contaminated by blood, blood products and infected needles is seen as related to the fact that these individuals are directly infected with T-cell tropic virus.
14. The first study was focused on 'resistant' individuals, making the demonstration of causality difficult.
15. Fauci's statement in 1996 (Fauci, 1996), may be compared to his insistence, three years earlier, on the importance of delicate physiological equilibrium in AIDS. (Fauci, 1993).
16. Fauci's affirmation that a protective gene may be absent from Africa is based on data of Samson *et al.*, who found that in causasian populations (selected on the basis of their appearance and European patronymics) about 10% of individuals carried the mutant gene, while no single mutant gene was found in their blood samples of 124 Africans. (Samson, Libert, Doranz *et al.*, 1996). Fauci adds, however, that the existence of exposed, non-infected Gambian prostitutes may indicate the participation of other factors besides CCR-5 in the protection of selected populations.
17. The term 'the defended body' was coined by Roger Cooter and John Pickstone (Cooter and Pickstone, 2000). Other proposed 'bodies' were 'the historiographical body', 'the healthy body', 'the industrial body', 'the third world body', 'the temporal body', 'the reproductive body', 'the psychological body', 'the psychoanalytic body', 'the psychiatric body', 'the diseased body', 'the disabled body' 'the 'analysed body', 'the genetic body', the 'experimental body', 'the ethical body', and 'the dead body'. The immunological body was also discussed by Scott Gilbert (Gilbert, 1995).
18. Researchers continue to speculate, however, about the role of 'hereditary predispositions' in individual susceptibility to HIV infection (Lever, 1966, p. 27; Paxton *et al.*, 1996).

REFERENCES

G. Alkhatib, C. Combadiere, C.C. Border *et al.*, 'CC CKR-5, A RANTES, MIP 1-alpha and MIP 1-beta Receptor as a Fusion Cofactor for Macrophage-Tropic HIV-1' *Science* (1996) 272: 1955–1958.

M.S. Asher, 'Paradox Remains' *Nature* (1995) 375: 196.

P. Bagnarelli, S. Menzo, A Valenza *et al.*, 'Quantitative Molecular Monitoring of Human Deficiency Virus Type 1 Activity During Therapy With Specific Antiretroviral Compounds' *Journal of Clinical Microbiology* (1995) 33: 16–23.

D. Baltimore, M.B. Feinberg, 'HIV Revealed: Towards a Natural History of the Infection' *The New England Journal of Medicine* (1989) 321: 1673–1675.

M. Burnet, F. Fenner, *The Production of Antibodies* (Melbourne: Macmillan, 1949).

M. Burnet, *The Clonal Selection Theory of Acquired Immunity* (Cambridge: Cambridge University Press, 1959).

A. Cambrosio, P. Keating, 'A Matter of FACS: Constituting Novel Entities in Immunology' *Medical Anthropology Quaterly* (1992) 6: 362–84.

A. Carrel, 'Remote Results of the Transplanation of the Kidneys and the Spleen' *Journal of Experimental Medicine* (1910) 12: 146–150.

'The Chronology of AIDS Research' *Nature* (1987) 326: 435–436.

H. Choe, M. Farzan, Y. Sun *et al.*, 'The Beta-Chemokinine Receptors CCR-3 and CCR-5 Facilitate Infection by Primary HIV-1 Isolates' *Cell* (1996) 85: 1135–1148.

F. Cocchi, A.L. DeVico, A. Garziano-Demo *et al.*, 'Identification of RANTES, MIP Alpha and MIP 1-Beta as the Major HIV Suppressive Factors Produced by CD8+ T Cells' *Science* (1995) 270: 1811–1815.

J. Cohen, 'AIDS: The Mood is Uncertain' *Science* (1993) 260: 1254–1256.

J. Cohen, 'Researchers Air Alternative Views on How HIV Kills Cells' *Science* (1995) 269: 1044–1045.

J. Cohen, 'Results on New Anti-AIDS Drugs Bring Cautious Optimism' *Science* (271) 1996: 755–756.

'Concorde: MRC/ANRS Randomised Double-Blind Controlled Trial of Immediate and Deferred Zidovudine in Symptom-Free HIV Infection' *The Lancet* (1994) i: 871–881.

R. Cooter, J. Pickstone, '(eds) Medicine in the Twentieth Century, Harwood Academic Publishers, 2000 .

P. Cotton, 'Controversies Continue as Experts Ponder Zidovudine's Role in Early HIV Infection' *Journal of the American Medical Association* (1990) 263: 1605–1609.

P. Cotton, 'HIV Surrogate Markers Weighted' *Journal of the American Medical Association* (1991) 265(11): 1357–1359.

P. Cotton, 'Many Clues, Few Conclusions on AIDS' *Journal of the American Medical Association* (1994) 272: 753–756.

J.W. Curran, 'AIDS-Two Years Later' *The New England Journal of Medicine* (1983) 309(10): 609–610.

J.W. Curran, 'The CDC and the Investigation of the Epidemiology of AIDS', in C. Hannaway, V.A. Harden, J. Parascandola, (eds.), *AIDS and the Public Debate* (Amsterdam: IOS Press, 1995) pp. 19–28.

H. Deng, R. Liu, W. Ellmeier *et al.*, 'Identification of a Major Co-Receptor for Primary Isolates of HIV-1' *Nature* (1996) 381: 661–666.

J.F. Desforges, 'AIDS and the Preventive Treatment in Hemophilia' *The New England Journal of Medicine* (1983) 308: 94–95.

J.M. Diamond, 'AIDS: Infectious, Genetic or Both' *Nature* (1987) 328: 199–200.

T. Dragic, V.a Litwin, G.P. Allaway *et al.*, 'HIV-1 e Entry into CD4+ Cells is Mediated by the Chemokine Receptor CC-CKR-5' *Nature* (1996) 381: 667–673.

S.K. Dritz, 'Medical Aspects of Homosexuality' *New England Journal of Medicine* (1980) 302(8): 463–464.

D.D. Durack, 'Oportunistic Infections and Kaposi's Sarcoma in Homosexual Men' *New England Journal of Medicine* (1981) 305: 1465–1467.

P. Duesberg, 'HIV is Not the Cause of AIDS' *Science* (1988) 241: 514–517.

S.G. Epstein, *Impure Science: AIDS, Activism and the Politics of Knowledge* (Berkeley: California University Press, 1997).

P. Farmer, *AIDS and Accusation: Haiti and the Geography of Blame* (Berkeley: University of California Press, 1992).

A.S. Fauci, 'The Human Immunodeficiency Virus: Infectivity and Mechanims of Pathogenenesis' *Science* (1988) 239: 617–622.

A.S. Fauci, 'Multifactorial Nature of Human Immunodeficiency Virus Disease: Implications for Therapy' *Science* (1993) 262: 1011–1018.

A.S. Fauci, 'Resistance to HIV-1 Infection: Its in the Genes' *Nature Medicine* (1996) 2(9): 966–967.

E. Fee, D.M. Fox, (eds.) *AIDS, The Burdens of History* (Berkeley: University of California Press, 1988).

Y. Feng, C.C. Border, P.E. Kennedy *et al.*, 'HIV-Entry Co-Factor: Functional c-DNA Cloning of a Seven Transmembranase, G-Protein Coupled Receptor' *Science* (1996) 272: 872–877.

L. Fleck, *Genesis and Development of a Scientific Fact* (transl. F. Bradley and T. Trenn), (Chicago: University of Chicago Press, 1979 {1935}).

F. Fontenay, J.F. Chambon, 'La Conference de Washington: L'lndavir et le Ritonavir Relancent l'Intérêt pour le Trithérapies' *Journal du Sida* (1996) 82: 4–11.

M. Gottlieb, R. Schroff, H.M. Schanker *et al.*, '*Penumocysitis carinii* Pneumonia and Mucosal Candidiosis in Previously Healthy Homosexual Men' *The New England Journal of Medicine* (1981) 305: 1425–1431.

M. Grmek, *History of AIDS: Emergence and Origin of a Modern Pandemic* (translated by R. Maulitz and J. Duffin), (Princeton NJ: Princeton University Press, 1990).

M. Haas, 'The Need to Search for a Proximal Pathogenic Principle of Human AIDS' *Cancer Research* (1989) 49: 2184–2187.

R.B. Heberman, 'Adoptive immunotherapy with purified CD8 cells in HIV infection' *Seminars in Hematology* (1992) 29(2): 35–39.

D.D. Ho, T. Mougdil, M. Alam, 'Quantification of Human Deficiency virus type I in the blood of infected persons', *The New England Journal of Medicine* (1989) 321(34): 1621–1625.

D.D. Ho, 'Time to Hit HIV Early and Hard' *The New England Journal of Medicine* (1995) 333(7): 450–451.

D.D. Ho, A.U. Neumann, A.S. Perelson *et al.*, 'Rapid Turnover of Plasma Virions and CD4 Lymphocytes in HIV-I Infection' *Nature* (1995) 373: 123–126.

G.W. Hoffman, T.A. Kion, M.D. Grant, 'An Idiotypic Nework Model for AIDS Immunopathogenesis' *Proceedings of the National Academy of Sciences (USA)* (1991) 88: 3060–3064.

A.M. Johnston, M. Laga, 'Heterosexual Transmission of AIDS' *AIDS* (1988) 2, suppl. 2: s49–s56.

P. Keating, A. Cambrosio, 'Intralaboratory Life: Regulating Flow Cytometry', in J.P. Gaudillière and I. Löwy (eds.), *The Invisible Industrialist: Manufacturers and the Production of Scientific Knowledge* (London: Macmillan, 1998) pp. 250–285.

J. Klein, *Immunology: The Science of Self-Nonself Discrimination* (New York: John Wiley and Sons, 1982).

D.E. Koshland, 'The AIDS Issue' *Science* 1988: 239, 541.

M.M. Lederman, O.D. Ratnoff, J.J. Scillian *et al.*, 'Impaired Cell-Mediated Immunity in Patients with Classical Hemophila' *The New England Journal of Medicine* (1983) 308: 79–83.

A.M.L. Lever, *The Molecular Biology of HIV/AIDS* (New York: John Wiley, 1996).

J.A. Levy, J.L., Zingler, 'Acquired Immunodeficiency Syndrome is an Opportunistic Infection and Kaposi's Sarcoma Results from Secondary Immune Stimulation' *The Lancet* (1983) i: 78–80.

J.A. Levy, 'Mysteries of HIV: Challenges for Therapy and Prevention' *Nature* (1988) 333: 519–522.

R. Liu, W.A. Paxton, S. Choe *et al.*, 'Homozygous Defect in HIV-1 Co-Receptor Accounts for Resistance in Some Multiply-Exposed Individuals to HIV Infection' *Cell* (1996) 86: 367–377.

I. Löwy, 'The Immunological Construction of the Self', in A. Tauber (ed.) *Organism and the Origins of the Self* (Dordrech: Kluwer, 1991) pp. 43–75.

C. Mack, 'Acquired Immunodeficiency Syndrom Cause(s) Still Elusive' *Journal of the American Medical Association* (1982) 248(12): 1423–1426.

J. Maddox, 'AIDS Research Turned Upside Down' *Nature* (1991) 353: 297.

T.C. Marigan, 'What We Are Going to Do About AIDS and HTLV-III/LAV Infection?' *The New England Journal of Medicine* (1984) 311(20): 1311–1313.

M. Marmor, L. Laubenstein, D.C. Williams *et al.*, 'Risk Factors for Kaposi's Sarcoma in Homosexual Men' *The Lancet* (1982) i: 1082–1087.

J.L. Marx, 'New Disease Still Baffles Medical Community' *Science* (1982) 217: 618–620.

J.L. Marx, 'Clue Found to Cell Loss in AIDS' *Science* (1991) 254: 798–800.

U. Matur-Waugh, D. Mildvan, R.T. Seene, 'Follow Up at Four and Half Years on Homosexual Men with Generalized Lymphadenopathy' *The New England Journal of Medecine* (1985) 313(24): 1542–1543.

P.B. Medawar 'Immunity to homologous grafted skin' *British Journal of Experimental Pathology* (1946) 27: 15–24.

J.E. Menitove, R.H. Aster, J.T. Casper *et al.*, 'T-Lymphocytes Subpopulations in Patients with Classic Hemophilia Treated with Cryoprecipitates and Lyophilized Concentrates' *The New England Journal of Medicine* (1983) 308: 83–86.

A.M. Moulin, *Le Dernièr Langage de la Médicine: L'Immunologie de Pasteur au Sida* (Paris: PUF, 1992).

A.M. Moulin, 'Reversible History: Blood Transfusion and the Spread of AIDS in France', in C. Hannaway, V.A. Harden, J. Parascandola, (eds.) *AIDS and the Public Debate* (Amsterdam: IOS Press, 1995) pp. 170–184.

A.R. Moss, 'Prediction Who Will Progress to AIDS' *British Medical Journal* (1988) 27: 1067–1068.

A.R. Moss, P. Bachetti, 'Natural History of HIV Infection' *AIDS* (1989) 3: 55–61.

M.A. Novak, 'AIDS Pathogenesis: From Models to Viral Dynamics in Patients' *Journal of AIDS and Human Retrovirology* (1995) 10 (Suppl. 1): s1–s5.

J.E. Osborn, 'Sounding Board. The AIDS Epidemics: Multidisciplinary Trouble' *The New England Journal of Medicine* (1986) 314(12): 779–782.

G. Pantaleo, O.G. Cohen, D.J. Schwartzentruber *et al.*, 'Pathogenic Insights from Studies of Lymphoid Tissue from HIV-Infected Individuals' *Journal of AIDS and Human Retrovirology* (1995) 10 (Suppl. 1): s6–s14.

J.W. Pape, 'Prévention et Implications Sociales de l'Infection au VIH: Comparison Entre Pays Industrialisés et Pays en Voie de Development' *Rétrovirus* (1989) 2: 134–143.

W.A. Paxton *et al.*, 'Relative Resistance to HIV Infection of CD4 Lymphocytes From Persons Who Remain Uninfected Despite Multiple High-Risk Sexual Exposures' *Nature Medicine* (1996) 2: 412–417.

J. Pepin, F.A. Plummer, R.C. Burnham *et al.*, 'The Interaction of AIDS Infection and Other Sexually Transmitted Diseases: An Oportunity for Intervention' *AIDS* (1989) 3: 3–9.

P. Piot, J.K. Kreiss, J.O. Ndynia-Achola *et al.*, 'Editorial Review: Heterosexual Transmission of AIDS' *AIDS* (1987) 1: 199–206.

T.C. Quinn, P. Piot, J.B. McComick *et al.*, 'Serologic and Immunologic Studies of Patients with AIDS in North America and Africa' *Journal of the American Medical Association* (1987) 257(19): 2617–2621.

X. Rey-Coquis (interview with), *Le Journal de la MGEN*, (May 1996) 41–42.

C. Rosenberg, 'Disease and Social Order in America: Perceptions and Expectations' *The Milbank Quaterly* (1986) 64(1): 34–55.

C. Rosenberg, 'What Is an Epidemic? AIDS in Historical Perspective' *Dedalus* (1989) 118(2): 1–17.

M. Samson, F. Libert, B.J. Doranz *et al.*, 'Resistance to HIV-1 Infection in Caucasian Individuals Bearing Mutant Alleles of the CCR-5 Chemokine Receptor Gene' *Nature* (1996) 382: 722–725.

F.P. Segal, C. Lopez, G.S. Hammer *et al.*, 'Severe Acquired Immunodeficiency in Male Homosexuals Manifested by Chronic Ulcerative *Herpes simplex* Lessions' *The New England Journal of Medicine* (1981) 305: 1439–1444.

M. Seligman, L. Chess, J.L. Fahey *et al.*, 'AIDS—Immunological Reevaluation' *The New England Journal of Medicine* (1984) 311: 1288–1292.

M. Setbon, *Pouvoirs Contre le Sida. De la Transfusion Sanguine au Dépistage: Décisions et Pratiques en France, Grande Bretagne et Suède* (Paris: Seuil, 1993).

B. Seytre, *Sida, les Sécrets d'une Polémique: Recherche, Intérets Financiers et Médias* (Paris: Presses Universitaires de France, 1993).

P.A. Treichler, 'AIDS, Homophobia and Biomedical Discourse: An Epidemics of Signification' *Cultural Studies* (1987) 1: 263–305.

J. Viera, E. Frank, T.J. Spira, S.H. Landesman, 'Acquired Immune Deficiency in Haitians' *The New England Journal of Medicine* (1983) 308: 125–128.

P. Volberding, 'AIDS-Variations on the Theme of Cellular Immune Deficiency' *Bulletin de l'Institut Pasteur* (1987) 85: 87–94.

X. Wei, S.K. Ghosh, M.E. Taylor *et al.*, 'Viral Dynamics in Human Deficiency Virus Type 1 Infection' *Nature* (1995) 373: 117–122.

R.A. Weiss, 'How Does HIV Cause AIDS?' *Science* (1993) 260: 1273–1279.

J.L. Zeigler, D.P. Sittes, 'Hypothesis: AIDS in an Autoimmune Disease Directed at the Immune System and Triggered by a Lymphotrophic Virus' *Clinical Immunology and Immunopathology* (1986) 41: 395–316.

CONCLUSION: WHAT WENT WRONG WITH MY GENES?

Jean-Pierre Revillard

For historians of biology and medicine, the second half of the XXth century will remain that of triumphant genetics. The impact of molecular genetics on biotechnologies with the production of recombinant plants and a series of new drugs, the nearly complete sequencing of human genome, and the identification of genetic alterations that account for a long list of inherited diseases are only a few examples of the major achievements of genetic research. Awareness of such rapid progress triggers ambivalent reactions in the society with a mixture of fears and fascination. In societies where the myth of progress is still dominant, medical applications of discoveries in human genetics offer fascinating prospects, including gene therapy, prevention of genetic diseases, predictive medicine, increased life span and prevention of disability associated with aging. That such progress could involve a more drastic control of human procreation or even deliberate human genetic selection, and could generate reactions of social exclusion supported by alleged scientific evidence is deliberately ignored. Expression of fears is restricted to a minority of 'ignorant' laymen who are regarded as resistant to progress. Yet the triumphant genetics raise major philosophical issues about identity, freedom and determinism, that were already approached in the most ancient expressions of cultural activities. Given the uncontrollable environmental events and the determinism imposed by the genes I inherited, is there anything left to my 'self'? Is 'I' an illusion or a combination of genetic background and environmental processes leaving room for a unique and undetermined way of reaction and creativity? Are the expressions of individual or collective achievements in architecture, arts or literature primarily genetically and socially determined or still unpredictable? And finally, in all the difficulties any individual encounters in his life, the most frustrating interrogation may become: what is wrong with my genes?

This volume on transmission is a timely and scholar contribution to these interrogations. The in-depth analysis of historical debates on the role of heredity and infectious agents in transmissible diseases enlights the contemporary analysis of host's *versus* pathogens factors in the development of transmissible diseases. It questions the historical dichotomy between monofactorial hereditary diseases (genetic deficiencies) and infectious diseases which may be controlled by antibiotics or vaccination. The origin of vaccination relies on the observation that individuals who survived a first infection were resistant to subsequent exposures to the same infectious agent. One may think that a similar observation made nowadays at the time of triumphant genetics would have led to the conclusion that such individuals were genetically resistant. In this cultural context the whole medical adventure of vaccination would not have been given the slightest chance to be initiated.

THE HOST-MICROORGANISMS ECOSYSTEMS: AN EVOLUTIONARY PRESSURE THAT GENERATES GENETIC DIVERSITY

Microorganisms are remarkable by their capacity to adapt to different environments. For those which grow in an animal host, survival—as a species—requires to fight efficiently the host defense mechanisms. This is achieved by regulation of gene expression, but mostly by genomic plasticity through multiple mechanisms including mutations, gene conversion, exchange of genetic material between microorganisms and between viruses and host's cells. This process of Darwinian selection is highly efficient because of the very short replication time of most microorganisms. The occurrence of resistance to antibiotics is the best documented example of adaptation within bacterial species, and resistance to antiviral chemotherapy indicates that the same process may apply to any new drug against infectious agents.

Because they are exposed to an infectious environment, pluricellular organisms have developed resistance mechanisms. In all animal species defense is primarily achieved by systems of preformed or induced molecules (e.g. stress proteins, complement, cytokines, kinines, lectins that bind sugar expressed by microorganisms) and by cells specialized in phagocytosis (macrophages, neutrophils) or cytotoxicity (natural killer cells). In vertebrates, the emergence of thymus and lymphoid cells along with that of three types of highly diversified molecules—products of the major histocompatibility complex (MHC), T cell receptors, B cell receptors or antibodies—defines the capacity to mount an adaptative immune response characterized by antigenic specificity and memory. MHC is a multigenic multiallelic system of codominant expression which represents so far the most highly diversified allelic system known among vertebrates. In humans, nearly 600 alleles of HLA genes have been characterized and new alleles are being identified. The molecular diversity of T and B cell receptors is achieved during lymphocyte differentiation by gene rearrangements, and in the case of B cell receptors, by somatic mutations. Hence the organism is able to produce antibodies against nearly all antigen epitopes, whereas the repertoire of non-self peptides recognized by each individual is highly restricted by the MHC. In other words within a single microorganism each individual recognizes a set of peptides defined by his/her own HLA alleles, and usually different from the sets of peptides recognized by other individuals. This restriction, along with other allelic polymorphisms, contributes to the heterogeneity of the clinical response to infectious agents among members of a human population. Exposure to the same infectious agent may result in an inapparent disease, demonstrated by the development of immunological memory, or in an acute disease with various levels of severity, from benign to lethal, or in a chronic disease. In any case, clinical expression and severity depend both on the direct pathogenicity of the microorganism (production of toxins, invasiveness, resistance to the host's immune response and subversion of defense mechanisms) and on the host's immune response that may account for lethal systemic acute inflammatory syndrome (e.g. septic shock) or for various chronic inflammatory disorders (e.g. hepatitis B and C).

The molecular diversity of environmental pathogens represents the major evolutionary pressure for the generation of interindividual allelic diversity among members of the host's species. This assumption relies on convergent historical and experimental evidences. The deliberate dissemination of myxomatosis in Australia resulted in a transient high mortality among rabbits followed by a return to an equillibrium of the ecosystem by genetic adaptation, leading to the emergence of more resistant rabbits and less pathogenic viruses. Human ethnic isolates (e.g. the dutch colony in Surinam) could survive in a highly hostile environment during several generations but remained highly susceptible to common pathogens that account for a very low mortality in outbred populations. Experiments of genetic selection according to the level of antibody production in mice resulted in the selection of good producers that, despite increased resistance to several infections by extracellular pathogens, were highly susceptible to infections by intracellular parasites. Converging evidences suggest that sickle cell anemia developed from independent mutations in several areas around the Mediterranean, and that some of those mutations were selected because they conferred some resistance to malaria. Similarly, schistosomiasis stimulated the emergence of high IgE producers who may be prone to IgE-mediated hypersensitivity to a wide variety of environmental allergens. Altogether, observations collected over the last ten years in different fields of life sciences provide for the first time a coherent conceptual framework to account for the increasing allelic polymorphism among individuals. There is no doubt that factors linked to human 'civilization'—population migrations, urban concentration, worldwide rapid transportation, synthesis of drugs and new chemicals—will accelerate and profoundly alter the interplay between human genetic selection and the environment.

CANDIDATE INFECTIOUS AGENTS IN IDIOPATHIC DISEASES

According to the Koch-Henle postulates, a microorganism should be considered as causative of an infectious disease if: i) it can be regularly isolated from patients, ii) but not from patients with other diseases, and iii) the isolated microorganism can reproduce the disease in susceptible hosts. In fact these postulates should be reconsidered because they exclude a wide variety of diseases that depend at least in part on infectious agents for their initiation or progression. For instance the Epstein-Barr virus, an herpes virus which causes infectious mononucleosis, is a major cofactor for the development of two neoplastic diseases, Burkitt's lymphoma and nasopharyngeal carcinoma. The human herpes virus 8 has been associated with Kaposi's sarcoma and multiple myeloma, HTLV-1 with T cell leukemia and spastic paresis. The list of idiopathic inflammatory diseases that may be triggered or maintained by an altered immune response to an infectious agent is extending rapidly.[1] While in some instances (e.g. *Helicobacter pylori* and duodenal ulcer) the causative relationship is strongly supported by the efficacy of antimicrobial therapy, in most cases the link is rather weak, because the microorganism is present in a large proportion of healthy subjects. Indirect evidence for a possible causative link may come from epidemiological studies demonstrating temporo-spatial distributions compatible with those of a transmissible disease (e.g. insulin dependent diabetes, multiple sclerosis, Crohn's disease . . .), or from the demonstration of an immune response to a superantigen or a peptide from the infectious agent.

Present difficulties in the search for infectious agents as a cause of idiopathic diseases reveal the limits of microbial taxonomy and mostly the weakness of the present definition of diseases. There are numerous examples of former diseases which turned out to be clinical syndromes common to various disease entities of different etiologies (e.g. pemphigus, reactive arthritis, chronic hepatitis induced by drugs, viruses or idiopathic autoimmunity, each type being associated with specific autoantibodies.[2]) In this respect it should be stressed that the search for causative agents should not be restricted to microorganisms but must be extended to drugs and chemicals.

A further step in the disintegration of the current clinical classification of diseases came with recent studies on Creutzfeld Jakob disease. Identification of prions as a new type of 'infectious' agents suggested that a structurally abnormal protein devoid of genetic information could account for a transmissible disease. Furthermore the description of inherited forms of the disease made the distinction between genetic and infectious diseases even more elusive. The boundary suffers from additional leakages. Hence an endogenous retrovirus was recently implicated as a possible factor in insulin dependent diabetes.[3] About 10% of the human genome is made of transposable elements which may code for a reverse transcriptase and contribute to the integration of viral RNA into the host cell genome. DNA from viral genome or from vaccines may persist for years as episomes or partially or fully integrated sequences. Transgenic mice that express a non pathogenic spumavirus develop progressive myopathy and encephalopathy,[4] whereas mice expressing the A subunit of cholera toxin present with pituitary hyperplasia, hyperthyroidism and gigantism.[5] Hence the possibility that microorganisms could integrate part of their genetic information into the host's germinal cells, as first suggested by the study of mouse mammary tumor retroviruses (MMTV), no longer belongs to fiction science.

THE SEARCH FOR SUSCEPTIBILITY GENES: MEDICAL AND ETHICAL PROBLEMS

In recent years efforts toward identification of genetic defects that characterize inherited monofactorial diseases have been so successful that Pasteur's aphorism 'one microbe, one disease, one vaccine' could be replaced by 'one gene, one disease'. All debates on ethical issues in genetics are surprisingly focused on monofactorial diseases and often limited to diagnosis, screening among populations, consequences on human procreation and gene therapy, not to mention the forbidden but recurrent fantasy about genetic manipulation of germ-line cells. This is obviously an oversimplified view of current human genetic research. With the study of HLA and diseases initiated more than 30 years ago, immunologists were the first to identify certain HLA alleles as risk factors for autoimmune or inflammatory diseases. The most popular example is the association of some HLA-B27 alleles with spondylarthropathies. Such associations provided a framework that led to the most advertised concept of 'predictive medicine' proposed by J. Dausset and others. In fact information on those genetic risk factors has not so far permitted to propose any adequate prevention. Moreover only a minority (about 2–5%) of HLA-B27 positive persons will suffer from spondylarthropathy, but the knowledge of this genetic risk by the 'patient to be' often entails a major psychological burden, causing repeated costly medical examinations. Finally availability of such information to employers, insurance companies, or social security raises major ethical issues as well as justified criticisms against a health system that fails to treat pain and to ensure adequate care to the poor but extends the population of patients to healthy individuals. George Orwell's Big Brother[6] is playing Doctor Knock—the famous quack in Jules Romains' novel[7]—to the greatest satisfaction of those who make a living from unacceptable drifts in medical research.

The definition of a disease needs to be re-visited in view of the environmental pressure towards increased inter-individual genetic diversity discussed above. It comes as no surprise that the best documented allelic diversity deals with genes that control natural and acquired immune responses (cytokines, cytokine receptors, MHC proteins, intracellular signalling molecules, complement . . .). Because of this genetic diversity, the distribution of 'normal' values of many immunological parameters is so large that it is difficult to determine a cut-off level between normal and pathological values. Hence auto-antibodies are present in all healthy individuals. Furthermore the immune system is highly redundant so that some genetic defects (e.g. C4 genes) may remain asymptomatic whereas others (e.g. common γ chain of cytokine receptors, *btk* tyrosine kinase in B cells) induce severe clinical manifestations.

We are the actors and the witnesses of a major conceptual revolution in medicine, which bears some analogies to the Renaissance period. Recent biological discoveries question our present classification of diseases, reconsider the relative roles of inherited *versus* environmental factors, and lead to a revision of the mere definition of self, with major predictable changes in our health care systems. The emergence of a medicine of (and for) individual patients instead of syndromes entails introduction of a new level of complexity in medical decision-making. The type of care that was already provided by good general practionners applying their common sense without evidence-based scientific grounds may become scientifically justified. The obvious risk in this revolution is the fascination of power afforded by improved scientific knowledge and its use as a domination tool for promoting a mythic better future. History tells us that the myth of progress may generate so much oppression and so many wars that it could be more lethal than our infectious environment.

NOTES

1. See for review P. Berche. 'L'origine infectieuse de certaines maladies idiopathiques.' *Médecine/ Sciences* (1998) 4: 395–97.
2. J.P. Revillard/ASSIM. *Immunologie*, De Boeck Université, Bruxelles, 1998.
3. B. Conrad, RN Weissmahr, J. Böni, R. Arcari, J. Schüpbach, B. Mach. 'A human endogenous retroviral superantigen as candidate autoimmune gene in type I diabetes.' *Cell*, (1997) 90: 303–13.
4. A. Saïb, J. Periès, H. Dithé. 'Recent insights into the biology of the human foamy virus.' *Trends in Microbiology* (1995) 3: 173–78.
5. F.H. Burton, K.W. Hasel, F.E. Bloom, J.G. Sutcliffe. 'Pituitary hyperplasia and gigantism in mice caused by a cholera toxin transgene.' *Nature* (1991) 350: 74–77.
6. G. Orwell. *1984*, London, 1949.
7. J. Romains. *Knock ou le Triomphe de la Médecine*. NRF, Paris, 1923.

SUBJECT INDEX

Medical Research Council 273, 292, 297, 323

AUTHOR INDEX

V

W

Y

Z